THE ORIGINS AND
EARLY DEVELOPMENT OF
THE HEAVY CHEMICAL
INDUSTRY IN FRANCE

THE ORIGINS AND
EARLY DEVELOPMENT OF
THE HEAVY CHEMICAL
INDUSTRY IN FRANCE

BY

JOHN GRAHAM SMITH

CLARENDON PRESS · OXFORD
1979

Oxford University Press, Walton Street, Oxford OX2 6DP

OXFORD   LONDON   GLASGOW
NEW YORK   TORONTO   MELBOURNE   WELLINGTON
KUALA LUMPUR   SINGAPORE   JAKARTA   HONG KONG   TOKYO
DELHI   BOMBAY   CALCUTTA   MADRAS   KARACHI
NAIROBI   DAR ES SALAAM   CAPE TOWN

British Library Cataloguing in Publication Data

Smith, John Graham
    The origins and early development of the heavy
    chemical industry in France.
    1. Chemical industries – France – History
    I. Title
    338.4′7′66100944   HD9653.5   78–41129

    ISBN 0–19–858136–X

Typeset by CCC, printed and bound in Great Britain by William Clowes (Beccles) Limited,
Beccles and London

PREFACE

LAVOISIER'S generation saw not only the emergence of a new chemical science but also the beginnings of the modern chemical industry, and in both developments France played a prominent part. Yet while the scientific achievements of the French chemists have received much attention, the parallel industrial developments in that country have been comparatively neglected. It is to help fill this gap that the present book has been written. Beginning in the 1760s, when such chemical manufacture as existed in France was a small and scattered activity of obscure artisans, the book traces the rise of the industry over the following half-century. It describes how in response to the growing demand from chemical consumers, to the new opportunities offered by an expanding chemistry, and to the special economic pressures of the Revolutionary and Napoleonic Wars, there arose a large-scale industry widely considered by contemporaries to constitute one of the most striking fields of industrial advance of the period. The book focuses particularly on the three central developments that were to provide the basis for the heavy chemical industry of the nineteenth century: the manufacture of sulphuric acid by the lead-chamber process, the application of chlorine as a bleaching agent for textiles, and the manufacture of soda by the process of Nicolas Leblanc. Since I write primarily as a historian of technology, I have been especially concerned to trace the evolution of manufacturing methods, to explore the understanding which underlay them, and to consider the industry's relations with the science of the day. From this point of view the book is offered in part as a contribution to the continuing debate on the relations between science and technology in the Industrial Revolution. But my aim has been to write a history of the industry, not just an account of its technology, and to this end I have also sought to examine the economic context of the innovations described, to chart their diffusion, and to indicate something of their economic significance. I hope that the book might therefore prove of interest to students of economic and social history as well as to readers whose interests lie in the history of science and technology.

The materials for writing a history of this kind tend to be thinly scattered, and one of the consequent pleasures of researching the book has been that of working in so many different libraries and archives, whose staffs have been so consistently welcoming and helpful. To all the institutions mentioned in the following pages I offer my thanks. I cannot here name them all, but perhaps I might be forgiven if I single out for special mention the Archives Nationales, the Bibliothèque Nationale, and the British Museum (now the British Library), without whose resources the book could hardly have been attempted. I would also like to extend my thanks to several other libraries on whose resources I have drawn but which I have not had occasion to name elsewhere: the library of the Pharmaceutical Society of Great Britain, that of the Institute of Historical Research (University of London), the libraries of Leicester, Loughborough, and Nottingham universities, and particularly the university libraries of Leeds and Cambridge. To Cambridge University Library I am further indebted for Figs. 1.1, 2.2, 2.8, 2.9, 3.2, 3.4, 3.5, 3.6, 4.9, and 4.10. For permission to reproduce other illustrations I am grateful to the various institutions cited in the accompanying captions.

The book originated in a doctoral thesis completed some years ago at the University of Leeds, and it was then that much of the basic research for it was done. For the maintenance grant which made that research possible I would like to thank the Science Research Council, while for a bursary which gave me the opportunity of an extended period of study in France I am grateful to the Centre National de la Recherche Scientifique. More recently I have also had cause to be grateful for assistance with travelling expenses from the Department of Social Sciences, Loughborough University.

My debts to friends and colleagues have in part been noted at appropriate points in the text, but a few remain to be acknowledged here. To Dr Bill Smeaton I offer my thanks for his comments on the original thesis and for his help in various ways over the years. To Mr Mark Hudson I am grateful for reading the book in typescript and for saving me from a number of embarrassing slips and infelicities of style. Above all it is a great pleasure to have this opportunity of thanking Professor Maurice Crosland, my former research supervisor at Leeds, who from the beginning of my work in this field has been unfailing in his kind interest and generous advice. To his steady encouragement and example I am deeply indebted.

Loughborough,   J.G.S.
August 1979

CONTENTS

A NOTE ON WEIGHTS, MEASURES, CURRENCY, AND CALENDAR

Weights

Before the development of the metric system in the 1790s the basic unit of weight in France was the pound, which varied somewhat in size from one region to another. The pound most commonly employed was the pound *poids de marc* used in Paris, which weighed 489·5 grammes. Of the provincial variants the most important was the pound *poids de table* used at Marseilles (and elsewhere in Provence and the Languedoc); this weighed 388·5 g and was thus equivalent to about 0·8 pounds *poids de marc*. The English pound (avoirdupois), for comparison, weighs 454 g. References to pounds weight in the text are to French not English measure, unless otherwise stated or clearly implied by the context.

For the measurement of larger quantities the units commonly employed in the period of our study were the *quintal* (= 100 pounds), the *quintal métrique* (= 100 kg), and the *millier* (= 1000 pounds).

Measures

Prior to the metric system the chief unit of length was the foot, which like the pound was subject to some regional variation. The foot of primary importance was the *pied de roi*, used in Paris, which measured 0·325 metres. The English foot is a little shorter, at 0·305 m. References in the text to dimensions in feet are to French not English measure, unless otherwise stated or clearly implied by the context.

Currency

The basic unit of currency under the *ancien régime* was the *livre* (often popularly called the *franc*); it was divided into 20 *sous*, each of 12 *deniers*. Monetary reforms in the 1790s replaced the *livre* with a new unit to which the name *franc* was now officially given. The new *franc* was divided into 100 *centimes* and was virtually identical in value to the old *livre*.

For purposes of comparison with English costs and prices we have throughout our discussion assumed an exchange rate of £1 = 24 *livres* (or *francs*).

Calendar

During the Revolution, France officially abandoned the use of the familiar Gregorian calendar in favour of a specially devised Republican calendar, whose Year I began on 22 September 1792. In the new calendar, weeks were replaced by ten-day periods called *décades*, and new months were defined, each of thirty days. The Revolutionary calendar remained in force until 1 January 1806, when there was an official return to the Gregorian system. In the present text we have for the most part silently substituted Gregorian for Republican dates, so as to spare the reader unnecessary distraction.

ABBREVIATIONS

AAS	Archives de l'Académie des Sciences, Paris
ADBR	Archives Départementales des Bouches-du-Rhône, Marseilles
ADH	Archives Départementales de l'Hérault, Montpellier
ADR	Archives Départementales du Rhône, Lyons
ADSM	Archives Départementales de la Seine-Maritime, Rouen
AML	Archives Municipales de Lyon
AMM	Archives Municipales de Marseille
AN	Archives Nationales, Paris
AS	Archives de la Seine, Paris
BM	British Museum, London (now the British Library)
BML	Bibliothèque Municipale de Lyon
BMM	Bibliothèque Municipale de Marseille
BMR	Bibliothèque Municipale de Rouen
BN	Bibliothèque Nationale, Paris
BUM	Bibliothèque Universitaire de Montpellier, Section de Médecine
CCM	Archives de la Chambre de Commerce de Marseille
CNAMA	Archives du Conservatoire National des Arts et Métiers, Paris
CNAMB	Bibliothèque du Conservatoire National des Arts et Métiers
INPI	Archives de l'Institut National de la Propriété Industrielle, Paris
MC	Minutier Central (Archives Nationales)
PP	Archives de la Préfecture de Police, Paris
SG	Archives de la Compagnie de Saint-Gobain, Neuilly
SRL	Science Reference Library, London (formerly the Patent Office Library)

1

THE DISTILLERS OF AQUA FORTIS

WHEN towards 1773 Demachy wrote his volume on the chemical arts for the Academy of Sciences' great series on the trades and industries of France, he took as his basis and title the art of the distillers of aqua fortis. Among the sparsely scattered artisans then engaged in chemical production of one kind or another, it was the distillers of aqua fortis, he observed, who were the most numerous, the most widely established, and the most important for the magnitude of their production, and who alone formed a significant identifiable group. It is consequently with the activities of these distillers that we might here appropriately begin, as chiefly constituting France's chemical 'industry' at the beginning of our period.

As the name implies, such distillers had as their primary concern the production of aqua fortis, or nitric acid, the first of the mineral acids to have come into appreciable commercial demand, and a material important for a variety of leading trades by the middle of the eighteenth century. The acid was required by dyers, for example, who employed it particularly to produce scarlet red with cochineal, one of the colonial dyestuffs whose use in the eighteenth century was greatly growing. Hatters needed the acid for their traditional 'secret' composition (a solution of mercury in aqua fortis), with which they rendered hairs proper for felting by making them curl. The use of the acid by gold- and silversmiths, and by refiners, was again of some importance, while it found lesser uses among furriers for degreasing and colouring skins, among brass and copper founders for cleaning their metal, and among etchers. Besides making aqua fortis itself, the distillers produced other materials, too—most commonly perhaps 'spirit of salt' (hydrochloric acid)—materials generally related to their primary product in being similarly prepared by distillation, or obtained as by-products from the distillation residues.

It was probably in Paris that the distillers were oldest established,[1] for already in 1639 we find the trade sufficiently practised in the capital to be brought under the statutory surveillance and jurisdiction of the *Cour des monnaies* (the authorities of the Mint were anxious at the possible use of aqua fortis to thin the coinage). In the seventeenth century the trade of the chemical distiller was not clearly distinguished from that of the distiller of *eaux-de-vie*, both activities being embraced by a single guild; but by the beginning of the eighteenth century the distillers of acids had emerged as a separate body, composing a small guild of their own under the name 'distillateurs en chymie', working under the continuing jurisdiction of the *Cour des monnaies*. Practice of the art was restricted, in theory at least, to those who were received master distillers after practical proof of their competence. By the time Demachy wrote, this guild organization had disappeared, abolished in 1746, but the distillers themselves continued, perhaps some six or eight of them, making the capital almost certainly the main centre of chemical production in France. It was probably also in Paris that the trade offered the most extensive range of products, with some distillers qualified to produce pharmaceutical materials as well as materials for trade use. Charlard, for example, from whom Demachy derived much of his information, was a master apothecary and a man of some chemical learning. Besides nitric and hydrochloric acids, it appears from Demachy that they distilled vinegar and alcohol (for spirit lamps and varnishes), and prepared a variety of aromatic and flavouring essences employed in pharmacy, perfumery, patisserie, and the confection of liqueurs. By-products obtained from the residues included a highly reputed cement, an oxide of iron employed as a polish by glass-makers, and various chemical salts.

Outside the capital it is doubtful whether the trade was significantly established before the eighteenth century, and even after the middle of the century distillers seem to have still been few and far between. In the north, Demachy mentioned distillers at Amiens, Roubaix, and Lille. The workshop at Amiens had presumably arisen since 1719, when we find a dyer in the town (Dehée), in the absence of any local distiller, setting up furnaces to make aqua fortis for himself, only to be ordered to desist since his guild did not have the right to conduct such a trade.[2] At Lille production of the acid evidently began towards 1750, for in that year a certain widow Schoutteten, a goldsmith by trade, announced that 'having for several years had the secret of the composition and making of aqua fortis', she had been preparing it in a country house some miles from the town, for the use of dyers as well (presumably) as for herself. The good widow was authorized in 1750 to establish a workshop in the town itself to make it in larger quantity, the acid previously having been brought to Lille from Paris or Holland. The subsequent conduct of production at Lille does not seem to have been continuous, however: in 1761 it was reported in Brussels that the distiller at Lille had recently ceased working,[3] and two new ventures there in the 1770s—one of them by an apothecary—were both short-lived. In the south of France, four distillers were reported at Montpellier in 1744 (two in 1768), and the third quarter of the century also presents odd distillers at Gignac, Clermont-Lodève, and Carcassonne.[4] The distiller working at Carcassonne in 1754 was said to have established himself there about twenty years earlier to supply aqua fortis to local dyers in what was the largest woollens centre in the province; he also sold his products in Roussillon and Catalonia. The most important centre of production in the south was the County of Avignon,

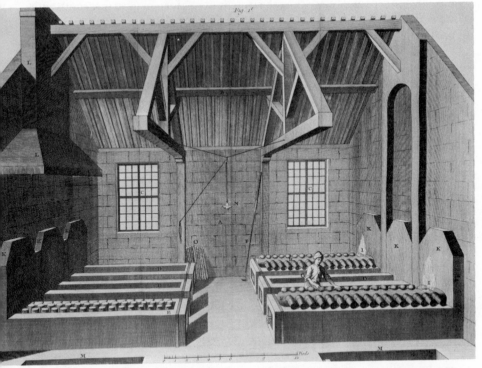

Fig. 1.1. Workshop for distillation of aqua fortis. (Demachy (1773), Pl. 1.)

where in 1781 there were five or more distillers: three at Avignon and two at Carpentras.[5] At this time (until 1791) the County of Avignon was not under French rule but a papal territory, and this gave distillers there the advantage of cheaper saltpetre. The distillers of the County were said to sell aqua fortis to the value of 200 000 francs a year—perhaps equivalent to some 200 000 pounds weight—to France, Spain, and Piedmont.

The work of the distillers of aqua fortis was more an artisanal than a truly industrial activity. The apparatus employed had scarcely changed in centuries. Aqua fortis itself was generally made by distilling a mixture of saltpetre and clay (or sometimes copperas) in stoneware or glass retorts, heated in rows on galley furnaces, a commercial output thus being achieved simply by multiplying laboratory-scale apparatus. A distiller's output of aqua fortis (probably by far his largest product) would be a few tens of thousands of pounds a year. The workshop of Charlard, one of the prominent distillers in the capital, could produce about 30 000 pounds.[6] That illustrated by Demachy (Fig. 1.1) can be reckoned to have had a likely capacity of some 45 000 pounds. The five or more distillers in the County of Avignon were said in 1781 to be annually distilling 3300 *quintaux* of saltpetre (equivalent to

275 000 Paris pounds, assuming the weight to be given in local measure), and we would expect this to have resulted in a roughly equal weight of acid. The distillers there together employed about forty people.

If the distillation of aqua fortis was the first branch of chemical production to acquire any great significance in France, it was not long to retain its pre-eminent position. Even as Demachy wrote, a new manufacture was beginning to arise which was soon destined to assume the focal role, and to become the basis for the development of the heavy chemical industry of the nineteenth century. This was the manufacture of sulphuric acid by the lead-chamber process.

NOTES

1. On the Paris distillers, and their rather complicated guild history, see particularly Scagliola [1943]; also Clacquesin (1900).
2. AN, F^{12*} 65, pp. 267–8, 346–7.
3. *Journal de commerce*, July 1761, p. 145.
4. Dutil (1911), 623; ADH, C2741.
5. Dubled (1972).
6. Estimated from description in Bellot *et al.* [1774].

2

THE GROWTH OF SULPHURIC ACID MANUFACTURE

In the early eighteenth century there were two processes in use by which sulphuric acid was made. One was by distilling green vitriol (ferrous sulphate) and collecting the acid vapours evolved, a method which gave the acid its early name 'oil of vitriol'. Acid was made in this way for use in the industrial arts, but it was expensive and its use was very limited. The second method was by burning sulphur and condensing the acid fumes produced; the simplest procedure was to suspend over the sulphur a glass bell, whence the process was known as the bell method. This was the way in which apothecaries had long prepared for their medicaments an acid known as spirit of sulphur, or oil of sulphur, which had at first been supposed to differ from oil of vitriol but which by the end of the seventeenth century was recognized to be essentially the same. The yield was extremely low, and if the acid made from vitriol was expensive that made from sulphur was even more so. It was nevertheless a variant of this old apothecaries' process which was to provide the basis for the development of the modern industry. The amendment which transformed its viability was the addition to the sulphur of a little saltpetre (potassium nitrate), whose action was not to be properly understood until after 1800 but whose practical effect was greatly to increase the chemical efficiency of the process. The contribution which saltpetre could make to the process was known for a good many years before being put to any significant commercial use. It was indicated in 1697, for example, in the popular chemistry text by Lemery, which described the production of the acid by burning sulphur and saltpetre over water in a large covered pot. Lemery, however, continued to prefer the older method employing sulphur alone, for the use of saltpetre introduced into the product some nitric acid impurity. In this he seems to have been followed by apothecaries generally. The first to take up the saltpetre process in a major way was the English quack doctor Joshuah Ward, whose venture is generally regarded as marking the beginnings of the modern industry.

In about 1736 Ward established a works at Twickenham, where he operated the process on a commercial scale by conducting the combustion in large glass globes, laid out in rows on a sand bath for easy servicing. The practice was to put a little water into the bottom of each globe and then introduce into the neck an iron dish bearing the combustion mixture. The mixture was ignited and the globe was stoppered to prevent escape of the acid fumes. A series of charges was burned until the water had become sufficiently acidulated for it to be worth drawing off and boiling down to the concentrated state in which it was sold. Ward is said to have reduced the price of the acid from a figure of between 1s. 6d. and 2s. 6d. an ounce to a similar price per pound, so opening the way for that extension in use which was to make sulphuric acid the most prolifically valuable of all industrial chemicals. He evidently produced in quite large quantity, but his apparatus, being of glass, was still essentially small in scale. It was therefore a further major landmark when glass globes came to be replaced by chambers of lead, the use of lead eliminating the size limit imposed by glass, and resulting in due course in the vast lead-chamber installations of the nineteenth century. Leaden chambers were first introduced by John Roebuck and Samuel Garbett when they began making the acid in Birmingham in 1746. Three years later these same manufacturers set up a more important works at Prestonpans—on the east coast of Scotland—which besides supplying local markets also exported to the Continent, and acquired international fame. Following their entry into the field the price of the acid in Britain quickly dropped still further, to about 4d. a pound. They naturally endeavoured to keep their method secret, but rival lead-chamber plants began to appear in Britain from about 1756, and multiplied fairly rapidly from the 1770s.

While these developments were taking place in Britain, France at first remained content to import her acid. It was still an item of only small consumption, and indeed in 1774 an observer in Rouen was to recall that a first venture at large-scale production in France, nearly twenty years earlier, had foundered partly on the insufficiency of the market.[1] The significance of the acid was growing, however, a mark of this being the Government's action in 1762 in reducing the import duty levied: though this was in part to remove an anomaly in the tariff, it was also on the grounds that the acid was needed by dyers and was not made in France.[2] In the 1760s the rise in consumption was such as to begin to make manufacture in France worthwhile.

Demand for sulphuric acid came in the first instance from dyers and cotton printers. Dyers employed it for producing on wool the shades known as Saxon blue and Saxon green, developed in Saxony in about 1740 and requiring a solution of indigo in the concentrated acid. Its use in cotton printing was the more important, the acid here finding two early applications.[3] One was for the dilute acid bath in which pieces were washed before printing to remove materials left by the bleacher. The second, more specialized but again important, was in the production of the blue-based prints which were one of the main lines. These were made by the so-called resist method, in which a pattern was printed on to the cotton using a dye-resistant composition, and the piece was then dipped in a blue vat to give a patterned fabric in which the parts protected by the 'resist' remained white. A sulphuric acid bath was used to remove the resist after the operation.

Consumption of sulphuric acid by cotton printers grew rapidly with the remarkable development of that industry from about 1760. Printed cottons had made their appearance in France in the seventeenth century, when they were imported from India as products of native craft. Such was their popularity that imitations had begun to be made in France, but this development had been arrested in 1686 when both importation and manufacture were prohibited in order to protect the home silk and woollen industries. For many years this prohibition was enforced quite strictly, until in the mid-eighteenth century printed cottons, supplied by contraband and by a number of illegal French works, again began to come into fashion, and a violent controversy ended in 1759 with the lifting of the ban. Numerous works then sprang up in many parts of the country. By 1785 they numbered a hundred or more, many of them quite sizeable enterprises for the period, for this branch of manufacture was one of the leading elements in the development of factory organization in the French textiles industry. One of the main early centres of production was Rouen, where the number of works rose from 2 in 1763 to 30 in 1774.[4] It is therefore not surprising to find that it was in Rouen and its region that the manufacture of sulphuric acid first developed.

Holker at Rouen

The credit for first introducing sulphuric acid manufacture into France belongs to the Englishman John Holker,[5] a colourful figure who had fled to the Continent to save his neck after fighting in the Jacobite uprising of 1745. Holker's original ambition had been to set himself up in the Lancashire textiles industry and to this end he had spent the earlier 1740s gaining experience in Manchester. On finding himself in exile he resolved to make his career in France, and in particular in Rouen, which at that time was the centre of a notable though still small-scale industry, producing chiefly linen goods, with some wool and less cotton. Holker's plans soon attracted Government interest in the person of Daniel Trudaine, France's effective director-general of commerce, and in 1752 he was able to establish in Rouen a works for the production of cotton cloth and velveteen, using English methods and equipment, and enjoying Government support and the title *manufacture royale*. Trudaine, increasingly impressed, in 1755 secured Holker's appointment as an inspector of manufactures in the specially created post of *Inspecteur général des manufactures étrangères*, based in Rouen but with country-wide responsibilities. This gave Holker the duty of promoting the introduction and diffusion of foreign manufacturing techniques, and he was to play a leading part in the development of France's textiles industry. As both manufacturer and inspector of manufactures he was obviously well placed to appreciate the needs of the industry in Rouen, which by the early 1760s included important dyeworks and the first of the printed-cotton establishments.

In his capacity as inspector of manufactures Holker in October 1764 recommended to Trudaine the project of a certain Brown, an elderly tax official (*inspecteur général des aides*) from Brussels. Brown was proposing to leave his Brussels post in order to establish in France a works for the manufacture of various chemicals, provided he was granted certain favours by the Government.[6] It is interesting that this followed closely upon a venture in Brussels of a similar kind, by another Englishman, Thomas Murry, a man of uncertain technical competence and shady moral character who had left England in 1759 after embezzling from his employers.[7] Murry's occupation in England had been as commercial agent for a number of manufactories of copperas (ferrous sulphate), and he had also had some involvement in a sulphuric acid works, apparently in London. Arriving in Brussels he had launched an enterprise for the manufacture of aqua fortis and sulphuric acid (in glass globes), with the production of copperas also planned in the longer term. By 1762 he had secured the financial backing of the Austrian government (the ruling power) and his works had become a State manufactory. It was destined in fact to fail, and in 1770 Murry abandoned his debts in Brussels and headed for France, where we shall meet him again later. In 1764, though, it must have appeared a very promising undertaking, and it was no doubt Murry's example which was the inspiration of Brown. Evidence is lacking of any closer connection.[8] Like Murry, Brown intended to produce copperas, aqua fortis, and sulphuric acid.

Brown's project was welcomed in Government quarters by Trudaine de Montigny, now increasingly taking over his father's duties. A report by his technical adviser, the chemist Hellot, cautiously approved of the scheme, mainly for its intention to make copperas, for Hellot quite failed to appreciate the growing industrial significance of sulphuric acid. Holker's advocacy was more positive, and it was probably his interest which was chiefly instrumental in securing two *arrêts* of the *Conseil d'état*, on 3 February 1765, authorizing the formation of a works and granting various favours of the kind commonly accorded to notable new industrial enterprises, including the promise of premiums on output and an annual bounty of 2000*l.* to Brown. The works was to be at Fécamp, on the Normandy coast, where Brown had found plentiful supplies of pyrites for copperas manufacture.

From the start things did not go smoothly. In June 1766 Holker reported that various difficulties had prevented the establishment of a works at Fécamp, but that he had now persuaded Brown to set up a works in Rouen instead. By now, too, Brown had changed his associates, and a new *arrêt* of 18 June transferred to Brown, Garvey, Norris, & Cie the privileges originally granted to Brown, Connely, and Tierney (or Terney).[9] It seems, moreover, that the intended method of making sulphuric acid had also changed. An original intention to use lead chambers is suggested by provision in the *arrêt* of February 1765 of 1500*l.* to bring over from England a workman able to join lead without solder, whereas in June 1766 arrangements were being made to obtain glass globes, from the glassworks at Maucomble. Nevertheless, Holker could report that the associates had now acquired a site and he was hopeful that production would begin before winter.

In the event Brown's venture never did come to fruition. The exact reason is not clear, but he evidently gave Holker cause to feel aggrieved, for two years later Holker ruefully reminded Trudaine de Montigny how he had fallen 'victim' to Brown, adding that the affair had cost him dear in every respect. Holker, however, had become too enthusiastic about the project to be willing to see it dropped. He therefore took it up himself, and by the beginning of 1768 was again petitioning Trudaine de Montigny, this time on his own behalf. A significant change was that whereas sulphuric acid had been only a subsidiary part of Brown's original plan, it had now become the essential concern.[10]

It was Holker's son, also called John, who handled the technical aspects of the new project, a task for which he was admirably fitted. Holker had long cherished the hope that his son would follow in his own footsteps, a hope equally shared by the Trudaines, and so young Holker had received a very careful education with this end in view.[11] For three years, from 1764, he studied in Paris under some of the leading men of science of the day, his studies including chemistry with Rouelle, Cadet, and others; in the summers his father arranged practical instruction for him in the various branches of local industry, and these preparations were to meet with a satisfactory conclusion when towards the end of 1768 Holker *fils* was appointed inspector of manufactures, assistant to his father. Even before this appointment, however, young Holker was in 1767 sent by Trudaine de Montigny on a mission in Picardy and Flanders. The journey took him as far as Brussels, and there he acquired knowledge of the technique for making sulphuric acid, apparently buying the 'secret' from Brown.[12] On his return he succeeded in making three or four pounds of acid himself, using two small globes, and in April 1768 he had the gratification of hearing his samples pronounced very good by the chemist Montigny, of the Academy of Sciences, to whom Trudaine de Montigny had referred them for a report.

By this time Holker *père* was submitting his plans and requests for Government favour. In what seems to have been the final plan, presented on 31 March 1768, he envisaged the production of 67 000 pounds of oil of vitriol a year, which he estimated on one occasion to be a half, on another the whole, of France's consumption. The main details of this plan are worth outlining, as they make clear the impressive scale of the enterprise. The acid was to be made in two sheds, each 360 feet long by 24 feet wide, and each containing 400 globes, $3\frac{1}{2}$ feet in diameter, arranged in four rows; it was reckoned that one globe could yield 7 pounds of oil of vitriol a month. There were in addition to be two workshops, each 130 feet long by 27 feet wide, in which the dilute acid was to be concentrated in 120 smaller globes, and where it seems aqua fortis was also to be made. Another shed was to house a horse-driven mill for grinding the materials. The capital required for site, buildings, and equipment was estimated at just over 90 000*l*. This Holker seems to have raised partly by private loans and partly by finding partners.

Negotiations with Trudaine de Montigny culminated on 24 September 1768 in two *arrêts* of the *Conseil d'état*, authorizing the establishment of a works for the production of oil of vitriol, aqua fortis, and 'autres drogues accessoires', and granting a number of privileges.[13] Holker was permitted to import, duty-free, up to 30 000 pounds of saltpetre a year; he was authorized to send his products throughout the kingdom and abroad free of duties; for a period of ten years he was promised a premium of up to 3000*l*. a year, at the rate of 10*l*. for every 100 pounds of oil of vitriol sold; and he was awarded a sum of 6000*l*. to cover preliminary expenses and to enable him to bring over from England one or two glass-blowers to instruct French workers. There were also such standard subsidiary favours as exemption from militia and other public duties for six of the workers, and tax concessions for Holker himself.

In anticipation of Government protection Holker had already in May bought a site in the faubourg Saint-Sever, to the south of Rouen, where his cotton works was also situated.[14] By Christmas he could report that the buildings were completed and globes were being made. The French glass-blowers were having some difficulty with these, breaking two for every good one made, but Holker was expecting not to require an English worker after all.[15] The commencement of manufacture was a little delayed, partly it seems by fierce hostility to the enterprise from local residents, who tried to obstruct the registration of Holker's letters patent at the Rouen *parlement*,[16] but production began in the summer and the first sale was made on 7 August 1769, appropriately enough to Oberkampf, of Jouy near Paris, the most celebrated of printed-cotton manufacturers.[17] In its first year the works sold over 45 000 pounds of acid, not only in Rouen and Paris, but as far afield as Lille, Nantes, Bordeaux, and Marseilles.

The success of the enterprise was assured by further Government protection in 1770. On 11 June the import duty on sulphuric acid was raised from 3*l*. 15*s*. per 100 pounds to 15*l*., which with the standard surcharge levied on all goods formed in reality a duty of about 20*l*.[18] This was a measure for which Holker had petitioned from the outset and it gave him a decided advantage: British acid, which in 1768 had cost 15–16*s*. a pound in Rouen, was in 1773 said to sell in France at 20–1*s*. (100–5*l*. per 100 pounds), while Holker's sold at 14–15*s*. (70–5*l*. per 100 pounds).[19] Holker also received assistance in the form of a Government loan. His request for such a loan in 1768 had been unavailing, but Trudaine de Montigny came speedily to his rescue in November 1770, supplying a sum of 20 000*l*.

within a week when Holker was embarrassed by the need to repay a private lender.[20]

In 1771 and 1772 the works was enlarged, and output then rose to 130–140 000 pounds in each of the next three years; in 1773 there were fourteen workers, of whom four worked through the night.[21] The changes in 1772 were evidently considerable, for manufacture was interrupted for over four months. They almost certainly involved the introduction of lead chambers. Holker *fils* had in 1769 been on an industrial espionage mission to England, visiting manufactories of many kinds, including those of sulphuric acid, aqua fortis, copperas, and alum; he had followed this in 1770 with a similar mission to Scotland.[22] It seems that on these trips he acquired information

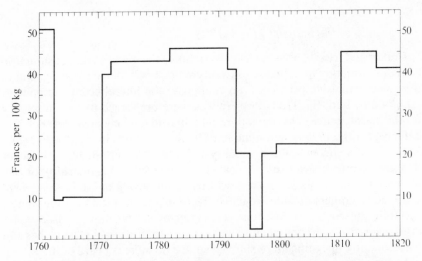

Fig. 2.1. Import duty on oil of vitriol, 1760–1820. We have here plotted not the nominal rate but the duty actually charged, i.e. including the standard surcharges (of up to 50 per cent) which were imposed at various times on all goods. (Sources: AN, F^{12}1506, [8]; Magnien (1786), vol. 1, p. lvi, and vol. 2, 248–9; id. (1815); *Tarif*, vol. 2, pp. 784–91.)

on the use of lead chambers, and there is a story that he learned the secrets of manufacture by penetrating Roebuck's works disguised as a workman.[23] Growth continued in the later 1770s. An *arrêt* of 4 December 1775 increased the duty-free saltpetre allowance from 30 000 to 70 000 pounds a year when Holker *fils* announced that he intended to double the size of his works, so buoyant was the market.[24] In the late 1780s and early 1790s the works was producing up to 300 000 pounds of oil of vitriol a year.[25]

Though established by the two Holkers the works from the outset went under the name of Chatel & Cie. Chatel was the resident director and the Holkers left the day-to-day running of the works in his hands. They both had other duties, of course, as inspectors of manufactures. The career of Holker *fils*, moreover, took a dramatic new turn in 1777 when, with France on the brink of entering the War of American Independence, he was sent to America as a secret diplomatic agent by the Foreign Minister Vergennes. In America he acquired responsibility for supplies to the French navy in American ports, and became consul-general at Philadelphia. This promising new career was shattered by a scandal in 1781, but he settled in America and spent most of the rest of his life there.[26] The Saint-Sever works remained under the direction of Chatel until 1792, when Chatel left to establish a works of his own. It was then run by Jean Holker (grandson of the original John), whom we shall meet later as a leading figure in the early nineteenth-century industry.

The growth of the industry, 1773–1793

Holker had tried to secure exclusive rights to the manufacture of sulphuric acid in Normandy for a period of twenty years, but had been unsuccessful in this. Such exclusive privileges ran counter to the liberal economic policies espoused by both the Trudaines. He was therefore forced to rely on secrecy for the maintenance of his monopoly, and this did not long prove an adequate defence, for the 1770s saw a number of further ventures in the Rouen area.

The first is encountered in January 1773, when a certain Jean Le Coeur and the brothers Jean and Louis Guyot formed a partnership for the manufacture and sale of 'essence de vitriol', proposing to establish a works a league (2½ miles) outside Rouen.[27] The Guyots in question were almost certainly those encountered four years later in the position of *Apothicaire-distillateur du roi*.[28] Their venture was perhaps for the use of glass globes and was evidently insignificant or short-lived, for no more is heard of it.

Far more important was the lead-chamber plant constructed in 1774 by Fleury, a manufacturer of printed cottons at Eauplet just outside Rouen.[29] Fleury had apparently acquired knowledge of the use of lead chambers by the British and the Dutch, and in view of the large sum he was spending on sulphuric acid decided to make it for himself. To this end he erected a chamber with a base of 10 feet by 12 and a height of 14 feet, built of leaden sheets soldered together and supported by some kind of wooden framework, as in later practice. Some of its features appear rather awkward: in shape it was of elliptical cross-section and narrowed towards top and bottom, while one-third of the chamber was buried in the ground, presumably for rigidity. Nevertheless, by the middle of 1774, after trials with various modes of operation, Fleury had succeeded in making several thousand pounds of acid. He was not to become a manufacturer of any note and indeed may not long have persevered with his enterprise. His venture did, though, play a significant part in the diffusion of the lead-chamber process in France, for Fleury made no secret of his methods. Since his aim was to make acid for private use rather than public sale, he had no reason to wish to obstruct manufacture by others, and so he proudly showed his chamber to De la Follie, a local businessman and amateur of science, who was a leading figure in the town's academy and agricultural society, and a man particularly known for his interest in the applications of science to industry.[30] De la Follie at first could scarcely believe that such a chamber would withstand the acid and not fall to pieces. Once convinced of its efficacy, however, it was only natural that he should seek to spread knowledge of so useful an innovation.

De la Follie accordingly described the chamber, suggesting certain improvements, in a memoir he presented to the Rouen Academy, a summary

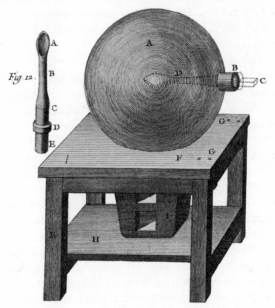

Fig. 2.2. Production of sulphuric acid by glass globe method. As portrayed by Demachy (1773), Pl. 5. The globe was heated on a sand bath to raise steam, and the mixture was burned on a stoneware spoon affixed to the stopper.

of which was included by the secretary in his report to the annual public meeting in August. Later in the year, De la Follie published a detailed article in the widely circulated journal *Observations sur la physique*, and this evidently aroused interest for he then received letters from a number of people who had been inspired to carry out trials themselves, with varying success. He published again on the same subject three years later. De la Follie, moreover, sent advance details of his 1774 article to Demachy, who had described production in glass globes in his *L'Art du distillateur d'eaux-fortes*, of 1773. Demachy now included details of the lead chamber method in his *L'Art du distillateur liquoriste*, published in 1775, and he also added an account to his earlier work when a new edition came out at Neuchâtel in 1780. Knowledge of the method was soon sufficiently general for Chaptal to speak in 1782 of 'the lead chamber, of whose construction few chemists are today ignorant'.[31]

It seems to have been De la Follie's 1774 article which primarily inspired the next lead-chamber plant near Rouen, that constructed in 1777 by Étienne Anfrye, a printed-cotton manufacturer at Déville.[32] Anfrye succeeded also in engaging an experienced worker, a certain Guillaume Lolivray, who by an agreement of 4 February 1777 undertook to work for three years at the manufacture of oil of vitriol and aqua fortis 'and to teach the secret of it to the said Anfrier *père*'.[33] It must have been Anfrye's chamber that De la Follie described in his 1777 article. The chamber marked a clear advance on the pioneering essay by Fleury, standing 22 ft. high and with a base of 144 sq. ft. It was perhaps raised above the ground, as was to become standard practice, and the floor was sloped towards one corner so that the acid could be drawn off through a tap, an improvement which De la Follie had suggested in 1774. By 1778 Anfrye had more than one chamber and was intending to produce for commerce as well as his own needs. For some fifteen years, indeed, his works did compete successfully with Holker, before closing during the Revolution, in about 1793–4. It was much smaller than Holker's works, however, producing in 1787 some 60 000 pounds of acid a year.[34]

Anfrye's example was almost immediately followed by Louis Stourme, another printed-cotton manufacturer at Déville, specializing in blue-based prints.[35] Stourme was stirred into action by Holker's patent inability to meet the growing demand. For six weeks in the summer of 1777 Stourme's printed-cotton works lacked acid, while repeated approaches to Holker proved unavailing. Having therefore conceived the idea of making his own, Stourme discussed his plans with De la Follie. He was also said to have obtained knowledge of Holker's methods from workers at the Saint-Sever factory.[36] By the middle of 1778 he had established a plant at Déville, like Anfrye intending to produce not only for his own consumption but also for sale. In fact Stourme went bankrupt in 1779, but his sulphuric acid works continued.[37] It is probably this works which one finds in 1781 in the hands of Jean Henry Carstens, and the works was subsequently run for some years by a certain Petou, before being acquired by Forestier, an established producer of aqua fortis and other chemicals in Rouen, who ran it until its final closure during the Revolution. It was again a small works, Petou in 1787 making 50 000 pounds of acid a year there.[38]

Besides De la Follie, another member of the Rouen scientific community to contribute to this growth of acid manufacture in the area was the scientific-instrument maker and mechanician Scanegatty. Scanegatty was again an active member of the Rouen Academy and a prominent local figure: in the 1780s he gave lecture courses on physics and mechanics in specially built premises on the quai de Paris, and attracted a visit from Arthur Young during the latter's travels through France.[39] It was Scanegatty who constructed the water pump when De la Follie wished to try the effect of spraying water into a chamber in the mid-1770s, and he was subsequently responsible for designing chambers, and also furnaces for concentrating the acid. In 1792 he was given an award of the highest class (6000*l.*) under the Revolutionary Government's policy of rewarding meritorious inventors.[40] The citation spoke of his work in many fields, and particularly of his contribution to the growth of sulphuric acid manufacture. He was credited with having supervised the construction of two sulphuric acid works (those of Anfrye and Stourme?), and one of his laudable deeds had been the donation, probably in 1788, of a model lead chamber to the royal collection of machines and models on public display at the Hôtel de Mortagne in Paris. This has come down to us in the collections of the *Conservatoire des arts et métiers*, and is shown in Plate 2.

At about the same time as Anfrye and Stourme were starting manufacture, there began to appear the first lead-chamber plants outside the immediate vicinity of Rouen. In 1778 there was established the famous works at Javel, near Paris, to which we shall return later. And in the same year a plant was built near Honfleur by Auvray, Payenneville, & Cie, a group of merchants from Rouen.[41] Auvray and his associates, surprisingly, were unaware of De la Follie's article, and consequently suffered at the hands of Thomas Murry, now arrived to continue in France the ignominious career on which he had embarked in Brussels. The company had established a copperas works at Honfleur in 1774, with Murry's assistance presumably, and this seems to have gone well enough; but when Murry induced them to invest further, in the manufacture of sulphuric acid, the result was abortive and occasioned considerable losses. It was probably a glass-globe venture. By May 1778 they had instead constructed large lead chambers on the same site, and were denouncing Murry as an 'ignorant foreigner'. The company was successfully making both copperas and sulphuric acid in 1782,[42] but seems to have disappeared by the early 1790s, perhaps as a result of the Revolution. As for Murry, he now moved to Amiens, where on 30 August 1781 he joined in partnership with De la Haye *frères* and Durand, merchants in the town, to establish a works for the manufacture of sulphuric acid, copperas, and aqua fortis. They refused him capital, however, after hearing of the Honfleur episode, whereupon Murry, with engaging hypocrisy, sued them for breach of contract. Litigation continued until 1787 when the case finally went against him.[43]

In the 1780s and early 1790s manufacture quickly became established in other parts of France. An obvious stimulus to local production was the difficulty and cost of transport. Costs can be illustrated from a list drawn up by Holker in 1779, giving the cost per bottle of gross weight 120–50 pounds, containing 95–120 pounds of acid.[44] Transport by sea to Nantes cost 7–8*l.* and to Marseilles 10–11*l.* Overland transport of course was dearer: 13–14*l.* to

Nantes, 14–16*l.* to Lyons, 20–2*l.* to Geneva. Since the price at the factory of this acid in the 1780s would be about 50–60*l.*, transport could clearly add fairly significantly to the basic price. Also significant was the risk of breakage *en route*, and indeed land carriers were reluctant to take the acid at all because of its dangerous nature.

Works now appeared in the south of France, most notably in the vicinity of Montpellier. A first factory was begun there in 1780, by Pitou, Caizac, & Cie, a group including Montpellier merchants and also associates in Paris.[45] It seems to have been at a farmhouse near the country residence of the local worthy Bonnier d'Alco ('au Mas de Vaneau, à côté d'Alco'), presumably somewhere in the region of the present rue d'Alco, a couple of miles west of the town. The works was to produce not only sulphuric acid but also other mineral acids and salts for use in pharmacy, in the mints, in foundries, and above all in dyeing. More important, though, was the next works to be established at Montpellier, that set up by the chemist Chaptal towards 1785. This was to become one of the best known in the country and it will receive separate discussion shortly. Close on Chaptal's heels, Rey and Cros launched a third Montpellier concern, which seems to have been of no particular distinction, though they evidently had considerable resources, also exploiting coal mines and a glass-works at le Bousquet-d'Orb.[46] The venture of Pitou, Caizac, & Cie was probably short-lived, for it does not figure in a local industrial survey of July 1788.[47] A further sulphuric acid producer there encountered, however, is a Sr Bertieu, also described as a manufacturer of cream of tartar.

At Nîmes, a works was established in 1782 by Lapenne to produce sulphuric and nitric acids and mineral salts.[48] Lapenne described his works as being 'mounted on a large scale, like those of England and Holland', and he was sufficiently ambitious to petition (unsuccessfully) for the title *manufacture royale*. In August 1782 he had a single lead chamber, with another larger one under construction, and there was also a sizeable building housing galley and other furnaces. At Agen, in the south-west, a works was formed in 1786 by a certain Guillebaut.[49] At Marseilles oil of vitriol was said as early as 1773 to be being made by Johannot and Malvesine, but in view of the date this was probably only a glass-globe venture and no more is known of it.[50] More significant was the lead-chamber works created there in 1787 by Janvier & Cie, who also set up an important factory at Lyons soon afterwards (see below). A second works to appear at Lyons in the early 1790s was formed by the Javel company of Paris.

Manufacture also began in further towns in the north-west. A works was established at Lille in 1783 by an Italian called Valentino, who had come to the town four years earlier as apothecary at the military hospital.[51] Though without money himself, Valentino succeeded in obtaining finance from a company of backers. He was evidently a man of enterprise and some chemical ability: in 1785 he began to give lectures on chemistry, covering the applications to dyeing and pharmacy, and in the same year he was one of the founders of the *Collège des Philalèthes*, a society which concerned itself with the improvement of manufactures and the useful arts. By 1794 his works had three lead chambers, 18–20 feet wide and totalling 120 feet in length, with a production capacity of up to 70 000 pounds of concentrated acid a year;[52] nitric acid was also made. Another town to acquire a works was Saint-Quentin, where a plant was begun in 1793 (or 1796) by Dupuis, head of one of the town's most notable and oldest established bleaching works.[53] Manufacture began, too, at Nantes in Brittany. A venture there mentioned by Demachy in 1773 probably only used glass globes and is not known of further, but a lead-chamber works was established some time between 1786 and 1794 by Athénas.[54] Like Valentino, Athénas had the background of an apothecary, and he manifested wide-ranging interests in science, agriculture, and industry. He had moved to Nantes from Paris in a pioneering attempt to establish soda manufacture there in the early 1780s, but this had failed by 1786 (see Ch. 4). He had next set up a dye-works in Nantes, and after this he had created a mobile distillery to tour the local vine-growing regions by river boat, a venture which collapsed after a series of poor harvests. It was then that he formed his sulphuric acid plant, financed by a group of backers ('à l'aide d'actionnaires'). Athénas became quite a prominent local figure: director of the mint in Nantes 1795–1817, secretary of the chamber of commerce 1803–29, and a founder and leading member of the *Institut départemental*, the local scientific and literary society, from 1797.

Finally, we may note the appearance of two further works in Normandy. At Rouen, in 1792, Jacques Chatel left the direction of Holker's concern and set up in partnership with Anthony Garvey and A. D. Bunel, raising a works on the river at Lescure, in the commune of Amfreville-la-mi-Voie, a league and a half (about 4 miles) outside the town.[55] This was much larger than the various enterprises launched at Rouen in the 1770s, and had a capacity approaching 300 000 pounds of concentrated acid a year. It was said indeed to be virtually identical to Holker's concern: in 1794 it possessed two lead chambers each 50 feet long by 20 wide, and another four half this size, and there were also four workshops (containing 68 furnaces, each with a retort) for concentrating the acid. The other new Normandy works was at Honfleur and was the creation of an English immigrant, Édouard Chamberlain. In the 1780s Chamberlain had been prominent in developing copperas manufacture in Picardy, and he had then moved to Honfleur, where in 1791 he set up a works at la Rivière to produce copperas from the local pyrites. It was as an adjunct to this that he erected à lead-chamber plant at the beginning of 1793. In 1794 he claimed to have two chambers, each 50 feet long, 20 feet wide, and 20 feet high (two years later just a single chamber, of these same dimensions).[56]

Of the various enterprises which we have now mentioned, three were of outstanding importance and merit fuller discussion: those of the Javel company near Paris, Chaptal at Montpellier, and Janvier & Cie at Lyons and Marseilles.

The works at Javel

The leading figure in the establishment of the Javel works was a certain Léonard Alban. Alban's background is regrettably obscure,[57] but he was a man known for his competence in chemical matters to the Paris inspector of manufactures Watier de la Conté, who encouraged him in his plan to attempt sulphuric acid manufacture and may indeed have first suggested it.[58] For financial backing Alban approached Jean Baptiste Peeters, described as a *bourgeois* of Paris, and after satisfactory trials Peeters joined him in partnership on 10 August 1776.[59] Peeters promised capital of 25 000*l.* and was to be in charge of business affairs, while Alban contributed simply his skills and was to be technical director. A works for the manufacture of sulphuric acid and aqua fortis was then installed in the village of Épinay, near Paris, but did not long remain there. Complaints from local residents about acid fumes led to its removal in the first months of 1778 to an isolated spot on the banks of the Seine, known as 'le moulin de Javelle' after the old windmill which stood there.[60] The factory was to remain on this site for over a century.

The works established at Javel was far more considerable than the modest partnership agreement on which it was based would suggest. Additional capital was provided by Jean Baptiste Buffault, Papillon de la Ferté, and Bourboulon de Boneuil, a wealthy trio who had had some interest in the venture from the start, for they were joint owners of the property at Épinay which Peeters and Alban had rented for their first establishment. It was they, in fact, rather than Peeters, who were to be the chief capitalists behind the Javel enterprise. The total capital involved we do not know, but it was of the order of hundreds of thousands of *livres* (Bourboulon and de la Ferté were said by the astronomer Lalande to have provided 100 000 *écus*, i.e. 300 000*l.*).[61] Buffault was Peeters's brother-in-law, and was a financier and tax official with the position of *Receveur général des domaines, dons et octrois* in the town of Paris. He had formerly been a merchant in Paris, dealing in silken fabrics and other luxury goods, and though he had now withdrawn from active participation he still owned that business and had 400 000*l.* invested in it.[62] De la Ferté (who claimed kinship with Molière) held the important Court office of *Intendant des menus-plaisirs du roi*, responsible for the organization of royal ceremonies, festivities, and entertainments. He was also a man with scientific and technical interests: in 1772 he contributed funds towards Lavoisier's burning-glass experiments; in 1778 he was one of the business directors of the *Compagnie des eaux*, formed by Périer to erect the first Watt-type steam engine in France for the water supply of Paris; and in the 1780s he wrote books on architecture, geography, mathematics, and astronomy.[63] Bourboulon held the reversion of a second *intendance des menus-plaisirs*, and was treasurer-general to the Countess of Artois. His position in the household of the Count of Artois (younger brother of the King) was no doubt of assistance in securing for the Javel firm the Count's protection, granted in 1780 by a *brevet* of 7 September. This was in recognition of their achievements and by way of encouragement for the future, and it permitted them to style their works 'Manufacture de Monseigneur Comte d'Artois, pour les acides & sels minéraux'.[64] There is no indication that Artois made any financial contribution, but even a nominal association with royalty was to be valued for the status it brought.

The company's primary product was sulphuric acid, which they, like Holker, supplied to a country-wide market. Towards 1780 the acid was being made in two lead chambers, fitted with the curious luxury of glass windows to light the interior.[65] Another early product was aqua fortis, whose manufacture they brought to such a point that it won preference for use in assaying in the nation's mints over that prepared in the refineries of the Paris Mint itself, which had hitherto exclusively been used. In this they were encouraged by the chemist Sage, who held a chair in mineralogy with accommodation at the Mint. Sage, in fact, had close relations with the Javel works in its early days, one of the company speaking of 'M. Sage . . . to whose knowledge and friendship we owe so much'.[66] The company also applied their sulphuric acid to the production of a range of other heavy chemicals,[67] and this is one of the most notable features of the concern. They were the first manufacturers to make serious efforts to develop an integrated series of products based on sulphuric acid. They began by producing green and blue vitriols, presumably by simply dissolving iron and copper in sulphuric acid. They then turned their attention to the less straightforward task of making alum, and succeeded in manufacturing a synthetic product by digesting clay in sulphuric acid; this was to become a major part of their activities. Hydrochloric acid was made from about 1780, presumably from sulphuric acid and salt, and when in 1786–7 Berthollet made known his discovery of a new means of bleaching by using chlorine, the Javel manufacturers were quick to seize the opportunity of increasing their sales by marketing hydrochloric acid complete with directions for its use in the preparation of this new bleaching agent. Moreover, recognizing the inconvenience for bleachers of having to make chlorine themselves, the company also developed and marketed a bleaching liquor for direct use under the name *eau de Javel* (see Ch. 3). The production of hydrochloric acid yielded sulphate of soda as a by-product, only some of which could be sold as Glauber's salt. The company therefore gave thought to its conversion into soda, and subsequently carried on this manufacture to some extent. They also developed processes for the manufacture of a so-called 'white lead' and 'verdigris' (probably in fact chlorides of lead and copper). Finally, we can perhaps see as a by-product of their manufactures the interest they developed in ballooning in the mid-1780s: they were said to be reckoning on filling their balloons for nothing, and so were presumably using hydrogen released in the preparation of mineral salts. Capitalizing on the public enthusiasm for this new pursuit following the pioneer ascents in 1783 of the Montgolfier brothers and of Charles, the Javel firm in 1785 announced a public course in balloon handling, for 400 subscribers, to be held at the works during the summer.[68] A little earlier, in March 1784, Alban's associate director at the works, Vallet, had supervised the filling of the balloon for the first ascent of J. P. Blanchard,[69] and Vallet's subsequent experiments at Javel win him a modest place among the footnotes of aeronautical history as one of the first pioneers of the airscrew for propulsion.

The Javel firm acquired an especial celebrity and its example was an inspiration to several of those beginning chemical ventures elsewhere in the 1780s. The firm also spawned two further works in a more direct fashion. In the first place, two of the Javel associates—Bourboulon and Vallet[70]—in 1787 crossed the Channel and set up a works near Liverpool. In the event this was to be short-lived, but the two did contribute considerably to the development of chlorine bleaching in England. The second offshoot of Javel was longer lasting and was at Lyons, where a subsidiary works was formed towards 1793, in the district of la Guillotière. This was solely or largely for the manufacture of sulphuric acid and was much less important than the parent works.

Chaptal at Montpellier

If Alban remains a somewhat shadowy figure, no such obscurity enveils Chaptal, who is remembered as a prominent member of that outstanding generation of French chemists which grew up at the time of Lavoisier.[71] He is noteworthy as one of Lavoisier's earliest supporters, and author of a highly popular textbook of general chemistry first published in 1790. He was not, however, a pure chemist of the first rank and it is on his contributions to applied chemistry that his claim to scientific fame chiefly rests. His extensive publications in this field mark him out from all other early manufacturers, including the Javel directors, as being not only a manufacturer but also an industrial scientist, and as such a figure more characteristic of the nineteenth than of the eighteenth century. His most important publication was an applied chemistry text in four volumes—*Chimie appliquée aux arts* (1807)— which is a work of strikingly more modern appearance than its only notable predecessor, that by Demachy thirty-four years earlier, clearly reflecting both the industrial advances and the revolution in chemical science of the intervening years.

Chaptal's education had been provided for by a rich uncle. Since the uncle was a leading physician in Montpellier, with ambitions for his nephew in the same profession, the boy had naturally enough studied at that town's famous

Fig. 2.3. Ballooning at Javel, 1785. The enclosure and principal buildings depicted are presumably those of the Javel chemical works. (From Alban and Vallet (1785). The original pencil drawing is in BN Estampes, Ib, tome 2, f° 90.) (Bibliothèque Nationale)

medical faculty. Young Chaptal, however, was less certain than his uncle of the attractions of a medical career, and after qualifying as a doctor of medicine at the end of 1776, he slipped away to Paris to widen his horizons. It was there that he began to develop a particular interest in chemistry, studying the subject under Bucquet, Mitouard, and Sage, and forming what was to be a lifelong friendship with Berthollet. He was soon presented with an admirable opportunity to exploit his new interest, when in 1780 the president of the *États de Languedoc* invited him to return to Montpellier in expectation of a chair.

The *États de Languedoc* were one of France's few remaining provincial assemblies, enjoying some measure of autonomy in economic affairs and holding a notable record for their promotion of the province's economic development. The plan to create a chair in chemistry was a manifestation of this, inspired partly at least by a belief in the practical utility which might be expected of public lectures in the subject. This was a belief which Chaptal fully shared, and his inaugural lecture of December 1780 was filled with sentiments which delighted his illustrious audience (among them the assembled *États*), and which Chaptal was to reiterate many times in his later writings. He deplored the gulf which existed between the academic man of science and the layman, spoke of the need for the useful knowledge of the sciences to be rendered accessible to the people, and stressed the importance of an understanding of chemistry in particular for the development of many of the province's industrial and agricultural productions. Chaptal was a gifted lecturer and the ensuing course proved highly successful: he continued to teach it until interrupted by the events of the Revolution in 1792. Attendance reached some three or four hundred and pressure of numbers required its removal from the rooms of Montpellier's *Société royale des sciences* to a special 'Hôtel des cours de physique et de chimie', which Chaptal shared with a less distinguished course in physics. His success so pleased the *États* that in their meeting of 1787 they created for him a second chair in Toulouse, where the course attracted even larger audiences than in Montpellier.

Chaptal's early efforts were by no means confined to the lecture room and the laboratory. He made practical studies of the region's products and resources, and with his appointment in June 1784 as an honorary inspector of mines he undertook official visits to mines and workshops, noting their state and advising on improvements. He served as adviser on such matters to the *États de Languedoc*, and later, in his autobiographical sketch, he claimed to have been a virtual minister of trades and agriculture to the province. The experience must have served him well when he became Napoleon's minister of the interior after the Revolution.

Chaptal's concern for the economic prosperity of his home province was one factor in his decision to establish there a chemical works, his aim being to supply local industry with the materials for which it was then largely dependent on Britain, Holland, and the new factories in the north of France. While in Paris he had no doubt heard with interest of the venture then arising at Javel, and he made admiring references to this works in 1781. With his marriage in that same year he gained the resources he needed to form a similar enterprise of his own, in the form of his wife's dowry of 70 000*l*., and a wedding present of 120 000*l*. from his uncle; at the same time he acquired extensive family connections in the trade and industry of southern France.[72] In November 1782 Chaptal bought a site at la Paille, to the south-west of Montpellier, where construction and installation of the works was to take several years, indicative of the care bestowed on its planning and on preliminary research. In September 1784 he took on Étienne Bérard,[73] his twenty-year-old laboratory assistant and former student, to direct the works; and regular production evidently began the following spring, for that is when the first sales are recorded.[74] The benevolent regard of the regional administration for the venture was demonstrated in April, when the *Intendant de Languedoc* sent out for official distribution copies of a printed prospectus detailing the factory's products, and asked his subordinates to recommend the new works to apothecaries, drysalters, dyers, and other buyers of chemicals.[75] Sales remained small in 1785, but in 1786 the works began to acquire a firmer appearance, production increasing threefold in the course of that year.[76] Further expansion was made possible when Chaptal's uncle died in November 1787, leaving him a legacy of 300 000*l*., and by 1788 business was such as to justify Chaptal in taking on a further associate, Étienne Martin, to supervise the commercial side of the concern.

At la Paille Chaptal developed a range of manufactures wider than that at Javel. The prospectus of April 1785 listed some twenty products, though these were largely pharmaceutical and therefore made in only a fairly small way. It was a works distinguished in its early years more for the range of its activities and the intelligence of its direction than for sheer magnitude of output. The main early product was aqua fortis, probably made in the traditional way, with sulphuric acid an important second: sales in 1788 were perhaps of the order of 40 000–50 000 pounds of each.[77] The sulphuric acid is said to have been made in just a single chamber until the end of 1788, when a second was added; by 1791 there seem to have been at least three, and by 1799 there were four.[78] Other heavy chemicals whose manufacture was successively undertaken during the factory's early years included hydrochloric acid, alum, copperas and other sulphates, sal ammoniac, and white lead. There was also a pottery and a foundry, producing laboratory and workshop equipment both for use in the factory itself and for outside sale.

By the end of the 1780s Chaptal's diverse activities had earned him a remarkable reputation. In 1788, at the age of 31, he was awarded letters of nobility and the decoration of the order of Saint-Michel by the King, in recognition of his services. This was at the instigation of the *États de Languedoc*, who the following year themselves voted him a sum of 50 000*l*. in token of their gratitude (never actually paid because of the Revolution). His reputation, moreover, was not confined to France. The Spanish Government in 1788 offered him a handsome salary, along with loans and other privileges, if he would transfer to Spain the works he had established so successfully at la Paille; and in the 1790s he received further invitations from the Queen of Naples and from George Washington, then President of the United States.

Janvier & Cie at Lyons and Marseilles

The most impressive of the early French chemical concerns was that of Janvier & Cie, with works at Lyons and Marseilles. This was also, however, one of the shortest lived, for it fell victim to Revolutionary violence in 1793, and it has consequently been almost completely overlooked by historians.

Of the firm's two works, that at Marseilles was the first to be established.[79] Janvier & Cie sought Government authorization for this towards the end of 1786, announcing that a major works was intended, comparable to those at Rouen, Javel, and Montpellier. Construction began the following year on a site outside the town's porte de Saint-Victor. By the end of 1787 the plant included a large lead chamber, said to have cost over 40 000l., and the manufacture of sulphuric acid had been satisfactorily mastered; it was intended to extend manufacture to include nitric and hydrochloric acids and alum. By the end of 1788 the establishment of the works had absorbed some 200 000l. There was apparently a significant local market for the products among dyers and others, and a particular attraction of Marseilles as a place of manufacture was the ready availability of raw materials, especially Sicilian sulphur. The company could also count on deriving some benefit from Marseilles' special status as a 'free port'. This was a status intended to facilitate the port's trade, and it meant that foreign goods were permitted to enter Marseilles without payment of entry duty, the duty becoming payable only if and when they passed on into the interior of France. Janvier & Cie could thus obtain their sulphur duty-free, and this would be of clear advantage for such acid as they sold in the town or exported. As for acid which they sent into the interior, they probably expected this to attract a moderate duty, designed to put it on an equal footing with that made in other French works, this being the way in which Marseilles industry was generally treated. In the event, however, they encountered an unforeseen obstacle in this respect, for despite their protestations (supported by the chamber of commerce), the customs authorities insisted on charging the full import duty on their acid, as on that of foreign manufacture, and this the firm described as prohibitive. They were still complaining in June 1789 and it is not known whether they ever obtained any concession. It may be that this difficulty was one factor in their development of a second works, this time in the French interior.

It was this second works, just outside Lyons, in the industrial quarter of la Guillotière, which was to become the company's main factory. Exactly when operations at Lyons began is not clear. The existence of a plant there was mentioned in July 1790,[80] but the creation of a major works probably came only after 1791, following property purchases that year at la Guillotière to the value of over 100 000*l*. The largest of these was the purchase on 18 August, for 72 000*l*., of extensive monastic buildings and gardens formerly belonging to the Picputian order.[81] It was there, presumably, near the parish church (formerly the church of the Picpus), that the works was installed.

By 1793 the company had built up at la Guillotière a flourishing and broadly based enterprise, employing (they said) 'a multitude of workers of all kinds'.[82] It supplied chemicals to the region's dyers, cotton printers, paper works, wallpaper manufacturers, and hat and button trades. It also exported. Sulphuric acid was produced in three 'immense' lead chambers, whose output was given as 1200 pounds a day; the capacity of the plant can thus be conservatively estimated at 250 000–360 000 pounds of concentrated acid a year (depending on whether chamber acid or concentrated acid was intended by the daily figure). Other products were nitric and hydrochloric acids, iron and copper sulphates, alum, sal ammoniac, acetic acid (vinegar), lead acetate, and white lead. There was also a lead foundry of which the firm was very proud (producing lead sheet measuring 40 ft. by 6 ft.), a colour workshop (making Prussian blue and other colours), a sulphur refinery, a pharmaceutical workshop (still in course of installation), and a chemical laboratory 'where all the precious discoveries of this science were repeated and perfected'.[83] Besides the workshops already in use it was intended that others should be successively added, plans being already prepared and locations designated. The administrative head of the Rhône Department spoke glowingly of the works as 'this superb manufactory, unique in France and perhaps in Europe, for its unity, its extent, and its perfection in all chemical products'. At the same time the company's Marseilles factory was also continuing to work, producing sulphuric and nitric acids and copper sulphate, while serving too as a repository for imported raw materials destined for Lyons.

The company had a very large capital of 1 200 000*l*., supplied by forty associates, both French and foreign; it had been agreed, moreover, that for the first ten years profits were not to be paid out but were to be used for the further development of the concern. It must have been decidedly the best financed of all the enterprises we have so far mentioned, and indeed it is doubtful whether it was matched by any other chemical firm in France before the rise of the Leblanc soda industry in about 1810. To draw another comparison, it is not until towards 1830 that one meets with concerns of such a size in the silk industry at Lyons.[84] The directors of the enterprise—of whose skill and intelligence the authorities in Lyons spoke highly—were a certain Janvier and Jean François Vincent. Janvier is otherwise unknown, but the Vincent in question was probably the Vincent who in 1779 had been planning to set up a sulphuric acid works in Grenoble, with the backing of two local apothecaries (Delange and Perret), and who described himself then as a teacher of chemistry in Grenoble. That project was apparently not realized and in 1781 Vincent had been reportedly intending to establish a works in Lyons instead.[85]

Protection of the industry

It will be recalled that in establishing his pioneer works, Holker received Government support in the form of a number of special privileges, including for instance the payment of a premium on his acid sales. Such privileges, of course, put him at an advantage over other producers when these began to arise in the later 1770s, and Holker himself seems to have been so imprudent as to draw attention to this fact (if we can believe his opponents): alarmed at the prospect of competition, he was said to have warned that he would put any rivals out of business by capitalizing on his privileges to undercut their prices. Naturally enough the nascent rival manufacturers in Normandy complained to the *Contrôle général des finances* in 1777–8 of Holker's unfair position, and sought to share the most important of the privileges he enjoyed.[86] In 1778 the new Javel manufacturers, too, sought privileges identical to those granted to Holker, including the title 'manufacture royale'; and in 1779 a petition from Vincent, of Grenoble, requested a variety of privileges for the works he was proposing to establish in that town, including exclusive manufacturing rights for the region and again the title of 'manufacture royale'.[87] The Government's response to these various petitions was to conclude that the new industry could now be considered safely implanted in France, and that consequently no further measures of special encouragement were needed. Rather than multiplying privileges, therefore, it was resolved that those which Holker enjoyed should be terminated. As it happened, Holker's premium on sales came to an end anyway in September 1778, with the expiry of its ten-year period; his other privileges were ended in January 1781.[88] One privilege, however, was continued and extended to the industry in general by an *arrêt* of 12 October 1780: this was exemption for all manufacturers from payment of duties on the circulation of their acid inside the kingdom (the country still retaining, until the Revolution, an antiquated system of internal duties).[89] Besides this, of course, the industry continued to enjoy a more important kind of support in the form of the protective tariff against imports: at 22*l*.10*s*. per 100 pounds this was virtually prohibitive in the 1780s.

The protective tariff did not please everyone. It was criticized from the consumer's point of view in 1788 by the dyers and cloth finishers of Picardy, who complained at the higher price charged by French acid-producers than by their counterparts in Britain.[90] Sulphuric acid was indeed rather dearer at first in France than in Britain, though by the early 1790s the gap had largely closed. In Britain, Dossie in 1758 indicated a price of fourpence a pound, and other figures encountered up to the early 1790s all range between threepence-halfpenny and fourpence-halfpenny (c. 38–48l. per 100 pounds French measure), the variation depending partly on the credit given.[91] In France, Holker's acid was said in 1773 to sell at 70–5l. as we have seen. By 1782, however, his price had fallen to 50l., for immediate payment, or 55l. with six months' credit. In 1789 the Javel works advertised at 50l. with six months' credit. And in the early 1790s, according to a source in Honfleur, the price in general was 40–50l.[92]

If the price of the acid was higher in France, manufacturers' costs were probably higher there too. Certainly one cost which drew complaints was that of saltpetre. It was an unfortunate coincidence for the chemical industry that this basic raw material should also have been a material of prime military importance, entering as it did to the extent of 75 per cent into the composition of gunpowder. For this reason saltpetre had long attracted the special interest of the Government, and this was particularly the case after 1775. Responsibility for the supply of saltpetre and gunpowder had previously been farmed out by the State to a private company, but the Seven Years War had highlighted the defects of this arrangement, and in 1775 the system was completely changed. Saltpetre and gunpowder supply now became a State industry, administered by a *Régie des poudres*, and in the interests of national security the new *Régie* made determined efforts to end the country's heavy dependence on saltpetre from India, by expanding its production at home. The programme was very successful. Unfortunately, though, it remained distinctly more expensive to make saltpetre in France than to import it, and so to protect home production the Government found it necessary rigidly to prohibit private importation, and enforce the *Régie*'s monopolistic control over the supply of saltpetre to all users. Thus, whereas in 1768 Holker had been permitted to import saltpetre (duty-free) as one of his privileges, after 1775 the Government was no longer prepared to countenance private importation by others. Chemical manufacturers were obliged to buy from the *Régie*, at prices fixed by the Government.

This restriction brought particular complaint from Chaptal and from the Javel manufacturers. After an unsuccessful petition in 1787, Chaptal did win an apparent concession at the beginning of 1789, when the King promised that steps would be taken 'to assure him of the supply either of national saltpetre, or of foreign saltpetre, on the most advantageous terms possible', but the promise was almost certainly without effect. It mystified the authorities in Paris, for it was emphatically not the reply which they had intended should be sent, and they decided therefore simply to take no action.[93] The Javel manufacturers petitioned for freedom to import from the very creation of their works in 1778, but obtained not even an illusory success. They then took up the issue with renewed hope in the first years of the Revolution, when France's tariff system was being fundamentally reformed.[94] Their petitions now claimed that the high price of saltpetre had always been the greatest single hindrance on their works: it was restricting the development of new branches of manufacture, such as alum and soda, by inflating the cost of the necessary sulphuric acid, and it put French manufacturers at a disadvantage compared with other countries. Their demands won a somewhat unsatisfactory first concession on 28 September 1791, with a provision that the *Régie des poudres* would sell crude saltpetre to chemical manufacturers at cost price. Their real demand was finally met on 31 May 1792, when the import ban was lifted, allowing saltpetre to be brought in on payment of a modest duty.[95] The measure was of little benefit, however. Already in 1792 the outbreak of war was making imported saltpetre difficult to obtain and increasingly expensive, and by 1794 foreign trade was totally dislocated. The situation can scarcely have begun to recover before the prohibition of saltpetre imports was reimposed on 30 August 1797.[96]

It can be seen from Fig. 2.4 that between 1784 and 1792 saltpetre was indeed rather dearer in France than in Britain. Assuming its price in Britain to have been on average about 70 per cent of that in France, it can be estimated that the extra cost incurred by French manufacturers would amount to about 5–10 per cent of their acid's selling price.

The state of the industry in the early 1790s

We have now traced the growth of the industry into the early years of the Revolution and it will be useful at this stage to pause briefly and take stock of the progress it had made. The year 1793 provides a convenient vantage point, since the industry had then reached a peak in its development, prior to the disruptive impact of the Revolutionary War. The number of lead-chamber works had grown steadily since the late 1770s: the establishment of twenty has been described above, and though the subsequent history of some of these is unknown we can be fairly certain that at least fifteen existed in 1793. The output of the works ranged from perhaps 50 000 to 300 000 pounds of concentrated acid a year (or its equivalent in chamber acid). Most or all of the works made at least one or two subsidiary products as well, while the largest firms were extensive integrated concerns producing a considerable number. The capital sums involved in these early ventures ranged from tens of thousands of *livres* (probably) for the smaller works, to rather over a million for the largest firm, that of Janvier & Cie. They therefore covered the greater part of the spectrum formed by industrial enterprises in general at this time: towards the end of the *ancien régime*, we are told, these varied from small concerns with less than 75 000*l.*, common in textiles, paper, and glass, to large firms with over 300 000*l.* (and very large with over a million) found particularly in mining and metallurgy.[97] In one exceptional instance, that of Chaptal, the new chemical venture owed its capital to the private fortune of the chemist himself. In a number of other cases a works arose when an existing textiles concern decided to add sulphuric acid manufacture to its activities, the cotton printers at Rouen, and the bleacher Dupuis at Saint-

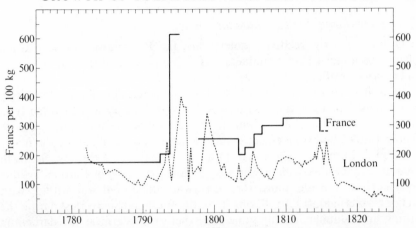

Fig. 2.4. Saltpetre prices in France and England, 1775–1825. The French price shown up to 1793 is that charged by the State for the grade known as *deuxième cuite*, which was of about 90 per cent purity and which seems to have been the grade generally used in sulphuric acid manufacture. The State also sold a fully refined saltpetre, and until 1795 a crude (70 per cent) product. In parts of France the price prior to the Revolution was sometimes up to about 20 per cent higher than the Paris price shown here, owing to regional differences in the pound weight (Jourdan *et al.*, vol. 26, 163). From 1797 only two grades were sold: industrial grade (>85 per cent), which is that plotted here, and fully refined. After 1815 acid manufacturers generally imported their saltpetre instead of buying from the State (see p. 86). For England we have plotted the London price—including duty—of rough Indian saltpetre, which seems to have been of about 90 per cent purity (*Arch. parl.*, 2nd series, vol. 23, 54). We have assumed £1 = 24 fr. and have taken no account of fluctuating exchange rates during the Revolutionary and Napoleonic Wars; the graph should therefore be regarded as offering a rough comparison only. French prices are from: Jourdan *et al.*, vol. 23, 180–5; *Arch. parl.*, 1st series, vol. 44, 392–3, and vol. 74, 590; *Moniteur*, 20 germinal an III (9 April 1795); Conseil d'état (1809) [giving prices 1797–1809]; AN, AF^{IV}395, and AF^{IV}523.
London prices are from: Tooke and Newmarch (1838–57), vol. 2, 409.

Quentin being examples of this. Perhaps most commonly works were established by specially formed partnerships, often when a man with technical expertise and ambitions to manufacture secured backing from a group of capitalists: Alban, Athénas, Valentino, and probably Holker, Vincent, and Chatel are such figures.

By the early 1780s France's new industry could produce more than enough sulphuric acid to meet her needs, trade figures showing an almost constant small surplus of exports over imports after 1781.[98] The country's total production capacity by 1793 was probably of the order of $1\frac{1}{2}$–2 million pounds a year, reckoned as concentrated acid. Her consumption had obviously risen considerably since 1768, when Holker's highest estimate had put it at some 130 000 pounds a year.

Disruption during the Revolutionary Wars

During its early, relatively quiet years, the Revolution had no marked effect on sulphuric acid manufacture. The industry could not fail, however, to be seriously affected by the war crisis of 1793–4.

In the first half of 1793 the position of the young Republic grew increasingly precarious. The execution of Louis XVI on 21 January was followed by a dangerous escalation of the war, as Austria, Prussia, and Sardinia, with whom France had been engaged since mid-1792, were joined by the major remaining European powers: Britain and Holland in February, Spain in March. France was soon at war with almost the whole of Europe. To this menace from outside, moreover, there was added civil war within, with major revolts against Paris flaring up in the West and South-East during the spring and summer. At the same time there was alarm at the darkening economic situation, as shortages pushed up prices and the value of the paper currency shrank. In response to the desperate situation an emergency war-government came into being. The rule of the country effectively passed into the hands of one of the committees of the elected Convention, the great Committee of Public Safety, which took definitive shape in the summer months of 1793 and proceeded to rule with growing centralized control and dictatorial powers for the following year, its policies imposed by that strategy of coercion and intimidation known as 'the Terror'. Despite its notorious image in the popular imagination ever since, the stern rule of the Committee of Public Safety during this critical year was undoubtedly the salvation of the Republic.

Perhaps the most remarkable aspect of the Committee's achievement was its mobilization of the economy in support of the war effort. In the course of 1793 mass conscription raised the strength of France's armies from 200 000 to some 750 000 men, the largest armed force ever yet raised and one posing unprecedented problems for its supply. The imperious need to feed, clothe, equip, and arm this vast multitude forced the Revolutionary government to swallow its libertarian economic principles and impose on the economy a degree of centralized control far surpassing anything known under the *ancien régime*. The country's entire resources became subject to requisition. The celebrated *levée en masse* of 23 August 1793, besides conscripting into the armies all able-bodied and unmarried men between the ages of eighteen and twenty-five, declared the whole population to be liable for war work of one kind or another. Along with this there came the requisitioning of material and industrial resources. Goods and equipment of all kinds were compulsorily purchased (or just taken) as needed. Mines, iron and textiles works, tanneries, and shoemakers shops, were required to work on State orders for the supply of military requirements, while to meet the particularly pressing need for arms and munitions, many new works and workshops were set up at the Committee's behest, turning out sabres, bayonets, muskets, artillery, and gunpowder in quantities which astounded Europe. Besides this military programme the Committee also made some endeavour to facilitate and control the provisioning of the civilian population, but its efforts in this sphere were much less forceful. Its direction of the economy was essentially for the benefit of the nation's embattled forces.

The Committee's urgent development of gunpowder production[99] directly affected the chemical industry since both gunpowder and chemical manufacture required sulphur and saltpetre as fundamental raw materials, and supplies, moreover, were problematical. Sulphur was a material for which France was entirely dependent on imports, chiefly from the Italian countries, and these imports were now obstructed by the war. As for saltpetre, although this was made in France under the direction of the *Régie des poudres*, the output which had satisfied the country's normal needs was far from sufficient in the present war crisis. Between mid-1793 and mid-1794 the Committee of Public Safety therefore introduced a succession of emergency measures regarding these materials, that aspect which concerns us here being the reservation of available supplies almost exclusively for gunpowder production. A decree of 21 September 1793 declared that all the sulphur in France was to be at the disposal of the *Conseil exécutif provisoire*, and in January and February 1794 the requisitioning of large stocks was ordered from merchants in Marseilles, Bordeaux and Nantes.[100] The same decree provided that, munitions apart, the *Régie des poudres* was to supply saltpetre only to the mints and to military hospitals and pharmacies. Since the *Régie* had monopolistic control over saltpetre production, French saltpetre was consequently denied to chemical manufacturers.

With the interruption of foreign trade, and the commandeering of existing supplies for war purposes, chemical manufacturers found it difficult or impossible in 1794 to obtain sulphur and saltpetre, particularly the latter. The Committee of Public Safety, in an *arrêté* of 12 May 1794, reiterated that saltpetre could not be supplied for chemical manufacture, though it did permit manufacturers to make whatever use they could of the rejected mother liquors from saltpetre production.[101] In September 1794 the Committee's *Commission des armes* reported that manufacturers of aqua fortis and sulphuric acid had long been demanding saltpetre to allow them to continue their work;[102] their demands were to no avail, of course, save in certain exceptional cases where limited quantities were released to allow production for Government use. In November 1794 the sulphuric acid works in Paris, Rouen, and Honfleur were said by a frustrated customer (Carny) to have long been reduced to a state of almost total inaction for want of saltpetre.[103] By the end of the year, of the four Rouen works the two small establishments at Déville were both completely idle; Jean Holker's factory was working at only a very modest level, while Chatel's works was similarly operating at reduced capacity and was making no concentrated acid, only chamber acid, through lack of coal.[104] The position was the same elsewhere. In Lille work had almost totally ceased for want of saltpetre.[105] In Nantes Athénas abandoned manufacture as a result of the lack of materials and the revolutionary disturbances.[106]

Shortage of raw materials was the basic factor tending to disable the industry, but for some works this difficulty was compounded with others arising from particular local and personal circumstances. In Rouen, for example, Jean Holker had to leave his works in the hands of a subordinate when he found himself conscripted into the navy (he was authorized to return in January 1795, after petitioning the Government).[107] In Montpellier, Chaptal shut down his factory during his own enforced absence, and the concern suffered badly.[108] Chaptal had taken an active part in the southern insurrection against Paris in mid-1793, and with the quashing of that rebellion had been obliged to go into hiding in the Cévennes mountains. At the end of the year, the Committee of Public Safety, overlooking his transgression on account of his potential utility, called upon him to assist in its emergency development of saltpetre production, and put him in charge of this programme in the Midi; the following April he was summoned to Paris to take part in the administration of the programme at national level. Chaptal had probably closed his works by the end of 1793, and it remained closed until 1795. Materials and workers were diverted to saltpetre production, as were his two associates Bérard and Martin. When Chaptal eventually returned to the works in March 1795, he described it as a cadaver, its workshops dilapidated and drained of materials by requisitioning. He wrote that the episode almost ruined him, and estimated at 500 000 fr. his losses resulting from the closure, and from the concurrent monetary crisis.

The most serious casualty, however, was the Lyons-based firm of Janvier & Cie, which fell a direct victim to revolutionary violence.[109] Lyons was one of the chief rebel centres in the insurrections of 1793, and when the town finally succumbed to republican forces in the autumn fierce repression ensued, including some of the most notorious mass executions of the Revolution. One of the firm's directors, Vincent, was now executed as a 'rebel', and Janvier himself was lucky to escape a similar fate; others of the company were also said to have perished, and the rest dispersed. It seems that the firm's misfortunes stemmed in part from their foresight in having stocked up with raw materials before the effects of the war began to bite; this laid them open to a charge of hoarding, which a law of 26 July 1793[110] made a capital offence. The works was plundered and stripped of materials for the supply of the armies: Dubois-Crancé, the *représentant en mission* directing the siege of Lyons, wrote to the Convention of the importance of the capture in this respect. Janvier and Caminet later claimed that over 200 *milliers* (200 000 pounds) of copper, 400 of lead, 500 of sulphur, 30 of saltpetre, 300 of coal, 500 *ânées* (c. 46 500 litres) of vinegar, an enormous quantity of manufactured goods, boilers, tools, horses, carts, fodder, provisions of all kinds, furnishings, and a library of nearly 4000 volumes, were all taken, some with, some without, requisition orders. Their buildings both at Lyons and at Marseilles were confiscated by the local authorities.

By the end of 1794 sulphuric acid manufacture in France was almost at a standstill. Acid needed by French industry had to be imported, and to facilitate imports the entry duty was cut on 31 January 1795 to one-tenth its former level, in line with a general lowering of duties on necessary goods. As a result of scarcity the price of the acid rose well above its pre-war level of 8–10 *sous* a pound; it is said to have reached a peak of 40 *sous*,[111] though one should beware of attaching any precise significance to this figure in view of the collapse of the paper currency at this time.

The crisis facing the Republic had passed its peak, however, by the late summer of 1794. On the military front, France's victory at Fleurus in June marked the beginning of a turn in her fortunes, and prospects of peace were soon coming into view. In 1795 the First Coalition against France largely collapsed, when peace was signed with Tuscany, Prussia, Holland, and Spain in the spring and summer; the only major powers with whom France then remained at war were Austria (who made peace in October 1797) and Britain. At home, the easing of the military tension marked by Fleurus was accompanied by mounting political dissension, both within and without the Committee of Public Safety, culminating in the fall of Robespierre and his party at the end of July. Reaction against the Terror and the dictatorship of the Committee immediately followed. The Committee was reorganized and deprived of its supreme powers, and the end of 1794 saw the beginnings of a return to a more liberal economic régime with the dismantling of the major economic controls instituted during the war crisis. Attention was turning to the restoration of France's trading relations and the resurrection of her industry, though it must be mentioned that the lifting of Government controls brought major new problems to hinder this, in the form of severe inflation and the precipitous collapse of the paper *assignats*.

For the chemical industry, as for other branches of the economy, the attempt at recovery now began. Manufacturers who had been taken away from their works by other duties during the height of the war were able to return in the early months of 1795: the examples of Holker and Chaptal have been mentioned above. The supply of raw materials continued to be difficult, however, and recovery of the industry was to take several years. With the State gunpowder administration still not selling saltpetre, the Javel manufacturers secured authorization in December 1794 to import from neutral Switzerland, though they complained in advance of the difficulties and expense this would involve.[112] After the general freeing of trade from the grip of Government control in the new year, importation presumably became the recourse of other manufacturers too. The situation must have greatly improved a year later with the resumption of supplies by the *Régie*—now renamed the *Agence des poudres*—the Government having decided in April 1796 to sell up to a million pounds of saltpetre to meet the needs of trade and industry.[113] Even this, though, was by way of an extraordinary measure, and sales were not re-established on a regular basis until August 1797.[114] As for sulphur, there are signs that here, too, supplies were not without difficulty. In October 1796, for instance, Chatel of Rouen informed the Ministry of the Interior that since his regular means of supply had come to an end in 1794 he had been able to continue manufacture only with such odd batches of sulphur as he had procured from individuals, and he asked whether the State could sell him any quantity whatever; happily the State was able to oblige.[115] In Year V of the Revolutionary calendar (September 1796–August 1797), France continued to make quite heavy imports of sulphuric acid, buying acid to the value of 443 000 fr. from Britain, Holland, and Denmark (equivalent perhaps to something like 300 000–350 000 pounds).[116] Still in June 1797, the British were said in Honfleur to be selling acid in considerable quantities at three times its pre-war price.[117]

Some works did not recover from the Revolutionary disruptions, most notably those of Janvier & Cie. In 1795 the surviving members of this company set about trying to obtain compensation for their losses, with the hope initially of re-establishing their factories; but though they fairly soon secured the return of their buildings, the collection of documentary evidence for an official estimate of their other losses was to take several years. The compensation eventually claimed, as evaluated by local experts named by the prefect in Lyons, amounted to 1 627 329*l.* The claim reached the Minister of the Interior—now Chaptal—in August 1802. He promptly decided that it was not his province and passed it on to the Minister of War.[118] We may doubt whether compensation was ever paid. Apart from this Lyons firm, the two small works at Déville near Rouen also failed to reopen, and there may have been others. The majority of works, however, do seem to have survived (certainly nine, and probably eleven, can be affirmed to have continued after the Revolution), and soon further works were being established. By 1800 the industry had probably recovered its lost ground and was again steadily growing.

Before we proceed to trace the growth of the industry in the Napoleonic period, we might at this point consider the widening uses to which the acid was being put.

The uses of sulphuric acid

The most important users of sulphuric acid until the end of the eighteenth century seem to have remained the dyers and cotton printers whose demand had first called the industry into being. Other significant early users, commonly mentioned from the 1770s, were hat-makers, and the acid also found early application in the preparation of leathers: its use by curriers was mentioned in 1777 by Alban and Peeters, it was introduced into tanning in 1794 by Séguin, and by the early nineteenth century it was being used by tawers and morocco-leather makers. From about 1790 a major new outlet began to open up in textiles, following Berthollet's introduction of the technique of bleaching with chlorine. Bleachers now used the acid in preparing their chlorine liquors, while at the same time they were also beginning to adopt it in place of sour milk for the acid baths which they employed as cleansing agents. By the beginning of the nineteenth century, minor applications of the acid included its use in generating carbon dioxide for the newly developing manufacture of aerated mineral waters, and its use in the extraction of phosphorus from bones, for the early types of ignition device developed from the 1780s as forerunners of the modern match. Of greater economic significance, probably, was its application to the purification of vegetable oils. This was suggested by Thenard in 1801, and was developed commercially to supply oil for the new Argand-type lamps that were then coming into use. A further new role for the acid came with Napoleon's creation of a beet sugar industry in France, in the years following 1810: the acid was employed in the extraction process for the purification of the beet juice. Somewhat related to this was its application to the saccharification of starch, following the discovery in 1811 by the St Petersburg pharmacist G. S. C. Kirchhoff that when starch was boiled with sulphuric acid it yielded glucose. The discovery attracted considerable scientific interest and was taken up commercially in France (and elsewhere) for the production of syrups from potatoes, generally for subsequent fermentation to alcohol. The acid was also finding use in the early nineteenth century in metallurgy, for cleaning metal surfaces ('pickling'), prior to plating.

Within the heavy chemical industry itself the acid found great employment in a variety of other manufactures. It seems to have become fairly common in the early nineteenth century for sulphuric acid producers to employ perhaps half or more of their own output in making such further materials as alum, sulphates, and nitric and hydrochloric acids. A factor promoting such internal consumption was a tendency for the production of sulphuric acid to run ahead of the outside market. Chaptal in 1807 remarked on this as common experience, and it had already been the experience of the Javel manufacturers in the 1780s: in 1790 they observed that 'the consumption of vitriolic acid . . . is not very considerable in France, and it is a misfortune for factories of this kind', pointing, however, to the great potential outlets in chemical manufacture itself if the acid could be made cheaply enough.[119] By far the most important application to be made of the acid in this field came with the development of soda manufacture by the Leblanc process. This manufacture, whose growth is traced in Chapter 4, supplied synthetic soda to such industries as soap, glass, and textiles, in place of the vegetable ashes on which they had hitherto relied. It grew up with extraordinary rapidity in France from about 1809, partly as an addition to the miscellaneous manufactures carried on by existing acid works, but to a much greater extent as a separate and very considerable branch of industry in its own right. Being heavy consumers, the new Leblanc soda factories usually made their own acid as a subsidiary activity, and were responsible for a very brisk growth in France's sulphuric acid production. By the end of the Empire the use of the acid in soda manufacture had almost certainly outstripped all its other uses combined.

The outstanding utility of sulphuric acid as an industrial chemical was already clear by the end of the eighteenth century, when Fourcroy observed that 'in the manufactures and the arts, there are few substances which are so useful, or so frequently employed'.[120] By the 1820s it had become a commodity of such basic industrial importance that Dumas, foreshadowing a later more famous remark by Liebig, could take its consumption as a general indicator of the industrial state of a nation:

> If we possessed an exact table of the quantities of sulphuric acid consumed annually in various countries or at various periods, there is no doubt that this table would at the same time present the precise measure of the development of industry in general, for those periods or for those countries.[121]

The growth of manufacture after the Revolution, 1795–1815

Hopes for a general peace after the War of the First Coalition did not materialize. Apart from the brief Peace of Amiens in 1802–3, France was to remain constantly at war until 1815. There was no repetition of the munitions crisis of 1793–4, however, and sulphuric acid manufacture did not suffer any renewed disruption. The state of war did, nevertheless, cause sizeable increases in the cost of sulphur and saltpetre. The price of sulphur was high throughout the war, and rose to a peak of something like seven times its pre-war value between 1808 and 1810, as a result of Napoleon's Continental System. This famous policy, which sought to undermine Britain's economy by closing Continental markets to her commerce, came to affect sulphur supplies when the Milan Decree of 17 December 1807 extended the earlier prohibition on imports from Britain herself to those from countries under British occupation. This included Sicily, France's chief supplier of sulphur, and the consequent prohibition of Sicilian sulphur[122] lasted until 1810, when Napoleon's relaxation of his policy permitted a resumption of imports under special licence. Saltpetre rose less dramatically, but by 1809 the product sold by the State had reached a peak of almost double its pre-war price (Fig. 2.4). Chemical manufacturers did have the alternative of importing their saltpetre if they wished, on payment of a small duty of 6 fr. 12 per 100 kg, for they were granted this right in February 1800.[123] To what extent they availed themselves of the facility is doubtful, however, for in the years 1805–7 (the only years for which we have figures), imports in accordance with the provision were negligibly small; they came from Holland.[124] It seems that with Indian trade firmly in British hands, cheap saltpetre was not to be had in France even by importing. The rise in the cost of raw materials inevitably brought a considerable increase in the price of sulphuric acid (see p. 99), though this was mitigated by improvements in manufacture to be discussed later. Despite these disadvantageous circumstances, the production of sulphuric acid expanded greatly from the late 1790s, with the important extension of its use.

The industry continued to enjoy protection against imports. The entry duty was re-established at 20 fr. 40 per 100 kg in November 1796, equivalent to about a quarter of the acid's market price in the opening years of the new century. Further protection was afforded by the outright prohibition of acid from Britain. Already during the Revolutionary Wars, the French Government was adopting a policy of commercial warfare foreshadowing Napoleon's Continental System. In particular a decree of 31 October 1796 prohibited the importation of British products. At first sulphuric acid was excluded from this general prohibition on the grounds that its importation was then necessary, and a request by the Honfleur manufacturer Chamberlain that it be barred was rejected by the Government in 1798.[125] British acid was officially prohibited by 1802, however, although some was then still illegally entering the Belgian departments by means of false certificates of origin:[126] acid from Britain was evidently sufficiently cheap to make such introduction profitable despite the entry duty. On the whole the protective measures appear to have proved effective, and trade figures for the years after 1806 (we have no earlier figures) show imports to have been very small and inferior to exports.[127]

We shall now proceed to a survey of the state and growth of the industry after the Revolution. In general we shall include here only works in which sulphuric acid was the focal concern, deferring to Chapter 4 those which were essentially alkali works, where acid was made as a subsidiary to soda manufacture. It will be convenient to begin in the east and then proceed in a clockwise direction round the country.

It seems to have been only after the Revolution that sulphuric acid manufacture developed in the east of France. The first indication of production there is found in September 1794, when Koechlin and Marchais were mentioned as manufacturers of oil of vitriol at Mulhouse.[128] At this time Mulhouse was still an independent town allied to the Swiss Confederation, but it was to be incorporated into France in 1798. It was an extremely important textiles printing centre, and it may be that the Koechlin in question was a member of the Koechlin family prominent in that industry, and that the acid was made as a subsidiary to textiles printing. It may also be that manufacture was just being established, for Koechlin and Marchais were authorized to buy a large quantity of lead sheet from France, in exchange for aqua fortis. The works would seem to have been short-lived or commercially unimportant, for a statistical survey published in 1804 made no mention of it. The region was instead said to obtain its sulphuric acid from Javel, from

Fig. 2.5. France: Places mentioned in connection with sulphuric acid manufacture.

'Wenterthen' in Switzerland (no doubt Winterthur, where there had been a works since 1781), and from Nordhausen in Saxony.[129]

By 1798 there was a works at Nancy, established since the Revolution by Charles Léopold Mathieu.[130] This made sulphuric, nitric, and hydrochloric acids, sal ammoniac, and Prussian blue, and had an oven for making pottery and a furnace for the calcination of potash and 'the obtaining of soda by decomposition of the sulphate'. It was not a large works, though, its principal plant in 1798 consisting of a single lead chamber and six galley furnaces. It was perhaps this works which the *Almanach du commerce* listed at Nancy in 1816, under the name of Despaze Dufour.

More important was a works established at Strasbourg in about 1799 by Jean Geoffroy Oppermann and Edme Armand Lefebure. In 1806 their works was making 100 000 kg of oil of vitriol a year, most of which was exported over the Rhine; in 1810 they employed 25 workers. Lefebure was a military pharmacist employed in about 1801 as a chemical assistant at the Strasbourg *École de santé*, and subsequently (1804–24) assistant professor and secretary at the town's *École de pharmacie*.[131] A second, somewhat smaller works was set up at Strasbourg in about 1802–4 by F. B. Meunier & Cie. Meunier, the director, was a teacher at the town's medical faculty; he employed 7 workers in 1810. In 1811 these two works were said to be together producing 315 000 kg of sulphuric acid a year, and also soda (175 000 kg), alum (80 000 kg), nitric acid (166 000 kg) and tin salt (6000 kg), and so they were evidently not insubstantial enterprises for the period.[132]

A more ambitious manufacturer was Philippe Charles Kestner, whose ventures provide a striking illustration of the rapid expansion the chemical industry was undergoing at this time, as well as of the no less rapid reversals of fortune which some firms suffered.[133] Kestner began in 1806 by establishing a chemical works at la Chartreuse, near Strasbourg, which was evidently quite successful and employed 16 workers by 1810. Its early success encouraged him soon after its formation to set up a further works at Thann (where he made sulphuric acid, nitric acid, tartaric acid, and tin salt), and another at Roedelsheim, near Frankfurt, in Germany. Then in 1809–10 there came the great soda boom in Marseilles, and Kestner, with his brother Théodore, hastened to establish a works there for the manufacture of sulphuric acid and perhaps soda; they also conducted a works at Aix. The success of Kestner's ventures was short-lived, however, for they collapsed during the economic crisis of 1810–11. The difficulties of his Strasbourg, Thann, and Frankfurt works had led him to suspend payments at Strasbourg by January 1811, and on 28 January his failure was declared at the *Tribunal de commerce* in Marseilles. The Marseilles works was then sold for the payment of his creditors and its subsequent fate is unknown,[134] as also is that of the plants at Strasbourg, Frankfurt, and Aix. The works at Thann did continue in production, under different ownership but Kestner's management, and Kestner was able to re-acquire it in 1816. In his hands, and later in the hands of his son Charles, it was then to grow into one of France's more important chemical concerns as the century wore on.

Still in the east, there were works by 1810 at Langres and at Couternon.[135] The works at Langres had recently been established in the former seminary there by Guillaume Guérinot, in company with a merchant from Rouen. It was said in 1810 to be making sulphuric acid (employing 12 workers) and copperas (employing 6). That at Couternon, near Dijon, had been formed by Vivent Lazare Chervaux and was said to be prospering, making sulphuric acid and soda with 12 workers. Another venture, begun by Isaac de Rivaz in 1808, at Thonon (by Lake Geneva), was of negligible economic significance but will be referred to later for its unusual processes.

Moving to Lyons[136] we find that the works established there by the Javel company survived the Revolution and worked into the nineteenth century under the continuing ownership of Buffault's sons. Its director was a certain Alban, who remarked in 1836 that he had been running it for some forty years; he was presumably related to Léonard Alban, the director at Javel. Alban's works was soon followed by a number of others, all of them similarly located on the left bank of the Rhône. Being as yet little developed, this was an area which particularly attracted chemical producers and other artisans, since they could there hope to pursue their trades relatively unmolested by complaining neighbours. The first of the new works was established by Oblin and Mariette, in a property which they purchased in the Béchevelin district of la Guillotière, about a mile north of Alban, in October 1803. Mariette was from Rouen and was presumably the Rouen acid manufacturer of that name who a year earlier had been thinking of setting up a works at Amiens.[137] By 1812 the firm was styled Oblin, Bouvier & Cie, Bouvier being Oblin's son-in-law. A mile or so north of Oblin, near the pont Morand in the district of les Brotteaux, a third works began making sulphuric acid in 1808. The manufacturers were Paris and Mussard, who had been making iron and copper sulphates there since 1806; one lead chamber came into operation in April 1808 and two more in 1810. A fourth sulphuric acid works, again on the left bank, is said by Tissier to have begun in 1814, but we have no information on its owners or its location. An abortive venture at Saint-Clair in 1815–16 will be referred to again later for its processes.

The Lyons concerns which we have just enumerated were of a fairly typical size for the period, none of them comparable with the great pre-Revolutionary enterprise of Janvier & Cie. That of Alban was reported by Tissier in 1822 to have three chambers with a combined capacity of 15 000 kg of acid a month (say 150 000 kg a year), about a quarter of which was applied to the manufacture of sulphates. Oblin's seems to have been only a small plant under the Empire, with just a single chamber. It was in financial difficulties in 1812 and was declared bankrupt in 1813 or 1814. After standing idle for some time it was bought and resurrected by Colin in 1815–16,[138] and it is probably to be identified with the works which Tissier in 1822 described as having a single chamber, producing up to 5000 kg of acid a month, which was then applied to the manufacture of mineral acids and other chemicals. The works of Paris and Mussard had a capacity of 15 000 kg of acid a month in 1810 but declined during the depression which immediately followed; if our identification is correct, it was said in 1822 to be selling 10 000 kg of acid a month, as well as producing large quantities of iron, copper, and magnesium sulphates. The anonymous fourth works in 1822 had two large chambers with a total monthly capacity of 15–20 000 kg, most of the acid being used internally for further manufactures. The third and fourth works just mentioned both seem to have disappeared by 1836 but those of Alban and Colin had a longer existence.[139] Alban's works was bought in 1834 by the chemical concern of Claude Perret, which had itself begun with the manufacture of soda at les Brotteaux in 1819. As part of Perret's burgeoning enterprise it was evidently still working in 1845, when the director was named as Michel Alban,[140] but its development was then beginning to be hindered by the construction of a fort nearby, and in due course it was to give way to further military building. The barracks which today occupies the site perpetuates the memory of the old acid works in its name, 'La Vitriolerie'. As for the works of Colin, that was to be developed in a major way from about 1827 following its acquisition by Estienne and Jalabert, and it continued to work until 1921 when its demolition greatly improved the neighbourhood.[141]

Continuing south, we find sulphuric acid being made at Carpentras in 1810 by François Caire, along with nitric acid and soda, though with 8 workers and a capital of 80 000 fr. this cannot have been a large concern.[142] We do not know whether it had any link with the distillers of aqua fortis who operated at Carpentras in the eighteenth century. In Marseilles the early nineteenth century presents no trace of Janvier's old concern, but sulphuric acid was now made by others, a particular feature here being its association with sulphur refining, one of the town's traditional industries. At a works dating from about 1796 in the rue du Coq, outside the porte de Rome, Jean Baptiste Michel made sulphuric acid and other chemicals and operated a sulphur refinery. The works was ordered to close in January 1804 following complaints from neighbours, and Michel is next encountered in 1806 simply as a sulphur refiner. In the later years of the Empire, however, he was again making sulphuric acid, at a works in the rue Perrier said to have begun in 1809.[143] Another works, outside the town at le Bas Canet, dated from 1808 and was exploited by François Augustin Porry, who also made copper sulphate in the rue Fauchier. This manufacturer was no doubt identical with (or related to) the Auguste Porry who sent samples of refined sulphur and copper sulphate to the 1806 exhibition in Paris, and who in turn presumably belonged to the firm of Porry, Bausse, & Cie, the largest sulphur refiners in Marseilles, with works in the rue Crudère.[144] Further discussion of chemical manufacture in Marseilles is best postponed to Chapter 4, since its major development came with the growth of the soda industry there from 1809.

Prior to the rise of the soda industry the most important centre of chemical manufacture on the south coast was Montpellier, where the three sulphuric acid works operating on the eve of the Revolution all seem to have continued into the nineteenth century. In 1816 a works making sulphuric acid and sulphates on the cours des Casernes was said to have been there for over 30 years, and so is probably to be identified as that established by Rey and Cros. It had been run for some time by a certain Valedau and was now in the hands of J. Laurent, his former associate: in 1816 it seems to have had just a single chamber. In the faubourg du Courreau there was a works which Jean Pierre Jaumes was said in 1813 to have had 'depuis longtems', and which from its unusual coupling of sulphuric acid manufacture with the production of cream of tartar would appear to have been that encountered under the name of Bertieu in 1788. It almost certainly closed in 1813, when Jaumes's landlord gave him notice to quit. By then a further new works had appeared, that of Delmas & Cie on the main road to Lodève (faubourg Saint-Dominique), which was established following prefectoral authorization in April 1807 and was continuing to operate after the Empire.[145] Easily the most important of the Montpellier works, however, was that of Chaptal at la Paille. We have already seen how Chaptal returned there after the Revolution to find it in a cadaverous condition. In reviving it and making good his losses he was greatly assisted by the happy circumstance of Spain's entry into war with Britain in October 1796. With characteristic acumen Chaptal seized the markets in Catalonia hitherto supplied by Britain, and finding himself without competitor put such a high price on his sulphuric acid and other chemicals that he claims to have made profits of 350 000 fr. in a year.[146] At the same time he was successfully resuming a teaching career in Montpellier, as professor of chemistry at the newly reorganized medical school. After his time in Paris, however, and with Montpellier no longer the town he had known, a provincial life could not satisfy him for long. In March 1798 he returned to the capital, there to settle, and the la Paille works was now ceded to his two former associates.[147] Following a notarized act of 31 December 1798 it went under the name of Bérard, Martin, & Cie, and in 1802 Chaptal's interest in the concern was finally terminated when by an act of 6 September it was acquired by Bérard and Martin for the sum of 103 770 fr. (with an additional 25 000 fr. for a house and other property). Martin withdrew at the end of 1808, leaving the works then in the hands of Étienne Bérard. It continued to be highly regarded for the intelligence of its direction, and was described by the prefect in 1810 as occupying very large premises which Bérard had considerably expanded; 25 workers were then employed and Bérard's fortune was estimated at 250 000 fr.[148] The works remained with Bérard and subsequently with his sons until 1861, when it closed, though still said to be prosperous, following the death of Henri Bérard.[149]

In the west, the *Almanach du commerce* by 1815 lists a works at Bordeaux, making sulphuric acid and copperas and going under the name of Le Cerf. At Nantes, the works of Athénas was revived after being closed during the Revolution, and in the opening years of the nineteenth century was said to be making some 100 000 kg of sulphuric acid a year, together with nitric and hydrochloric acids, although a local statistic indicates only four workers.[150] At Honfleur Chamberlain's works was continuing, and though his fortunes varied considerably over the years he was still making sulphuric acid, copperas, alum, and other products there in the early 1820s.[151] Local rivals appeared at the turn of the century, in a manner which Chamberlain described with some indignation in June 1801:

a *négociant* of le Havre, to whom I had entrusted the books of my manufactory and the secrets of my operations, in contempt of the customs of society and of the laws, to satisfy his cupidity, has hired my day-labourers to form with them oil of vitriol establishments following my processes in the towns of Honfleur and le Havre.[152]

Of the Honfleur venture here mentioned no more is known. The works at le Havre was presumably that known to have been established in Year VIII (1799–1800) at Ingouville, to the north of the town. The proprietor there in the later years of the Empire was J. Foubert, also described as 'receveur des bois de la marine'. It would seem to have been only a small works, said in 1811 to employ three workers and to consume 14 000 kg of sulphur a year (equivalent to an output in the region of 20–35 000 kg of concentrated acid), though since 1811 was the year of depression these figures may not reflect its full capacity. A second works near le Havre, at Sanvic, was formed in about 1808 by Augustin Olivier, and was of a similar size, said in 1811 to employ two workers and to consume 15 000 kg of sulphur.[153] In 1810 a works at Sainte-Marguerite near Dieppe was described as making sulphuric acid,[154] but it seems likely that this was an error based on misinterpretation of the term 'vitriol'; the main pursuit there, at all events, was the manufacture of iron sulphate. There were, however, manufacturers by 1810 at Laigle and at Rugles;[155] the *Almanach du commerce* for 1815 indicated more than one works at Laigle and named the Rugles manufacturer as Valette.

Further north, at Lille, the works founded by Valentino was still operating in the early nineteenth century. Now under the direction of Lachapelle, it was said in 1804 to be producing 44 000 kg of sulphuric acid a year with four workers, and to be finding an important new market among the chlorine bleachers of the department.[156] This works had perhaps disappeared by the 1820s, for when in 1825 Kuhlmann began his important concern at Lille it was in response to the difficulty local industry was then having in obtaining chemicals, buying sulphuric acid in Paris and Rouen.[157] More sizeable than the Lille works was that of Dupuis at Saint-Quentin, whose establishment in 1793/6 we have noted earlier. A statistical note of 1806 recorded that it was making sulphuric and hydrochloric acids and salts, and that its workshops were very large, with three lead chambers capable of producing 400 *milliers* (200 000 kg) of concentrated acid a year.[158] This may have been an exaggeration: in the mid-1820s the works was said to be selling up to 100 000 kg a year and to be employing 8–10 workers.[159] Apart from these continuing enterprises, a new factory was established in 1809 at Amiens, by a medical group consisting of a physician called Barbier in company with two surgeons (Routier and Jusse) and an apothecary (Dambresville). This was to survive until 1922, from 1847 as part of Kuhlmann's concern.[160]

Finally, we come to two of the most important centres: Rouen and Paris.

Rouen

The two major works at Rouen, those of Holker and Chatel, both survived the Revolutionary disruptions, though Holker's emerged somewhat diminished and that of Chatel was probably in like condition. Three of Holker's six chambers had collapsed and the state of the market dissuaded him from replacing them until 1809–10.[161] Chatel's works was by 1810 in the hands of his eldest son, who four years later was to be nominated to the *Conseil général des manufactures* in Paris as one of the department's leading industrialists. From the late 1790s Holker and Chatel were joined by a proliferation of small new plants, most of them along the main road to Caen. The first to appear was one erected by Forestier, whom we have already met as the exploiter of a works at Déville until its closure in 1794. His new works was installed in 1797 at 24 route de Caen, in the faubourg Saint-Sever, near the boundary with the commune of Petit-Quevilly. By 1804 it had been taken over by his

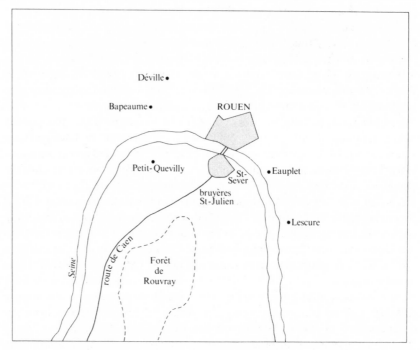

Fig. 2.6. The environs of Rouen. (Based on *Plan routier de la ville et des faubourgs de Rouen, avec ses environs*, Paris (1832).)

sometime associate Michel François Dubuc (Dubuc *le jeune*). It had only a single chamber until 1807 when a second seems to have been added.[162] In 1798 J. F. J. Mariette, eight years an employee in an existing works, set up on his own account next door to Forestier at 23 route de Caen. He again had a single chamber, producing 220 *milliers* of concentrated acid in his first four years (i.e. about 27 000 kg a year). The works was subsequently acquired by Foucard and Baril, probably in 1807.[163] Across the road from Forestier and Mariette, at 27–8 route de Caen, a works was established in Year X (1801–2) by Le Bertre & Cie, again with a single chamber.[164] In about 1804–5 manufacture was begun by Lefrançois *aîné, père & fils*, at Déville.[165] And at about the same time Forestier, after handing over his works on the route de Caen to Dubuc, added a small chamber to a works at 28 rue du Pré in the heart of the faubourg Saint-Sever, where he had for twenty years or so been making nitric acid and various salts.[166] By 1808 this works in turn had been acquired by Haag and Pelletan, who since early 1806 had been operating a pioneer Leblanc soda factory nearby (see p. 253). At the end of 1805 another plant appeared on the route de Caen, further out than those of Dubuc, Mariette, and Le Bertre, in the middle of the plaine de Quevilly (commune of Petit-Quevilly). It was the creation of Pierre Malétra and Delacroix (a merchant in Rouen).[167] These various enterprises all combined the manufacture of sulphuric acid with that of several other products, commonly nitric and hydrochloric acids, iron and copper sulphates, tin salt, and in one case artificial alum.

Steady growth gave way to dramatic expansion in 1809 and 1810, when the development of soda manufacture in France greatly increased the market for sulphuric acid. The existing manufacturers now hastily threw up additional chambers and further new works appeared.[168] At the beginning of 1809 the indefatigable Forestier established a works at 44 rue Benoît, in the faubourg Saint-Sever. At the end of the year Malétra left Delacroix and set up on his own near by his former partner. Simultaneously a further plant was erected in the same locality by Christophe Baril, apparently son of the Baril already mentioned. In 1808, according to Delahalle *oncle & neveu* (Le Bertre's backers), the various sulphuric acid makers in the department of the Seine-inférieure had possessed 12 chambers, producing 750 000 kg of acid a year.[169] By December 1809 an official list indicates 11 works in the region of Rouen, with 27 chambers, capable of producing the equivalent of 6592 kg of concentrated acid a day, in other words some 2 million kg a year (see Table 2.1). Nine of the chambers were still under construction, but these were duly completed early in 1810 and yet more were built: Malétra added a second chamber in May, and in the summer a further new venture was launched, on the bruyères Saint-Julien, by Louis Bazire in association with Charles Roulier and Jean Baptiste Dubuc (a spinner in the Sotteville district).[170] There were other projects, too, but these did not bear fruit since the local authority judged the sites unsuitable and refused authorization.[171] At first the bulk of the increased output was sent to the soda factories in Paris, but by the end of 1809 five of the Rouen acid manufacturers were making soda themselves, and after setting up special soda plants in the spring of 1810 it was claimed that they could employ all the acid made in Rouen.[172]

In the event this impressive expansion at Rouen proved ill-founded, for the town was particularly hard hit by the over-production to which France's soda boom quickly led. Soda manufacture in Rouen depended largely on sales to the soap-makers of Marseilles, and collapsed with the growth of a Leblanc industry around Marseilles itself in the course of 1810. The situation

TABLE 2.1
Sulphuric acid manufacturers at Rouen, 1809

Noms des propriétaires	Nombre des chambres de plomb		Dimensions des chambres			Produit journalier en huile de vitriol concentrée	
			Longueur	Largeur	Hauteur		
Mr Haag, [28] rue du Pré, fauxbourg St Sever	1		6 m	6 m	6 m	72 kg	
Mr Forestier, [44] rue Benoît, fauxbourg St Sever	1		25	9	6	300	
Mr Holker à St Sever [40–41 rue de Sotteville]	5	A	23	6	4		*(Two of these chambers still under construction)*
		B	id.	id.	id.	1500	
		C	25	6	3·70		
		D	33	7	5		
		E	id.	id.	id.		
Mrs Le Bertre & Cie, [27–28] route de Caen [f.b. St Sever]	2	A	11	6	4·70		*(One under construction)*
		B	23	7	5	500	
Mr Dubuc, [24] route de Caen [f.b. St Sever]	3	A	17	10	5		*(One newly completed)*
		B	23	13	2·70	1000	
		C	20	8	4·70		
Mrs Foucard & Barril, [23] route de Caen [f.b. St Sever]	3	A	16	6	3·25		*(One newly completed)*
		B	17	6·25	4	530	
		C	9	9	5		
Mr Christophe Barril, route de Caen, commune de Quevilly	1		16	7	4	150	*(Under construction)*
Mr Malêtras, Quevilly	1		16	7	5	240	*(Under construction)*
Mr Lacroix, Quevilly	3	A	10	5	5		*(One newly completed)*
		B	20	7	5	700	
		C	15	7	4		
Mrs Lefrançois aîné, père & fils, à Déville	4	A	17	7	7		*(One under construction)*
		B	id.	id.	id.	1000	
		C	id.	id.	id.		
		D	id.	id.	id.		
Mr Chatel, à l'Escure-lès-Rouen	3	A	17	7	6	600	
		B	id.	id.	id.		
		C	id.	id.	id.		

Observations: L'évaluation du produit journalier a été portée au dessous de ce qu'il est réellement, à raison de la quantité de soufre employé par chaque combustion, mais cette diminution du produit doit être maintenue, à raison des accidens, des réparations & des non-valeurs.

Note: from 'Tableau des fabriques d'huile de vitriol de l'arrondissement de Rouen', 14 Dec. 1809, in ADSM, 5 MP 1252, doss. 'Application … Le François'. The size of Forestier's chamber is elsewhere given as 53 × 25 × 18 ft. (5 MP 1252, [1]).

was aggravated, moreover, by the depression of 1811 which at the same time reduced other outlets for sulphuric acid. At the beginning of 1811 Pierre Malétra was hit by insolvency, his failure ('faillite') being declared at the town's *Tribunal de commerce* on 4 January; he was reported to have overreached himself in borrowing heavily to raise the 100 000 fr. establishment cost for his new works.[173] By November 1811 the works of Christophe Baril, of Haag, and even the historic enterprise of Holker had all closed, and Haag and Holker were said by the *Commissaire de police* to have left Rouen 'en faillite' (though they do not in fact figure in the official lists of business failures). Other manufacturers had dismantled redundant chambers, with the result that of the former 22 chambers in the faubourg Saint-Sever and at Petit-Quevilly, only 11 remained.[174]

Though the attempt to establish a major soda industry at Rouen thus proved abortive, the town nevertheless remained at the Restoration a notable chemical centre, supplying mineral acids and salts to the region's dyers and bleachers. Five of her chemical works are definitely known to have continued after the Empire, those of: Chatel *fils*; Dubuc *le jeune*; Malétra; Le Bertre & Cie (continuing under the name of Delahalle); and Lefrançois (continued under the Restoration by Lefrançois's widow in the name of Pouchet). Also probably still continuing was the works of Foucard & Baril, which appears to have been acquired towards 1819 by Gessard; and possibly also the works of Delacroix, for this still figured in the *Almanach du commerce* in 1819.[175] Of these various concerns, the one which was destined to acquire the greatest importance in the long term was that of Malétra. Despite its inauspicious beginnings in 1810, this had grown by 1830 to be employing 70–80 workers, and by 1890 its Petit-Quevilly plant covered over 60 acres and employed 700–800. It was then one of France's leading chemical enterprises, with factories elsewhere too. In the present century the firm joined with the electrochemical concern of Bozel in 1925 to become Bozel-Malétra, and in 1958 merged with the French Nobel company.[176]

Paris

In the Paris region the most important works at the beginning of the nineteenth century was one newly established there by Chaptal on his removal to the capital from Montpellier. Chaptal was in the capital by mid-March 1798, quickly slipping into place in its scientific life. In April he was named to fill the chemistry chair vacated by Berthollet at the *École polytechnique*, and in May he won the coveted distinction of election to the Academy of Sciences, now reconstituted as the First Class of the Institute. The vacancy at the Institute resulting from Bayen's death in February had probably been a major factor in his move. At the same time Chaptal was exploring the possibilities for chemical manufacture. On 19 September 1798 he bought from the State a property at les Ternes (commune of Neuilly) which had housed a Revolutionary munitions factory, and there, assisted it seems by a number of principal employees from Montpellier, he proceeded to install a chemical works.[177]

The political role which Chaptal assumed under the Consulate left him with little time for private business affairs. From December 1799 he was a hard-working member of the *Conseil d'état*, and then from 6 November 1800 was Minister of the Interior, responsible for virtually the whole of France's internal administration. The Ternes works was therefore looked after by Philippe Coustou, an old friend from Montpellier, who had probably played a part in the enterprise from the outset and with whom Chaptal formed a partnership on 1 January 1802.[178] The firm operated under the name Coustou & Cie. After resigning his ministerial post in August 1804, Chaptal was again able to devote attention to his works, but the business soon became more particularly the concern of his son, now in his twenties. On 25 August 1808 Chaptal *fils* formed a partnership with Amédée Barthélemy Berthollet, son of the great chemist, and the exploitation of the works now passed to the two youngsters. According to the terms of their agreement, Chaptal *fils* was to supervise the firm's business affairs, while Berthollet *fils* was to take up residence at les Ternes and direct manufacture.[179] In the following year the two partners also set up an important soda factory near Marseilles. Their partnership was to be cut short by the tragic death of Berthollet *fils* in 1810, and the two works were thereafter carried on by Chaptal *fils* alone. Chaptal *père*, after handing over the Ternes works to his son, still continued to take an active interest in the enterprise, particularly between 1809 and 1811 when his son was away for long periods in the south.

The products made at les Ternes under the Empire included sulphuric acid, alum, iron sulphate and acetate, tin salt, and oxalic acid. Nothing is known of the quantities produced but the concern was evidently fairly substantial. The capital brought to the 1808 partnership by Chaptal *fils*, who alone provided the funds, was 200 000 fr. This presumably covered the plant and working capital, but not the land and buildings at les Ternes, nor the firm's business premises in Paris, since these properties were rented by the partners from Chaptal *père*; they were evaluated in 1816 at 70 000 fr. for the Ternes property and 80 000 fr. for the house in the rue des Jeûneurs where the firm's offices were located.[180] Some glimpse of the character and scale of the works is provided by a description and plan drawn up in 1816.[181] These show the property at les Ternes to have amounted to some 8 acres, three-quarters of it occupied by gardens. With its vines and fruit trees, its flowers and lawns, it must have presented a curious appearance for a chemical works. The factory itself occupied an area of some 130 by 530 feet, and was housed in single- and double-storey stone buildings grouped around yards. Over a dozen buildings were devoted to workshops and stores, and there was also a laboratory, a house for the director, and living accommodation for the foreman and several workers. Regrettably, although the buildings are described in detail (down to the eight rabbit hutches), we are told nothing about the plant. A report by Deyeux in July 1811 spoke of the factory's 'vast lead chambers', but Chaptal himself had written in 1807 that he thought chambers with sides 20–5 feet long and 15 feet high to be best, which was not particularly large.[182] Of the number of workers again nothing is definitely known: a statistical note of January 1807 on employment at chemical works in the Paris region perhaps points to 10 workers at that time, but identification of the factories on the list is somewhat uncertain.[183]

The works continued under the Restoration. In March 1816, on the occasion of his marriage, Chaptal *fils* received the properties at les Ternes and in Paris as a gift from his father. Four months later he entered into partnership with Jean Holker (formerly of Rouen) and Jean Pierre Darcet, who were themselves running an important soda factory at la Folie, three or four miles to the west. The two works were then operated in conjunction and must have constituted the largest chemical enterprise in the north of France. This joint concern, and also the soda factory of Chaptal *fils* near Marseilles, will be considered more fully in Chapter 4.

The Paris region also still possessed the old Javel works, of course, though by the Empire its original founders had all disappeared. Bourboulon and Vallet had moved to England in 1787, as we have seen, and though Bourboulon returned to France in Year IV (1795–6) he did not resume an interest in the concern.[184] Buffault died in 1792 and de la Ferté was executed in 1794. The works then remained in the hands of Buffault's sons and Léonard Alban, who together formed a new partnership for its exploitation on 25 April 1797.[185] The death six years later of Alban, director of the factory since its foundation, removed the last member of the original company. The capital value of the concern was estimated in the partnership agreement of 1797 at 116 674*l.*, and an inventory of the works made in 1803 shows it to have included three lead chambers, each measuring $50 \times 14 \times 12$ ft., together with plant for concentrating the acid, a large number of pans and boilers for the manufacture of alum, a horse-driven mill, a laboratory, a house (presumably occupied by the director), and living quarters for workers.[186] The activity of the works was somewhat diminished after the Revolution, a local authority report of 1801 recording only eight workers, compared with twenty before the outbreak of the Revolutionary Wars.[187] After the death of Alban the direction of the works seems to have been undistinguished and it was now overshadowed in repute by Chaptal's enterprise. A report of June 1817 named the director as a certain Grivel and indicated only three workers.[188] Only after our period did it regain something of its former eminence. Annual entries in the *Almanach du commerce* show it to have been operated from the early 1820s by Lepelletier & Cie, and from the later 1830s by Lepelletier's son-in-law, Edmond Fouché-Lepelletier. In the latter's hands 'the celebrated works of Javel'[189] was a prize-winner at the 1851 exhibition in London, and was then said to be turning out 3 600 000 kg of concentrated sulphuric acid

Fig. 2.7. The environs of Paris. (Based on map in Chanlaire (1815), vol. 2.)

a year. Four years later it was there that Henri Sainte-Claire Deville was to conduct his first industrial trials for the manufacture of aluminium. The works was still in existence in the 1880s but had disappeared by 1912. The site—between the rue Leblanc and the rue Cauchy[190]—is today occupied by the car factory begun at Javel in 1915 by André Citroën.

The works at Javel and les Ternes were not the only sulphuric acid plants operating in the Paris region in the Napoleonic period. By 1807 the *Almanach du commerce* listed a further five manufacturers and it is interesting that two of these bore the name Alban. One was Pierre Alban, in the commune of Vaugirard, whose close relationship with the Javel director is clear from his being named guardian to Léonard's son when Léonard died. Pierre continued to figure in the *Almanach du commerce* until at least 1831 but was only a small manufacturer: three workers are indicated in the list of 1807 mentioned above, and two in the report of 1817. The other manufacturer with an Alban connection was the widow Simon, née Anne Alban, who was continuing a works in the commune of Issy which had been operated by her husband, Louis Simon, until his death in 1803. This again was not a large works, probably making little or nothing but sulphuric acid, and in 1803 possessing two chambers, one measuring $17 \times 13 \times 12$ ft. and the other an 18 ft. cube;[191] two or three workers are indicated in 1807 and two in 1817. More important was the works established in about 1797, in the hamlet of la Glacière (at Petit-Gentilly), by Marc, Costel, & Cie,[192] who are described in the *Almanach du commerce* as also having a textile printing works there. Besides sulphuric acid the firm made iron and copper sulphates and 'jaune minéral', sending all these products to the 1806 exhibition; from the list of 1807 they seem to have then employed ten workers. Later, during the soda boom, they also made soda for a while. When the company was dissolved in 1816 it consisted of seven members, among them C. C. H. Marc (a physician), J. V. Costel (a pharmacist), and M. J. J. Dizé (former partner of Nicolas Leblanc), but it is not known whether it had been so constituted from the outset.[193] The works was subsequently carried on by E. Bodin, Jolly *jeune* & Cie, who in 1817 were making 75 000 kg of sulphuric acid and 70 000 kg of iron sulphate with eight workers. The other two works operating by 1807 were those of Reuflet Duhameau at Issy (continued under the Restoration by the widow Duhameau, with two workers in 1817), and Bourrelier at Vaugirard (not mentioned in the 1817 report but figuring in the *Almanach du commerce* until at least 1820). In addition to these seven works in the immediate vicinity of Paris, there was also a sulphuric acid plant by 1812 at Pontoise, about twenty miles away; the manufacturer there was Cartier, who also made *eau de Javel* and oxalic acid.[194]

The size of the industry at the Restoration

We have now traced the formation of some sixty sulphuric acid works in France, from Holker's pioneer venture of 1769 to the end of the Napoleonic era in 1815. There were no doubt a few others too, which we have missed. Recognition of works is sometimes hindered by vague or ambiguous description: a works described as making 'vitriol', for example, was usually producing iron sulphate, but its product may occasionally have been sulphuric acid; and one described simply as manufacturing 'mineral acids' may or may not have numbered sulphuric acid among them. We have included here only works in which it seems explicitly clear that sulphuric acid was produced. From the summary of our sixty works in Table 2.2, it will be seen that twenty-three can be confidently affirmed to have still existed in 1815, and that the continuance of a further ten can be conjectured with varying degrees of probability. Fourteen works, on the other hand, had fairly definitely closed down by that date. On the remaining thirteen information is lacking or doubtful. It seems likely, in sum, that there were about 30–40 works operating in France at the Restoration with sulphuric acid manufacture as their focal concern. The number probably did not differ greatly from that in Britain, for Mactear records twenty-three works in England in 1820, to which we can perhaps add a further ten or a dozen in Scotland.[195]

TABLE 2.2
Sulphuric acid works established in France, 1769–1815

Place	Manufacturer	Date works established	Whether still existing in 1815
1. Rouen (Saint-Sever)	Holker	1769	No
2. Rouen (Eauplet)	Fleury	1774	No
3. Rouen (Déville)	Anfrye	1777	No
4. Paris (Javel)	Léonard Alban	1778	Yes
5. Rouen (Déville)	Stourme	1778	No
6. Honfleur	Auvray, Payenneville, & Cie	1778	No
7. Montpellier	Pitou, Caizac, & Cie	1780	Very probably not
8. Nîmes	Lapenne	1782	Not known
9. Lille	Valentino	1783	Not known
10. Montpellier (la Paille)	Chaptal	1785	Yes
11. Montpellier	Rey & Cros	c. 1785	Yes, very probably
12. Agen	Guillebaut	1786	Not known
13. Marseilles	Janvier & Cie	1787	No
14. Montpellier	Bertieu	By 1788	No
15. Lyons (la Guillotière)	Janvier & Cie	c. 1790	No
16. Rouen (Lescure)	Chatel	1792	Yes
17. Honfleur (la Rivière)	Chamberlain	1793	Yes
18. Saint-Quentin	Dupuis	1793/6	Yes
19. Nantes	Athénas	By 1794	Not known
20. Lyons (la Guillotière)	Buffault *frères*	By 1794	Yes
21. Mulhouse	Koechlin & Marchais	1794?	Not known
22. Marseilles	Michel	1796	No
23. Rouen (Saint-Sever)	Forestier	1797	Yes
24. Paris (la Glacière)	Marc, Costel, & Cie	c. 1797	Yes
25. Nancy	Mathieu	c. 1796–8	Yes, perhaps
26. Rouen (Saint-Sever)	Mariette	1798	Yes, very probably
27. Paris (les Ternes)	Chaptal	1798	Yes
28. Strasbourg	Oppermann & Lefebure	c. 1799	Not known
29. le Havre (Ingouville)	Foubert	1799–1800	Yes, perhaps
30. Honfleur	?	c. 1800	Not known
31. Rouen (Saint-Sever)	Le Bertre & Cie	1801–2	Yes
32. Paris (Issy)	Simon	By 1803	Yes
33. Paris (Vaugirard)	Pierre Alban	By 1803	Yes
34. Lyons (la Guillotière)	Oblin & Mariette	1803	Yes
35. Strasbourg	Meunier & Cie	c. 1802–4	Yes
36. Rouen (Déville)	Lefrançois	1804–5	Yes
37. Rouen (Saint-Sever)	Forestier	c. 1805	No
38. Rouen (Petit-Quevilly)	Delacroix & Malétra	1805	Yes, perhaps
39. Strasbourg (la Chartreuse)	Kestner	1806	Not known
40. Paris (Issy)	Duhameau	By 1807	Yes
41. Paris (Vaugirard)	Bourrelier	By 1807	Not known
42. Montpellier	Delmas & Cie	1807	Yes
43. Thonon	Rivaz	1808	No
44. Lyons (les Brotteaux)	Paris & Mussard	1808	Yes
45. Marseilles	Porry	1808	Yes
46. le Havre (Sanvic)	Olivier	c. 1808	Not known
47. Thann	Kestner	c. 1806–9	Yes

TABLE 2.2—*continued*

Place	Manufacturer	Date works established	Whether still existing in 1815
48. Amiens	Barbier *et al.*	1809	Yes
49. Marseilles	Michel	1809	Not known
50. Rouen (Saint-Sever)	Forestier	1809	Not known
51. Rouen (Petit-Quevilly)	Malétra	1809	Yes
52. Rouen (Petit-Quevilly)	Christophe Baril	1809	No
53. Langres	Guérinot	By 1810	Not known
54. Carpentras	Caire	By 1810	Yes, perhaps
55. Laigle	?	By 1810	Yes, perhaps
56. Rugles	Valette	By 1810	Yes, perhaps
57. Rouen (bruyères St-Julien)	Bazire *et al.*	1810	Probably not
58. Pontoise	Cartier	By 1812	Yes, probably
59. Lyons	?	1814	Yes
60. Bordeaux	Le Cerf	By 1815	Yes, perhaps

From what is known of early nineteenth-century production figures, the output of French plants appears to have been generally in the range 20 000–200 000 kg a year, reckoned as concentrated acid. Guessing at an average figure of about 100 000 kg, the combined production of our supposed 30–40 works can thus be estimated at 3–4 million kg. To arrive at a figure for France's total production we must add to this, of course, the large quantities of acid being made in the new soda factories, as a preliminary to the Leblanc process. Records for the soda industry's salt consumption can serve as a guide here, the use of 9 600 000 kg of salt in 1816 pointing to a sulphuric acid requirement in the region of 7 200 000 kg.[196] Allowing for some marginal overlap between the activities of soda factories and sulphuric acid works, an estimate of 9–11 million kg for France's total sulphuric acid production in 1816 would seem reasonable. The figure was rising quite rapidly with the expansion of soda manufacture. By 1818 that industry's acid consumption can be reckoned to have further increased by over 1·1 million kg, bringing total acid production to some 10–12 million kg. These estimates can be compared with those of contemporaries, such as Clément, who in September 1810 guessed at a production of over 10 million pounds (5 million kg), and Chaptal who in 1819 gave a figure of 20 million kg.[197] Chaptal's figure seems distinctly too high for the date at which he wrote, but production probably reached the level he indicated in the early 1820s. In his lecture course of 1825, Clément reckoned France's production at 22 million kg (on the basis of sulphur imports of 9–10 million kg).[198] Estimates of British production for comparison are scarce. In 1815, however, Parkes commented that consumption 'in these kingdoms' was upwards of 3000 tons a year (*c.* 3 million kg), while Hardie and Pratt have recently indicated a figure of the order of 10 000 tons for United Kingdom production in the early 1820s.[199] France had now evidently acquired a lead over Britain in the volume of her acid production, a fact which is unsurprising and plainly attributable to the heavy consumption by her new soda industry.

The technical development of manufacture: early lead-chamber theory

The lead-chamber process as generally worked until the end of the eighteenth century was simply a scaled-up version of the older method of production in glass globes. It consisted in burning a mixture of sulphur and saltpetre inside the closed chamber, and collecting the sulphuric acid fumes so produced in a layer of water on the chamber floor. Since the limited air of the chamber allowed only a few pounds to be burned at a time, it was necessary to work in an intermittent fashion. A small charge was burned; the chamber was then left to stand for a time, still closed, while the fumes condensed; and then it was ventilated, to prepare for a new combustion, by opening the door and waiting for the residual fumes to clear. This cycle of operations typically took some four or five hours, and it was carried on repeatedly over a period of several weeks until the water was sufficiently acidulated to be drawn off and concentrated for sale. We shall see that this rather crudely conceived and inefficiently executed method of production was to receive notable improvements at the beginning of the nineteenth century. It will be useful to begin our account of the industry's technical development, though, by considering early understanding of the chemistry of the process.

The early growth of the industry in France coincided with that transformation in chemistry itself which in the course of a generation was to establish the science on an essentially modern basis. The year 1772, for example, to which we have ascribed the first erection of lead chambers at Rouen, was also, to quote the title of Guerlac's well-known study, 'the crucial year' in the evolution of Lavoisier's thought, leading him into that programme of research which he himself predicted would effect a revolution in science. Of the major advances which chemistry made in the 1770s it is pertinent here to mention two in particular. First, the dramatic progress in the study of gases, as the old belief in the uniqueness of common air as the sole 'elastic fluid' gave way to recognition of a multitude of distinct aeriform species, whose isolation and characterization now proceeded apace. Second, Lavoisier's reinterpretation of combustion as consisting in the combination of the combustible with a part of the air, that part to which by 1779 he had given the name oxygen. The lead-chamber process, depending as it did on combustion and on gas reactions, was in due course to derive great enlightenment from these central advances in basic chemistry, though as we shall see it was to be several decades before anything like an adequate understanding of the process was arrived at.

When manufacture first began in France it was in terms of phlogiston theory that the process was understood. Since the mid-century the hypothetical substance 'phlogiston' had come to play a key role in chemical thought, particularly in the explanation of combustion. It was the presence of phlogiston in inflammable materials which was held responsible for their combustibility, and whose release was supposed to constitute the phenomenon of combustion. On this interpretation sulphur was a compound of sulphuric acid and phlogiston, and the manufacture of the acid consisted in its extraction from this combined state by the combustion process. Combustion in general had been known since ancient times to require air, of course, but the phlogistonists did not recognize air to be a chemical participant in combustion, allowing it merely some kind of subsidiary mechanical action. Only with the studies of Lavoisier was the air's chemical role elucidated. In his famous sealed note of October 1772, Lavoisier reported his findings that when sulphur (and similarly phosphorus) was burned in a determinate volume of air, not only did the air show a reduction in volume but at the same time the combustion product showed a gain in weight. From these observations, together with the analogous gain in weight of metals on calcination, he proceeded in the following years to elaborate his new theory of combustion as an oxidation reaction. With the new interpretation of combustion, sulphuric acid could be correctly understood as an oxidation product of sulphur.

The production of sulphuric acid by combustion of sulphur is not an altogether straightforward matter, however, and unfortunately Lavoisier's understanding remained seriously defective in some respects, and perpetuated certain earlier misconceptions which were to obstruct a proper comprehension until the early nineteenth century. We now know that when sulphur burns the resulting fumes are a mixture of sulphur dioxide and sulphur trioxide. The sulphur trioxide, on meeting with moisture in the air—or on dissolving in a body of water—forms sulphuric acid, and so from the combustion a certain quantity of sulphuric acid immediately results. The amount of acid so formed is very small, though, since overwhelmingly the major combustion product is not sulphur trioxide but the dioxide: the proportion of trioxide is commonly given as about 2–3 per cent, though it varies somewhat with combustion conditions, and a figure of 7 per cent has also been given.[200] It was on this very minor formation of the acid as a more or less direct combustion product that the old bell method of preparation chiefly depended, the acid being collected on the walls of the glass bell that was suspended over the burning sulphur. In some cases a little additional acid was formed in an indirect fashion from the sulphur dioxide: this would occur if along with the sulphur trioxide a part of the dioxide was retained in solution as sulphurous acid, for on subsequently standing in the air this would then gradually yield sulphuric acid by slow oxidation. By far the greater part of the sulphur dioxide resulting from the combustion simply escaped to the air, however, and was a pure loss. It can thus readily be appreciated why the old bell process was so highly unproductive and wasteful a method of preparation, giving a practical yield of the order of only 1 per cent of the theoretical figure for the quantity of sulphur burned.[201] As we know, it was the addition of saltpetre to the burning sulphur which transformed this primitive preparation into a viable method of large-scale production, by greatly improving the yield obtained. The improvement resulted from the fact that the combustion was now quickly followed by a second stage of reaction among the initial combustion products, the new reaction converting the sulphur dioxide to trioxide and hence to sulphuric acid. In bringing about this desirable effect, the saltpetre acted by decomposing to give certain gases which served as a catalyst to the process, in a manner which we shall consider more fully later. The practical consequence was an improvement in yield in early manufacture to about 30 per cent of the theoretical, and eventually in the nineteenth century to almost 100 per cent.

It should be clear from this account that the combustion of sulphur alone can never yield more than very small quantities of sulphuric acid, and that the much larger amounts obtained in the lead-chamber process depend on a special mechanism in which the saltpetre plays a crucial role. These basic points quite escaped eighteenth-century chemists, who laboured under misconceptions deriving primarily, it seems, from that very influential figure of the early century, Georg Stahl. Stahl's understanding of the combustion of sulphur had marked an advance in one respect, for he had correctly recognized that it yielded not simply 'spirit of sulphur', but in fact two distinct products, sulphur dioxide and trioxide (to use modern terminology). Unfortunately he failed to appreciate the further point that by far the greater product is always the dioxide. Rather, he regarded the dioxide and trioxide as alternatives, either of which might predominate depending on the combustion conditions. It subsequently became a notion firmly embedded in chemical lore that when sulphur burned rapidly, with a hot, lively flame, it yielded vitriolic acid (i.e. sulphur trioxide and hence sulphuric acid), whereas when combustion was slow and less vigorous there resulted 'volatile sulphureous acid' (i.e. sulphur dioxide and hence sulphurous acid). In terms of phlogiston theory these two supposed modes of burning were explained by the idea that brisk combustion effected a complete removal of the sulphur's phlogiston, leaving plain vitriolic acid, whereas feeble combustion removed only a part, resulting in the volatile acid.[202] When Lavoisier mounted his attack on phlogiston theory and reinterpreted the combustion in terms of oxidation, he retained this fallacious distinction between two modes of burning. He and his followers now supposed that the brisk burning of sulphur produced a full oxidation to sulphuric acid, while feeble burning effected only a partial oxidation, with sulphurous acid as the result. The idea is of course a very plausible one. It is nevertheless surprising that Lavoisier did not perceive his error, in view of his own studies of the combustion of sulphur in formulating his anti-phlogistic theories; it is not clear, indeed, how he obtained some of the experimental results he describes. Towards the end of his life he did recognize the combustion to be more problematical than he had originally admitted, but he seems always to have wrongly regarded sulphuric acid as a major direct product.[203]

With regard to the role of saltpetre in industrial production, eighteenth-century chemists generally held a view which was just as superficially plausible but just as fundamentally erroneous as their understanding of the combustion of sulphur. When formerly the acid had been prepared without saltpetre, by the old bell method, it had been recognized already in the seventeenth century that the process was rather inefficient. Chemists ascribed this inefficiency, however, to their own failure effectively to capture all the acid fumes which the combustion produced. In an attempt to condense them more completely, a variety of condensers were accordingly devised as alternatives to the bell. The difficulty appeared to lie in the fact that the burning of the sulphur and the collection of its fumes called for conflicting conditions: the sulphur required the open access of air in order to burn, while the use of open apparatus inevitably allowed fumes to escape. When industrial production of the acid developed, based on addition of saltpetre to the sulphur burned, the value of the saltpetre seemed to be that it allowed a resolution of these conflicting requirements. Saltpetre was known to be a supporter of combustion, and so it was thought that its contribution was simply to enable producers to burn their sulphur inside closed vessels, serving as a substitute for a free supply of air. The higher yield now obtained found a ready explanation in the more complete condensation that was obviously to be expected from operation in closed vessels. According to phlogiston theory the saltpetre performed its role by attracting phlogiston from the sulphur; according to the Lavoisian theory it acted by supplying oxygen for the combustion.[204]

This was the network of largely erroneous ideas in which early efforts to improve manufacture were grounded.

Attempts to make acid without saltpetre

When chemists gave thought to the improvement of manufacture, early theory conspired with economics to suggest that the primary objective should be the elimination of saltpetre from the process. Economically, although saltpetre was used in only small proportion (generally 10–20 parts to 100 of sulphur), its high price made it an important element in manufacturing costs: prior to the price fluctuations of the Revolutionary and Napoleonic Wars, the saltpetre can be reckoned to have accounted for some 17–43 per cent of the acid's market value, depending on the proportions used, while the sulphur accounted for some 14–20 per cent.[205] Theory, for its part, implied that since saltpetre served simply to support the combustion, to avoid its use one had only to find some alternative means of allowing the sulphur to satisfactorily burn. At the same time it implied that the heavy capital expense of lead chambers might also be avoidable, for the chambers seemed to act merely as crude condensers, and so one might reasonably hope to replace them by some type of condenser which would be more efficient and less costly. Endeavours to dispense with saltpetre were thus sometimes allied with 'improved' modes of condensation. Needless to say, in these two aims experimenters were equally misguided.

The most commonly attempted method of dispensing with saltpetre was by burning the sulphur in a current of air, so arranged as then to carry the acid fumes into the chamber or other condenser. It was probably a method of this kind for which Vincent sought an exclusive privilege in 1779 when he was proposing to establish a works in Grenoble. He claimed that his process used no saltpetre, instead 'the combustion of the sulphur takes place with all the assistance of the air which is necessary to it'.[206] The earliest detailed description of such a method that we have, however, is one appended by Struve in 1780 to the Neuchâtel edition of Demachy's *L'Art du distillateur d'eaux-fortes*. The combustion was here performed inside a stoneware globe maintained at a red heat on a furnace. Sulphur was continuously fed into the globe through a two-inch opening, which at the same time gave entry to a constant current of air, drawn in, our phlogistic author explains, by the rarefaction of the air inside. The air current then carried the resulting acid fumes through a condensing system consisting of a series of six globes, half-filled with water. These were heated on small furnaces, and the resulting steam condensed the fumes to such complete effect that the air emerged from the final globe entirely odourless. Struve claimed that he had often tried this method, with constant success. He also described an alternative arrangement in which the sulphur was burned in a similar way but a different method of condensation was used, the fumes this time being drawn through a body of water by means of a suction pump. Struve's account was reprinted in 1783 in a volume of the *Encyclopédie méthodique*, and his method was described again in 1786 by Guyton de Morveau, in a further volume of the same work. Guyton remarked that the replacement of saltpetre by an air current represented the highest degree of perfection in sulphuric acid production, and suggested the use of a furnace which he had recommended in 1782 for the calcination of zinc, and which was similar in principle to Struve's apparatus.[207]

An air current was probably also the basis of a method devised in the early 1780s by Hollenweger.[208] As one of France's early adventurers in the field of soda manufacture Hollenweger was proposing to make the acid as an associated activity. His request for an exclusive privilege for his projected soda factory led to an examination of his processes by the chemist Macquer, adviser to the *Bureau du commerce,* and Macquer was clearly impressed by the method of acid production. Referring to the apparatus, whose effect he had seen, for the extraction of vitriolic acid by burning sulphur without addition, he remarked in a report of July 1783 that it 'infinitely surpasses all that has been devised up to the present on this subject and . . . would alone merit a privilege or a distinguished reward'. A month later he again spoke of Hollenweger's 'excellent apparatus for procuring this acid in abundance, very pure and at the lowest possible cost'. The approval which such endeavours met with from leading scientific authorities is further illustrated by the hopes of Berthollet, in 1789, that the cost of his new chlorine bleaching process might be reduced by the improvement of sulphuric acid manufacture in this direction.[209]

Among others to concern themselves with the use of an air current was the inspector of manufactures Pajot Descharmes, who in about 1790 (probably) conducted experiments resembling those of Struve, later publishing them in his book on bleaching. More interestingly, a certain Du Porteau in 1790 published what he claimed to be the method of manufacture secretly employed by the British. His information came from a friend who had seen manufacture in Britain, and the method described depended on the replacement of saltpetre by an air current secured by 'a kind of ventilator furnace'; the acid fumes, after combining with steam, were condensed in a water-cooled leaden pipe. A. L. Brongniart, professor of chemistry at the *Muséum d'histoire naturelle*, also proposed a furnace for the production of acid without saltpetre in a memoir to the Institute in 1797. And another who seems to have given at least passing attention to the matter was the chemist and inventor Conté, for a manuscript jotting bearing his name depicts a crude experimental arrangement in which an air current was to be drawn over burning sulphur by a suction pump.[210]

The most notable experiments of which we have knowledge, however, are those which Chaptal conducted when he established his works at Montpellier in the mid-1780s. Already in 1782 Chaptal had remarked on the aim, conceived by some, of dispensing with saltpetre, but he himself was at that time sceptical about such hopes, pointing to the practical difficulties which would ensue if no saltpetre was used:

in that case one must introduce atmospheric air to make up for that given by the saltpetre in its deflagration; in that case one must lose vapours, obtain sulphurous acid, and add to all these drawbacks that which is worse still, of maintaining only with the greatest difficulty the combustion of the sulphur in a place soon filled with elastic vapours, which, forming an equilibrium with the outside air, will allow it only difficult access.[211]

Three years later, however, Chaptal was experimenting in just this direction and he published his results in 1789. The experiments were conducted on a manufacturing scale, the sulphur being burned in 'a kind of reverberatory furnace' with its chimney directed into a lead chamber; at first a chamber with a side of 30 ft. was used, later a smaller chamber 15–20 ft. square. In his first trial Chaptal burned sulphur in the furnace for seven consecutive days, until forced to stop by the mounting fumes, which made it impossible for the workers to tend the furnace and caused one of them to start bleeding at the nose. When he could get into the chamber Chaptal was surprised to find that although he had burned over half a ton of sulphur, and it had appeared to burn with great ease, the chamber water was scarcely acidulated. It was obvious from the skin on its surface that the sulphur had in large part simply sublimed. The experiment was repeated several times, with variations, but always gave the same result. Reasoning on the basis of the new Lavoisian theory of combustion, Chaptal concluded that it was necessary to reduce the air current, so as to facilitate fuller oxidation of the sulphur, an apparently paradoxical conclusion which is probably to be explained by a supposition on Chaptal's part that too rapid a current swept the vapourized sulphur from the flame before there had been time for appreciable oxidation. He therefore reduced the draught by fitting a damper to control the air intake and by enlarging the connection between furnace and chamber. Yet after burning more than a ton of sulphur over a period of thirty-three days (with several pauses necessitated by the fumes), he still found the water to be scarcely more acidulated than before, although sublimation was now avoided. He concluded that the sulphur had this time been converted almost entirely to sulphurous acid (sulphur dioxide), and he abandoned hope of making sulphuric acid in this way:

From that moment I renounced the hope of obtaining sulphuric acid by the oxygen of the atmosphere alone, and I recognized more than ever the necessity of the mixture of saltpetre.

While it was the air which seemed to offer the most obvious alternative to saltpetre, experimenters also considered other possibilities. Thus, the idea arose that one might use the oxygen contained in water, an idea clearly based on Lavoisier's celebrated demonstration in the early 1780s that water was not an element but a compound of oxygen and hydrogen, susceptible to decomposition by hot iron or charcoal. Du Porteau in 1790 hinted that the decomposition of water might be involved in the method of manufacture he described as being employed in Britain: the sulphureous fumes 'conveyed to the surface of boiling water . . . combine with the vapour of this water, and perhaps with its vital air [oxygen]'. And Socart-Château, a scientific amateur who visited Prestonpans in the course of travels in Britain, came away with the suspicion that it was by decomposition of water that acid was made in the famous works there. The aura of mystery surrounding the works, impenetrable behind its walls, encouraged him in his belief that a secret process was employed, which furnished the acid at only half the price of that made in France. The pharmacist Cadet de Gassicourt, in conveying Socart-Château's observations to the *Société d'encouragement* in 1802, asserted that according to experiments of his own, when water was dripped on to burning sulphur it was decomposed with the formation of sulphuric acid, and he suggested that this might be developed industrially. His views were challenged, however, by contrary evidence subsequently communicated to the Society by Guyton and by Vauquelin, members of its chemistry committee.[212]

Some particularly interesting researches were conducted by Chaptal, as a sequel to his fruitless trials with an air current. Those earlier studies had evidently convinced him that even in an air current saltpetre was still needed, as a supplementary source of oxygen to ensure the formation of sulphuric rather than merely sulphurous acid. He now experimented, therefore, with various other oxygen-bearing materials in search of some cheaper substitute to fulfil this role. Laboratory experiments and manufacturing trials showed, however, that neither water nor the several metallic oxides he tried were effective. Manganese dioxide did sensibly promote the combustion by its release of oxygen, but it nevertheless did not conduce to the formation of sulphuric acid, only sulphurous acid being formed in any significant quantity. Chaptal found, moreover, that even in pure oxygen gas, contrary to established belief, sulphur burned to give only sulphurous acid, although again combustion was plainly facilitated. The only materials which seemed conducive to sulphuric acid formation were saltpetre and (misleadingly) potassium chlorate, and of these saltpetre remained preferable. Chaptal recounted these experiments in his textbook of 1807,[213] but they had evidently been performed some years earlier for the results were mentioned by Cadet de Gassicourt in 1804.

The thoughts of early experimenters seem to have been chiefly dominated by the question of the air or oxygen supply, but another supposed factor in sulphuric acid formation which had long found some recognition was heat. We have already seen how the phlogistic school of chemists had contrasted the brisk, *hot* combustion which yielded sulphuric acid with the slow feeble combustion yielding the volatile sulphurous acid. Thus, when Struve burned his sulphur inside a red-hot globe, this was on the grounds that 'the more quickly the sulphur burns, the less sulphurous odour it has, and . . . the stronger the heat that one applies to the sulphur, the more quickly it is consumed'. The supposed importance of heat received particular emphasis at the beginning of the nineteenth century. When Chaptal's new experiments showed that even an abundant oxygen supply did not guarantee that sulphuric acid would be formed, it was in heat that he sought his explanation. He thought it significant that the only additives which had proved effective— saltpetre and potassium chlorate—not only provided oxygen but also generated a violent heat, and he now considered that heat was an essential requirement, and that the way to economize on saltpetre would therefore be to perform the combustion in strongly heated furnaces or pipes. Chaptal's views were endorsed by Cadet de Gassicourt, following the failure of some experiments of his own. Cadet had reflected on the inefficiency of the lead chamber as a condenser, and had been led to experiment with the possible alternative of drawing the fumes from the usual combustion mixture through a body of water by means of a suction apparatus, only to find, of course, that the product he obtained was sulphurous rather than sulphuric acid. After conversation with Chaptal he ascribed his failure to an insufficient heat in the combustion, and recommended as a likely remedy an arrangement communicated to him by Paul, of Geneva (presumably Nicolas Paul, the noted mechanician, who used sulphuric acid in his manufacture of aerated mineral waters).[214] This was an arrangement in which sulphur was burned inside a heated tube, in a current of air supplied by bellows, and Paul claimed that by linking the device with a small chamber of only 12 square feet ('12 pieds carrés') he could make as much acid as in an ordinary manufactory. Similar views on the importance of heat were also espoused by a certain Derrien, of Quimper. Some years earlier Derrien had experimented with the combustion of sulphur in an air current, dissolving the fumes by drawing them through water in a Woulfe's apparatus, but having observed a great deal of sulphurous acid in the product he in 1804 prescribed that to obtain sulphuric acid one should pass 'a mixture of sulphurous flame and atmospheric air' through an incandescent porcelain tube; whether he had actually tried this is not clear.[215] Perhaps the most intriguing experimenter in this vein was the chemical manufacturer Chamberlain, of Honfleur. In 1801 Chamberlain patented a method of manufacture in which saltpetre was to be replaced by a combination of oxygen gas and heat. There are some obscurities in his description and it is complicated by provision for various alternative modes of operation, but the central novelty consisted in burning sulphur in a heated furnace while directing into the furnace a stream of oxygen gas.[216] Chamberlain insisted particularly on the need for strong heat, and besides heating the furnace itself also applied heat to the iron pipe conveying the combustion fumes to the chamber, and to the glass tubes carrying the oxygen gas. He claimed by his methods to have obtained a yield of 34 900 lb. of acid, reckoned as concentrated, from 10 500 lb. of sulphur—over 100 per cent of the theoretical maximum!—and he complained that the Ministry's delay in granting his patent had allowed a former associate to steal his processes and set up on his own account. In 1822 Chamberlain still claimed to make acid without saltpetre (see p. 89).

The work of these various experimenters as they strove to make the acid without using saltpetre, though basically misguided, can be seen nevertheless to have contained some germs of truth, and it is likely that the arrangements they described did promote formation of the acid to some extent, albeit to a lesser degree than they hoped or supposed. In the first place we cannot dismiss as entirely fanciful the old distinction between brisk and slow combustion. An admittedly limited modern study has shown that the combustion temperature does affect the proportion of trioxide formed, with no detectable trace being produced at 305 °C but a rapidly rising quantity at temperatures above this; [217] the proportion of trioxide presumably remains very minor, however, even at high temperatures. In the second place we can conjecture that high temperatures might also in certain cases have yielded some acid by oxidation of the sulphur dioxide, with catalytic assistance from the walls of the apparatus. That acid might be formed in this way to a sufficient extent to be misleading is suggested by an experiment recommended by Fourcroy and Vauquelin in 1797 as a lecture demonstration to show the composition of sulphuric acid: a mixture of sulphur dioxide and oxygen was to be passed through a red hot glass tube, whereupon a very thick white smoke was said to result, condensible as the acid.[218] From the 1830s, of course, chemists began consciously to seek a manufacturing method based on contact catalysis, and this led eventually to the development of the modern 'contact process', employing the pre-eminent catalytic powers of platinum.

The theory of Clément and Desormes, 1806

By the beginning of the nineteenth century the idea that saltpetre served simply to maintain combustion, in place of a free air supply, could increasingly be seen to be inadequate. We have described how Chaptal in particular was led by his own experiments to a more specific view of its role: first to the view that it supplied the supplementary oxygen needed to form sulphuric rather than merely sulphurous acid; and then, when this in turn proved inadequate, to the view that it served to generate the necessary heat. These new interpretations, with their continuing fixation on the combustion conditions, were scarcely less erroneous than the old. It was to be the achievement of Clément and Desormes, in 1806, to break free from this fixation and to see that the true explanation lay not in the circumstances of the combustion itself, but in further reactions which succeeded the combustion.

Clément and Desormes were two bright young chemists with manufacturing interests of their own.[219] Desormes was a product of the *École polytechnique*. He had entered the school in 1794, at the age of seventeen, and after completing his studies had held a post there until 1804 as assistant to Guyton de Morveau. Clément, a year or two Desormes's junior, had taught himself chemistry in his spare time, at first in his home town of Dijon and then in the later 1790s in Paris. After three years in the capital, a lottery win had enabled him to abandon his job as a lawyer's clerk and devote his whole attention to science. His cultivation of relations with the brilliant young circle at the *École polytechnique* had brought him into a close and lasting friendship with Desormes, and in collaboration the two had begun to make a mark on the scientific scene with some notable researches on carbon monoxide in 1801–2. Their manufacturing interests were in the extraction of alum, for which they founded an important works in 1804 at Verberie, in the northern department of the Oise. They do not seem to have been engaged in sulphuric acid manufacture but were evidently familiar with lead-chamber practice. Clément was destined subsequently to have a varied and distinguished career in industrial chemistry, in 1819 becoming the first incumbent of the applied chemistry chair at the *Conservatoire des arts et métiers*, a post he was to hold for two decades. It is greatly to be regretted that, pre-eminently a man of action, he wrote no major work.

It was in a paper read to the First Class of the Institute in January 1806 that Clément and Desormes propounded the striking and imaginative new theory which still forms the basis of our understanding of the lead-chamber process.[220] Their paper began by dismissing existing explanations as being inconsistent with the evidence. Thus, the idea that saltpetre served as a source of oxygen was readily shown to be false by calculating the amount of oxygen it could actually yield. They were not in fact the first to make such a calculation. As early as 1782 Chaptal had given figures on the subject but ironically had missed their true significance: impressed by the 12 000 cubic inches of oxygen which he said each pound of saltpetre furnished for the combustion, he had not only failed to compare this with the quantity required by the hypothesis, but failed even to reflect on the comparative insignificance of those 12 000 cubic inches (7 cu. ft.) in a typical chamber space, for each pound of saltpetre burned, of at least 500 cu. ft. The chemist Berthollet by 1804 showed himself to be aware that the oxygen content of saltpetre fell greatly short of the needs, but mentioned the point only obliquely and took it as support for the rival heat theory.[221] It was Clément and Desormes who first explicitly developed the argument, and in doing so they provide us with an early example in chemical technology of the kind of quantitative reasoning which in the later eighteenth century had brought such triumphant successes in chemical science. As for the alternative view of saltpetre's role—that it acted by the heat it generated—this was challenged by Clément and Desormes as being inconsistent with the common industrial practice of mixing clay and water with the combustion mixture, for such additives had quite the reverse effect. Moreover, it was known (so they claimed, though how they did not say) that the combustion of sulphur at a temperature as high as 1000 °C gave no trace of sulphuric acid.

Their own theory rested on the novel supposition that the combustion yielded only sulphurous acid gas (sulphur dioxide) in the first instance, and that it was from this that sulphuric acid was then formed by subsequent oxidation. To explain how this occurred, they pointed to a fact which had been remarked on already in 1789 by the luckless Chaptal,[222] and which indeed must have been obvious to any manufacturer: that the deflagration of the saltpetre visibly filled the chamber with nitrous gases. 'This is the observation', they wrote, 'which affords the key to the true theory, and it is in following up the consequences that we find the clear explanation of the production of sulphuric acid.' Drawing upon the known properties of the gases involved, and on their own observation that sulphur dioxide and nitrogen dioxide spontaneously reacted when brought together, Clément and Desormes conceived a cyclical reaction scheme in which (using modern terminology) nitrogen dioxide oxidized sulphur dioxide while itself being reduced to nitric oxide in the process, and was then immediately re-formed from the nitric oxide by reaction with atmospheric oxygen present in the chamber. The nitrogen dioxide thus served as a constantly regenerated oxidizing agent. Though present in only small proportion, by its repeated action it was able to continue the oxidation until either all the sulphur dioxide or all the oxygen in the chamber was exhausted. Clément and Desormes's experiments revealed that the reaction between nitrogen dioxide and sulphur dioxide in fact occurred in stages, the two gases first forming a compound, observed as crystals, which was then decomposed by water. Their total theory can thus be represented as follows:

'nitrous acid gas'   +   'sulphurous acid gas'   →   unnamed compound
(nitrogen dioxide, NO_2)   (sulphur dioxide, SO_2)   (nitrosyl sulphuric acid, $NO.HSO_4$)

unnamed compound + water → sulphuric acid + 'nitrous oxide gas'
(nitric oxide, NO)

'nitrous oxide gas' + oxygen → 'nitrous acid gas'

That the conditions they supposed to exist in the lead chamber did indeed result in such a sequence of reactions they demonstrated by a very pleasing experiment, in which sulphur dioxide, air, and a small quantity of nitric oxide were fed into a glass vessel:

one sees the oxide grow red [forming nitrogen dioxide], and spread itself throughout the space; then clouds of white fumes roll across the vessel, and deposit themselves on the walls in shining, stellated crystals. These thick whirls of sulphuric acid are succeeded by clearness; and if at this moment one adds a little water, the crystals of acid dissolve with great heat, and the nitrous oxide gas, again becoming free, changes once more into the red acid, and the same phenomena begin anew until all the atmospheric oxygen is used up, or all the sulphurous acid burned. The remaining gases are just as we indicated in our conjectures; for the colour of the nitrous acid appears with almost all its initial intensity; after the complete operation there is no more odour of sulphurous acid, but much nitrogen, and oily sulphuric acid on the walls of the vessel.

This they considered a decisive demonstration of their theory, 'which is only a simple exposition of the facts'. From the time of Clément and Desormes to the present day, there has been much discussion of the finer details of the process, and indeed there is still some disagreement as to precisely what does go on inside the chambers,[223] but the general cyclical scheme which they proposed, and which can be represented in modern terms as

$$SO_2 + NO_2 + H_2O = H_2SO_4 + NO$$
$$NO + \tfrac{1}{2}O_2 = NO_2$$

remains the basis of the generally accepted interpretation.[224]

The work of Clément and Desormes is interesting as a particularly clear-cut early example of the enlightenment of technology by science, a happy instance of the application of scientific knowledge and skills to the investigation, explanation, and hence (in due course) amelioration of an industrial process. At the same time it illustrates the converse contribution which technological work could make to pure science, for the process they elucidated constituted the first well-characterized example of a catalytic reaction, and their theory was also in effect the first intermediate compound theory of catalysis. Clément and Desormes plainly recognized in the process the features we now consider definitive of catalysis, pointing out that the nitrous gases were not consumed but served rather as an 'instrument', and they predicted that further such reactions perhaps awaited discovery. We should add, however, that in the event the lead-chamber process was to play a less prominent part in the early history of catalysis than one might therefore have expected. The very success of Clément and Desormes in explaining the mechanism tended to diminish the apparent significance of the process, depriving it of the mystery which lay in the further (unexplained) instances of catalytic action subsequently discovered by others. It was to be these later discoveries—such as the contact ignition of gases by a platinum wire—which were to be chiefly responsible for shaping early ideas on catalysis when the phenomenon began to find general recognition in the 1820s.[225]

Clément and Desormes's paper received a wide circulation. After its presentation to the First Class of the Institute, it was published in September 1806 in the *Annales de chimie* and in the *Journal des mines,* and translations appeared the following year in Nicholson's *Journal of natural philosophy* and Gehlen's *Journal für die Chemie und Physik.* The new theory met with a cautious reception at first from some chemists. Thus, Berthollet and Guyton expressed certain reservations when they reported on the paper to the First Class, though they did recommend its publication in the *Journal des savans étrangers.*[226] On the other hand, support for the theory came from Gay-Lussac, who in 1807 brought forward fresh evidence and arguments in answer to some of the continuing doubts.[227] In the first place, Gay-Lussac further challenged the view, still held by Berthollet, that one of the principal factors involved was a high temperature. This view had seemed in keeping with accepted ideas on the general chemical behaviour of sulphurous and sulphuric acids, and was also related to Berthollet's theoretical notions regarding the manner in which the elements were there combined.[228] Gay-Lussac's own researches on the thermal behaviour of sulphates, however, led him to discover that sulphur trioxide was not after all stable at high temperatures, but decomposed, and this seemed clearly to imply that far from aiding formation of sulphuric acid, a high temperature would positively oppose it. In the second place, he sought to dispose of an objection to the new theory deriving from the experiment of Chaptal which had still yielded sulphuric acid when potassium chlorate was substituted for saltpetre. Gay-Lussac explained this by pointing to the known fact that in the presence of water sulphur dioxide was slowly oxidized by direct reaction with atmospheric oxygen, a reaction which he presumed to occur to some minor extent in lead chambers, alongside the more fundamentally important indirect reactions described by Clément and Desormes. It must have been by such direct oxidation, he argued, that sulphuric acid had been obtained in the old bell process, and it must have been by similar reaction that it was formed in Chaptal's experiment, the chlorate being essentially irrelevant, merely facilitating the combustion in an incidental fashion. Gay-Lussac thus further helped clarify the basis of sulphuric acid production, re-emphasizing that the acid was not formed as the direct combustion product, that this was always sulphurous acid, which might then yield the higher acid either through the intermediary action of nitrous gases, or by slower direct reaction with atmospheric oxygen.

After the initial doubts, the new theory seems to have quite rapidly won scientific acceptance (though occasionally accompanied by lingering traces of the older ideas). It was presumably only the late appearance of the theory which prevented its inclusion in Chaptal's *Chimie appliquée* (1807), and it does figure in such later chemistry texts as those of Bouillon-Lagrange (5th edn, 1812) and Thenard (1813). In Britain, too, the theory soon found adoption. Dalton was clearly delighted by it. He described it in the second part of his *New system of chemical philosophy* in 1810, referring to the 'excellent essay' of Clément and Desormes, and commenting that the theory 'has so very imposing an aspect, that it scarcely requires experiment to prove it'. He was moved to visit a factory near Bolton, to see the process for himself and to collect chamber gases for analysis, and in his account he briefly discussed some of the implications of the theory for the manufacture of the acid. The theory was also described by Humphry Davy, in his *Elements of chemical philosophy* of 1812; Davy presented a slightly amended mechanism, experiments of his own having indicated correctly that the initially formed compound of sulphur dioxide and nitrogen dioxide involved water too.

More difficult to gauge than the scientific reception of the theory is its industrial impact. Its authors certainly anticipated that it should lead to practical improvements, referring in passing to its likely influence on the size and shape of chambers, the conduct of the combustion, and above all on the consumption of saltpetre. In 1810, however, we find Clément saying that no major improvement had in fact resulted, and that many manufacturers had doubted the theory (see p. 82). Two years later, the Marseilles manufacturer Rougier, while remarking that the theory was 'known to all those who concern themselves with the physical sciences', added that 'This discovery has not, however, brought any change to the operations of manufacture, and it is still custom which directs the majority of manufacturers'. On the other hand, Chaptal in 1819 considered that 'The ingenious theory which MM. Clément and Desormes have given of this manufacture has contributed not a little to improve it'.[229] We shall see in due course how during the Napoleonic period the theory did have some direct impact in the industry with regard to the saving of saltpetre, suggesting new and more fruitful approaches to this old problem at the same time as it revealed the futility of those formerly tried; and the theory was also perhaps beginning to make less specific contributions through the new conceptual basis it provided for the general conduct of the chambers. The practical benefits of the new theory were to be felt more largely in the longer than in the shorter term, however, and the major manufacturing innovations which found general adoption in the early nineteenth century were essentially independent of it. It is to these advances that we must next turn our attention.

Improvements to manufacture by the intermittent method

Until the beginning of the nineteenth century there was no notable change in the method of working generally employed, and this was as inefficient in its results as it was simple in conception. The yield obtained seems to have commonly been only some 100 parts of concentrated acid for each 100 of crude sulphur burned, much the same as had been procured with glass globes.[230] A figure of 140 parts of acid cited by Bouillon-Lagrange in 1798–9[231] must have been untypically good, and yet was still barely half the theoretical. Calculation shows the theoretical yield in fact to be 287–313 parts of acid per 100 of sulphur, assuming an acid strength of 93–6 per cent and a sulphur purity of 90–5 per cent.[232] The causes of this substantial loss were various. A small part ($1\frac{1}{2}$–3 per cent) of the sulphur used can be reckoned to have ended up as potassium sulphate in the combustion residue. A considerably larger portion simply remained in the residue in an unburned state, since although manufacturers were aware of this loss they were unable to secure a complete combustion. Anselme Payen later remarked that they had thrown out as useless, residues still containing 25–30 per cent of unburned sulphur.[233] If anything, his figure is an understatement, for in 1794 a sulphur refinery in Rouen—established specifically to exploit such residues—was said to obtain from them a sulphur yield of 35 per cent.[234] From the Rouen data it can be estimated that something like 10–20 per cent of the sulphur employed left the chambers unburned. If we allow the production of the sulphuric acid itself to have consumed a little over 30 per cent, then we are still left with some 30–50 per cent of the sulphur unaccounted for. A small part of this was perhaps lost through sublimation. By far the greater part, however, presumably flew off in the form of sulphur dioxide gas, which we know to have emanated from early works in considerable quantities, to the annoyance of their neighbours. It would appear from our estimates that often no more than half the sulphur dioxide produced by the combustion was actually converted to sulphuric acid, the rest escaping, or being ignorantly released, into the atmosphere.

In the amelioration of manufacture it was Chaptal who was credited with having given the lead. Fourcroy in 1796 referred to the 'remarkable improvements' which Chaptal had achieved at his Montpellier works (see p. 77), and Cadet de Gassicourt in 1804 praised the example he had set in bringing down the price of the acid. For the present, however, we are not concerned with Chaptal's own methods, which he never fully disclosed. Rather we must examine the somewhat different improvements which

Chaptal's evident success elicited from the manufacturers around Rouen. The Rouen pharmacist Arvers later recalled that when Chaptal established his works near Paris in 1798, the Normandy manufacturers saw him to be making the acid far more cheaply than they could themselves, and being ignorant of his methods responded by bringing in improvements of their own.[235] These subsequently found widespread adoption.

One improvement which the Rouen manufacturers now made was in the manner of conducting the combustion. The usual method hitherto had been to burn the mixture on iron trays, which were conveyed into the chamber by a wheeled carriage running on rails through a door in the chamber side (see pl. 2). This arrangement is featured in the earliest known account of the process in France, that published by De la Follie in 1774, from whom we learn that it had been adopted by a Rouen manufacturer (Fleury) in imitation of British and Dutch practice. According to Arvers it was employed by Holker and others at Rouen until the end of the eighteenth century. It was also used at the Javel works, and is the method almost invariably indicated in early French writings.[236] Such carriages did, however, involve some trouble and expense, since the iron and wood used in their construction were soon corroded. Some manufacturers therefore preferred a simpler arrangement. Chaptal in his early years employed a fixed hearth built on the chamber floor, while in Britain it seems to have been the common practice to burn on iron trays supported above the water on pedestals.[237] These early methods were all essentially similar, and all shared the defect noted above of producing an incomplete combustion. An improvement encountered in some British works by the turn of the century was to burn the mixture on two or more trays arranged in tiers, so that the combustion below promoted that above. Such a method probably found some use in France, too, for Fourcroy in 1800 described a tiered array of three grid-irons, while Bouillon-Lagrange a little later spoke of a four-tiered array of grids or plates, introduced on a carriage.[238] The superior remedy adopted at Rouen, however, was to perform the combustion in a furnace built on to the chamber side and externally heated.

The use of a heated furnace was not a new idea. It had been advocated in 1774 by De la Follie, on the basis of trials by the manufacturer Fleury, though the end in view had at that time been different.[239] Fleury had been struck by the inconvenience of having to open the chamber door to introduce each new charge. If opened too soon there was a great loss of uncondensed vapours, and Fleury was impatient at the time he therefore had to wait before proceeding with the next charge. He was embarrassed, moreover, by the fact that even after allowing time for condensation, recharging was still accompanied by some release of acid vapours, which offended neighbours and damaged nearby trees. In an attempt both to avert the release of these fumes and to reduce the interval between charges, Fleury had devised the technique of burning the mixture in an external stove, from which the acid vapours passed into the chamber through a pipe; to facilitate combustion, the burning mixture was gently heated. It was thought that in this way one might avoid opening the chamber altogether, since to recharge one need only open the stove. De la Follie reported the method to have given encouraging first results, and since the combustion did not now heat the chamber a further benefit claimed was that condensation was faster. It seems clear, though, that in fact the method must have been seriously flawed by a failure properly to appreciate the need for air. In common with writers on the earlier glass-globe process,[240] De la Follie and Fleury were evidently unaware of the need for the chamber (or globe) to undergo ventilation between charges; nor does any deliberate provision seem to have been made for a supply of air from outside to the burning mixture in the stove. Presumably they supposed that the use of saltpetre made access of air redundant. It is no surprise therefore to learn from De la Follie's next article, three years later, that various people who had tried the use of a stove had found it unsuccessful, for the sulphur had tended to sublime instead of burning. De la Follie perceived something of the cause, though his phlogistic chemistry, of course, limited his understanding: he now stressed that an essential requirement for successful manufacture was renewal of the chamber air between charges. His reversion in this article to an account of operation by carriages seems to denote a tacit acceptance that since the chamber therefore had to be opened after all, the stove method had failed in its objective. He did not speak of the other advantages it might present.

It was thus only a generation later that combustion in furnaces was successfully introduced. Arvers tells us that it was then first adopted by Forestier. By 1809 its use at Rouen had become very general: a report of 30 October, for example, covering five of the six works in the faubourg Saint-Sever, described them all as having chambers with furnaces. [241] The furnaces now employed were built into the side of the chamber, or sometimes beneath its floor, and differed significantly from that of De la Follie in opening directly into the chamber, so that its air could freely support the combustion (see Figs. 2.8 and 2.9). Besides promoting the combustion by means of heat, the use of a furnace also permitted a measure of control, for the assisting fire could be regulated and the burning mixture could be raked over occasionally— by opening a suitable hole—so as to break up the crust whose formation hindered the mixture's burning. A more complete combustion was thus secured. As in the old way of working, it was at the expense of the confined air of the chamber that the mixture burned, and so operation was still intermittent, with ventilation between charges much as before.

A second innovation introduced at Rouen was the passage of steam into the chamber during manufacture. Again the idea was not new. The value of water vapour had been recognized already in the seventeenth century, when those wishing to make acid by the bell method were advised to work in damp weather since a greater yield was then obtained.[242] The deliberate generation of steam was suggested in 1682 by Hartman: some earlier writers on the bell method having prescribed burning the sulphur over a dish of water, Hartman now recommended that the porringer of sulphur be stood *in* the water, so that its heat might raise steam.[243] In the mid-eighteenth century, the bell process as described in Macquer's dictionary involved burning the sulphur over a dish of hot water, while the *Encyclopédie* spoke of passing a jet of steam into the bell during combustion.[244] When production of the acid in globes was developed, early accounts indicated that these should be heated, to raise steam from the water inside. It was thus simply extending an old principle when De la Follie in 1777 proposed the use of steam in lead chambers.[245] In his earlier paper, he had planned to speed condensation by spraying in cold water after burning each charge, but he had subsequently found this to have little effect. He now suggested therefore that steam should be introduced instead, by fitting the chamber with a large aeolipile (meaning, of course, not the familiar revolving toy, but the simple boiler and pipe which was used in the eighteenth century to provide a steam jet). Small-scale experiments in glass vessels, which De la Follie demonstrated before representatives of the local academy, convinced him that the method was sure of success, and he envisaged a doubling of the speed of manufacture and a reduction in the loss of acid fumes. He expected, too, that there would be some saving in saltpetre, for he believed the steam to play a chemical role in the combustion, aiding the acid's disengagement from the sulphur by combining with it upon its release.

Despite the simplicity of De la Follie's suggestion, and the high hopes he had of it, only at the beginning of the nineteenth century did steam actually come into use. Arvers ascribed its first adoption then to Chamberlain (at Honfleur), and it was evidently in common use at Rouen by 1809.[246] At first it was supposed that the steam simply helped to condense the acid vapours, but with the appearance of Clément and Desormes's theory in 1806, its deeper utility in promoting the chamber reactions began to be understood. Clément and Desormes themselves, in their classic paper, pointed to its useful mechanical effect in stirring up and mixing the reactant gases, and this became a common early view. By the 1820s Clément also taught that its action in warming the chambers was a further significant promotional factor. And towards the middle of the century, emphasis was being put on the steam's chemical role, in presenting water for the reactions in a widely dispersed form (another aspect already touched on in fact by Clément and Desormes). Whatever may have been the respective value of these various modes of action, it is clear that the adoption of steam must have contributed considerably to improving the efficiency of manufacture, effecting fuller and speedier conversion of the sulphur dioxide to sulphuric acid. Parnell in 1844 considered that:

The introduction of water into the chamber in the form of steam is one of the most important of the modern improvements in this manufacture, being the principal cause of the larger product and increased rapidity of the present process compared with the old method.[247]

Various other factors might be suggested as having also contributed to the gain in efficiency. Thus, one cause of sulphur dioxide loss at many eighteenth-century plants was perhaps simple leakage of the chambers; certainly Arvers described this as having been the case at Rouen, and considered that one of the notable early-nineteenth-century improvements had been to render the chambers gas-tight (see p. 93). In a less direct way, the scientific enlightenment afforded by the theory of Clément and Desormes was probably also of some relevance to the reduction of sulphur dioxide loss. Though the two chemists did not discuss the implications of their work in this respect, it does seem to have led to a significant change in the way the sulphur dioxide was regarded. Hitherto, the gas had been considered a waste by-product of the combustion, which once formed constituted an inevitable loss; improvement was to be sought, it seemed, in the amendment of combustion conditions so as to produce less of it. These views are exemplified with particular clarity by an unidentified English writer, who in about 1799, describing manufacture at Bealey's works near Manchester, observed that the purpose of the saltpetre

is to produce such rapid inflammation of the sulphur as will be productive of sulphuric instead of sulphurous acid . . . Even with nitre [saltpetre] a great deal of blue flame is present, and consequently much sulphurous acid [sulphur dioxide] formed, and this sulphurous acid is all lost to the manufacturer, as it flies off in the operation . . . Hence . . . the manufacturer loses 40 parts in every 100 of sulphur. This 40 parts is converted into sulphurous acid, and is lost.[248]

It was a similar view of this gas which led Chaptal in 1789, and still in 1807, to regard its release on ventilating the chamber as a normal feature of manufacture.[249] The theory of Clément and Desormes revealed that, far from being a waste by-product, the sulphur dioxide was on the contrary the essential reactant in the process, whose oxidation was the very basis of manufacture, whose escape therefore was to be avoided, and whose release on ventilating the chamber indicated incomplete reaction and faulty conduct of the operation. We can take it as an indication of the advance made by the later years of the Empire, both in the conduct and in the comprehension of manufacture, that writers on the industry then describe the evacuated vapours as consisting essentially of nitrous gases, with appreciable sulphur dioxide present only exceptionally, in the case of an occasional faulty operation. Besides the contribution made by the use of steam, this improvement must also have depended on small changes in working detail of a kind too minute to be readily identified. One such is suggested in a report of 16 November 1811 by the *Jury médical* of the Seine-inférieure.[250] In finding the fumes then emitted by Rouen's acid works to give no grounds for justifiable complaint, this body asserted that one reason for the release of sulphur dioxide in former times had been the combustion of charges too large for the chamber, through failure to appreciate that it was on the air of the chamber that the oxidation of the sulphur entirely depended. Such limited data as is available from the eighteenth century, however, would suggest that in fact the chamber capacity was then commonly some three times that theoretically required for the charge burned, and figures from the early nineteenth century afford no evidence of a reduction in charge.[251] It would seem that the routine release of sulphur dioxide by the early industry had generally been due to shortcomings in the manner of working rather than to any absolute disproportion between charge and chamber size.

Apart from Clément and Desormes's theory, another scientific advance whose practical bearing we should consider was that made in the quantitative analysis of sulphuric acid. Lavoisier's revelation of the essential chemical nature of the acid as an oxidation product of sulphur led to much work on the determination of its quantitative composition, and in due course such analysis showed that 100 parts of sulphur should theoretically yield a little over 300 of concentrated acid, if production methods were perfect. In this way the results of analysis enabled manufacturers to measure the chemical efficiency of their operations in precise and absolute terms, and assess the room remaining for improvement. This role of analysis in chemical technology might be compared with that of thermodynamics in power technology, where it was the efficiency of engines which became determinable, and it is one of the significant ways in which the quantification of chemistry in the eighteenth century was able to assist manufacturing improvement in the nineteenth century.

It is tempting to present a picture of chemical manufacturers after Lavoisier, their eyes opened by science to the gross inefficiency of their ways, setting about improving production in pursuit of theoretical perfection, but we must beware of over-simplification and exaggeration. In the first place, the exact analysis of sulphuric acid proved to be surprisingly difficult. The nineteenth century opened with widely divergent results in circulation, erring by up to 75 per cent, and in fact the figures employed by Clément and Desormes in establishing their lead-chamber theory were in error by 30 per cent; the direction of error was consistently such as to underestimate the acid yield to be expected. Only towards 1810 was there a general acceptance by chemists of values close to the true ones, and only then could their results become a reliable guide for the manufacturer. In the second place, there are disappointingly few instances known of those concerned with the industry actually considering the theoretical yield. The earliest is that of the anonymous English writer mentioned above, who in about 1799 made intelligent comparisons of theoretical and practical yields on the basis of Kirwan's analyses, which happened to be fairly accurate. In France no example has been found before 1810, when Clément reckoned the theoretical yield at 329 parts of concentrated acid from 100 of sulphur (a reasonable figure), and observed that there was therefore still room for improvement in manufacture.[252] In the same year the Rouen manufacturer Pelletan compared his own yield with theoretical expectation, though without giving us exact details; it is an indication of how misleading the early analyses could be that Pelletan, presumably using outdated figures, believed that he obtained more acid than was theoretically possible and was led into speculations about the possible non-elementary nature of sulphur.[253] These examples were all private computations. In the published literature, references to the theoretical yield are not encountered until the 1820s.

A fairly detailed description of manufacture by the improved methods outlined above was included in a memoir published in 1812 by Rougier, a soda manufacturer near Marseilles. Since this is the earliest such description known, we shall conclude this section by giving an account of it here, as an

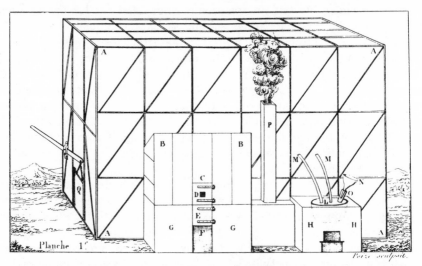

Fig. 2.8. Lead chamber with combustion furnace and steam boiler, 1812.
(From Rougier (1812), Pl. 1.)

illustration of the manner or working at the time of the Empire. Rougier describes the method as that practised at Rouen, and in some other places. His portrayal of the kind of chamber used can be seen in Fig. 2.8.

The coarsely powdered mixture of sulphur and saltpetre was spread out on an iron tray or pan in the combustion furnace, BB. Generally, rectangular trays measuring about $4\frac{1}{2}$ by 2 feet were employed, capable of taking up to some 50 kg of sulphur, with 4 or 5 kg of saltpetre. The charge actually burned depended, of course, on the size of the chamber and was to be determined by trial. The mixture was ignited by throwing in some burning coals or straw, and the charging door, C, was then closed and luted with clay. Once the whole surface had taken light, the fire beneath the burning mixture was lit, and then soon afterwards the fire under the boiler, L, was also lit, so that steam began to pass into the chamber through the pipes, MM. This steam replaced the layer of water formerly introduced, and the quantity to be passed in at each combustion was to be determined by trial. As the acid formed, it condensed on the cold chamber walls, dripping off the roof to produce a sound like rain. During the course of the combustion, the wooden stopper closing the hole D in the charging door was removed from time to time and the burning mixture was raked over to break up the crust which formed. It was also necessary, as the operation proceeded, to regulate the pressure inside the chamber so as to prevent damage. When the leads creaked, indicating a fall in pressure, the workman had to unstopper the hole and let in air. When, on the contrary, there was an excessive build-up of vapours inside, so that in the terminology of the workers the chamber was said to 'push', and white fumes began to escape from badly luted joints, it was necessary to slacken the fires. Some care was needed in handling the combustion to avoid sublimation of the sulphur, which would contaminate the acid; this would occur if insufficient air was admitted, and could also be caused by over-vigorous raking, and by use of insufficient saltpetre or saltpetre of poor quality. When all the sulphur had burned, so that no more flames appeared when the mixture was raked over, the chamber was left to stand for two or three hours while condensation proceeded. This was recognized to be complete when on hitting the charging door with a hammer the chamber resounded with an empty ring. The chamber was then ventilated by opening doors in two opposite sides. (To Rougier's account we might here add that this method of ventilation was the simplest but not the only one: at Rouen, vents in the chamber roof were sometimes employed as well, and occasionally chimneys, which might have a fire at the base to assist the draught. The best method in any particular case depended on the situation of the chamber.) On opening the chamber doors, the reddish-yellow nitrous vapours were seen to emerge. If these were suffocating, through the presence of sulphurous acid gas (sulphur dioxide), the workers said that the chamber was 'hard', and this was a sign that the operation had been badly conducted. But if they contained little or no sulphurous acid, so that men could work nearby without discomfort, the chamber was said to be good or 'sweet'. When the chamber had cleared, the combustion residues were removed and a new operation was begun. With a charge of about 50 kg, each combustion lasted about 12 hours. Acid of strength 50° Baumé (*c*. 63 per cent) collected on the chamber floor.

The development of the continuous mode of production

In parallel with the improvement of manufacture along the lines indicated above, the early nineteenth century also saw the development of a quite different mode of operation, by continuous as opposed to intermittent

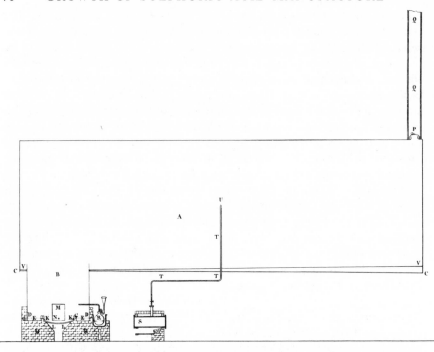

Fig. 2.9. Lead chamber, 1819. This chamber was built in 1819 to the design of Anselme Payen, and was for operation by the improved intermittent method. Sulphur was burned on the iron plate K (about 3 ft. in diameter), heated by a fire in L. Nitrous gases were generated separately, in the flask R, by reaction between nitric acid and molasses; oxalic acid was then extracted from the residues. Steam was passed into the chamber from the boiler S. After combustion and condensation the chamber was ventilated by opening the charging door M in the combustion compartment and two water valves P on top of the chamber, surmounted by wooden chimneys Q. (*Dict. tech.*, vol. 1 (1822), 126–9, and *Dict. tech. Atlas*, vol. 1 (1835), 'Arts chimiques', Pl. 3.)

combustion. This new method depended on supporting the combustion by means of an air current, instead of relying on the enclosed air of the chamber. The mixture was burned in an external furnace, supplied with air from outside, and while the resulting gases were poured constantly into the chamber at one end, those remaining after formation of the acid were steadily evacuated from the other. The perpetual draught through the plant eliminated the need to alternate the combustion with periods of ventilation, and so combustion could be carried on continuously. This method of working presented advantages which at length were to bring its total triumph over the intermittent plan. It is of particular interest, moreover, in having been probably the first continuous flow process to be employed in the chemical industry, an early foreshadowing of that general transition from batch-type to continuous processes which has so transformed the character of the industry since the late nineteenth century.

In the development of the continuous method a leading part was played by Chaptal, for it was he who pioneered the use of an air current. We have already discussed his experiments of the 1780s, in which by burning sulphur alone in a current of air he had hoped to make acid without saltpetre. Though he naturally did not succeed in this, he did thereafter become convinced of the value of this mode of combustion for the ordinary mixture, and he applied it to regular manufacture. In 1807 he wrote that before adopting this method he had for long followed the usual practice of burning inside the chamber, but in fact he seems to have had it in use with one chamber at least by 1789; this was an experimental chamber which he had built not of lead but of wood, with a protective coating of mastic, and his reason for burning the mixture outside was in this case to avoid melting the mastic (see p. 94). He evidently brought the method into general use at his works during the 1790s, and it was perhaps to this innovation that Fourcroy referred when in 1796 he remarked on Chaptal's manufacturing improvements:

The discoveries of the chemists on the nature of sulphuric acid, and the pneumatic theory to which they have lent such strength, have spread in the laboratories and workshops of the arts, where they have rendered great services. 'I avow', citizen Chaptal wrote to me some months ago, 'that if I have obtained some successes in my manufacturing operations, I owe them all to the application I have made of the principles of the new doctrine.' Indeed, it is according to these principles that this chemist has brought remarkable improvements to the manufacture of sulphuric acid.[254]

Apart from his use of an air current, we know virtually nothing about Chaptal's way of working, and it is not clear how far it was truly continuous. After the continuous method had come into widespread use its development was certainly ascribed to Chaptal by a number of writers. It was recorded as having originated at his Montpellier works by a local statistic of 1824, for instance, and J. E. Bérard, a son of Chaptal's partner in Montpellier, later spoke of continuous combustion as having been the basis for the firm's expansion during the 1790s, and for its healthy profits.[255] The continuous mode of operation would seem indeed an obvious corollary of combustion in an air current. It was not, however, an inevitable one, as appears from its use with closed chambers and intermittent working by Tennant in Scotland in 1807 (p. 100), and there is room for doubt about Chaptal's precise practices. His textbook of 1807 recommends the external furnace but contains no reference to continuous operation. That the method of working at his Paris works in 1810 was certainly in some degree intermittent seems clear from a remark to his son in a letter of 7 October:

I believe the new disposition of the chambers to be excellent, I believe indeed that the alternation of the combustion and the emission of steam is very good, but there is still something to be done so as not to lose the nitrous gas which escapes when one opens the chambers.[256]

Whatever Chaptal's own practice may have been, the continuous method rapidly became known in the industry after about 1810. It figures in the memoir of 1812 by Rougier, who considered it superior to intermittent combustion. Rougier described it as the method used in Paris, and that most generally followed (he was here perhaps thinking particularly of his home region around Marseilles). As typically worked in the 1810s,[257] the process employed a brick-built furnace—usually unheated—with a canopy of

Fig. 2.10. Lead chambers at Marseilles, 1823. The chamber on the left was for operation by the continuous method. A constant current of air entered through the small hole in the centre of the charging door, while the waste gases were steadily evacuated through the leaden chimney (4 ins. in diameter and 15–20 ft. high). The chamber on the right was for operation by the improved intermittent method, and besides the external stove for combustion of the mixture it can be seen to be fitted with a steam boiler. The stove illustrated is unusual in being for combustion on an unheated hearth, rather than on the heated iron plate which was more common with this mode of working. Periodical ventilation was by doors in the chamber side. (Péclet (1823), pp. 131–5 and Pl. 5.) (Bibliothèque Nationale)

masonry or lead communicating with the chamber through a wide pipe, or sometimes simply opening into the chamber directly. A small hole in the charging door gave admission to the steady current of air, while the waste gases continuously escaped from the far end of the chamber through a narrow chimney (perhaps 4 inches in diameter and 15 or 20 feet high). To capture any sulphuric acid escaping with the effluent gas, some manufacturers first passed the gas through a long leaden channel on top of the chamber, containing water; the acidulated water from this was then added to the chamber below. Unlike the improved intermittent method, the continuous process did not commonly employ steam, the chamber floor instead being initially covered with a layer of water, as in the old way of working; only after the 1820s did steam injection become a usual feature of continuous operation. Charges were burned in immediate succession until the acid on the chamber floor reached 35–45° Baumé (42–55 per cent).

A noteworthy elaboration already encountered in some works, and later to become standard, was the use of multiple chambers connected in series instead of single chambers worked individually. As early as the 1770s a pair of connected chambers had been employed at Javel, in conjunction then, of course, with intermittent combustion. The exact mode of operation there is unclear, but the mixture was burned on carriages and the vapours were transmitted from one chamber to the other through 'un guichet assez grand'; the arrangement was said to hasten condensation, so that the carriages could be recharged more frequently.[258] This was an isolated instance, however, and it was to be with the practice of continuous combustion that multiple chambers brought notable advantages. One of their earliest known uses in this context was towards 1815 at the Saint-Gobain soda works, whose methods were published in 1815 by Tassaert, the plant's director. Two chambers were employed, each 30 ft. long, 20 ft. wide, and 12 ft. high, and the gases passed from the first to the second through three pipes, the residual gases then finally escaping through a chimney. Burning at a rate of some 250 lb. of sulphur a day, with 13–15 per cent saltpetre, Tassaert claimed the very good yield of 283 parts of concentrated acid per 100 of sulphur. Methods at Saint-Gobain were still in a state of flux, though, and this multiple arrangement does not seem to have been retained. Evidence dating from between 1817 and 1824 suggests that the four chambers then operating by the continuous method were being individually worked. The effluent gases, however, were now passed through a 'réservoir supérieur, dit petite chambre', from which the water was subsequently run into the 'chambre principale' below, an arrangement presumably resembling the exit channels that we have previously mentioned.[259] Apart from the Saint-Gobain works, multiple chambers were probably also employed from 1813 at Darcet's soda factory near Paris, and they were used in the 1820s by Anselme Payen (see below). Towards the middle of the century the use of connected sets, usually of between two and five chambers, was to become standard practice. They gave a better yield than did the original single chambers. This was not simply through the provision of more chamber space, since it was generally considered that such a set worked more effectively than a giant single chamber of equivalent capacity. According to Lunge, the benefits resulted partly from the mixing of reactant gases upon their passage between chambers; partly from the promotion of liquid phase reactions, by the increase in wall area on which chamber mists could condense; and particularly from the cooling effect of the increased wall area, which prevented the heat generated by the reactions from producing detrimentally high temperatures.[260]

Although the continuous method was ultimately to displace intermittent operation, in the simple form in which it was at first generally worked it did not have a clear superiority. Each method had its own virtues and for many years the industry continued to employ them both. As far as yield was concerned, any advantage was on the side of intermittent working, since continuous chambers tended to lose some sulphur dioxide in the effluent gas. The difference was not very great, though: Payen in 1822 considered the continuous method typically to give 250–60 parts of concentrated acid from 100 of sulphur, while a particularly well handled intermittent process might yield up to 300. At Saint-Gobain, where continuous and intermittent methods were used side by side (and where manufacture was probably rather better than average), continuous combustion was reckoned in 1817 to give a yield equivalent to 269 parts of 95 per cent acid (using 11½ per cent saltpetre), while intermittent combustion gave 280 parts (using 8 per cent saltpetre).[261] One might have expected a lower yield to result also in a lower production capacity for continuous plants, but here there was a compensating factor in the ability of the continuous method to burn sulphur at a greater rate. In due course this was to give continuous chambers the advantage of a higher daily output, but at first there was probably little to choose between the methods on this score. Another advantage of the intermittent method was that it could produce a stronger chamber acid, by virtue of its use of steam. Intermittent producers commonly made acid of 45–50° Baumé (55–63 per cent), and sometimes 55° (70 per cent), the strength being simply governable by the quantity of steam admitted. Manufacturers using the continuous method brought their acid only to 35–45° (42–55 per cent), since when the aqueous layer on the chamber floor rose above this strength the process went less well. If the acid was to be concentrated for sale, that made by the intermittent method thus offered savings in fuel and associated costs. Such acid, moreover, was strong enough as it came from the chamber to be used directly for many purposes—such as soda manufacture—whereas that made continuously might require preliminary concentration in leaden pans. On the other hand, the intermittent method entailed significant fuel costs for the combustion furnace and steam boiler, and the iron furnace plates on which it burned the mixture needed frequent renewal. Intermittent operation also involved more labour: at Saint-Gobain it was reckoned in 1817 that to produce the factory's requirements entirely by the intermittent method would need four workers, compared with the two who sufficed while production was largely continuous. Finally, the chambers themselves had a more arduous time in the intermittent process, being subjected to the strain of constant pressure changes; continuously worked chambers were thus expected to have a longer life.

Among early manufacturers using the continuous method probably the most successful was Jean Holker, who indeed has been credited with its invention by some writers. This attribution was made particularly by the Rouen chemist Girardin, who wrote in the mid-nineteenth century, partly on information from Holker's son.[262] Girardin tells us that Holker devised the method at his Rouen works in about 1810, and we can perhaps take it as supportive evidence of Holker's concern with the method at that time, that continuous operation was an incidental feature of a manufacturing improvement patented in November 1810 by his partner Pelletan (see p. 86).[263] Whether or not Holker was really the first to employ the continuous principle is of little consequence: the important point is that he was evidently soon exploiting it with uncommon skill. Any use of it in his Rouen works can have been for only a short time in view of the works' closure in 1811, but in 1813 Holker introduced his methods into Darcet's soda factory at la Folie, near Paris, and there they were to enjoy great success. Darcet's nephew Grouvelle later wrote, no doubt with some exaggeration:

The process of M. Holker, employed for the first time in the factory at la Folie, near Nanterre, with its steady combustion of sulphur and the constancy of its system of supplying nitrous acid, its multiple chambers and the rational course of the production and condensation of the acid, carried this industry in a single bound to an acid production of 310 [parts] per 100 of sulphur, that is to say to theoretical perfection.[264]

It is possible that Holker's methods subsequently found use in the nearby works of Chaptal *fils*, at les Ternes, for as we have seen earlier, a partnership between Holker, Darcet, and Chaptal *fils* brought their works under joint control in 1816, with a pooling of manufacturing secrets. Holker's methods were probably later employed, moreover, at the soda works of Chaptal *fils* near Marseilles, for on 5 July 1820 the Marseilles firm acquired the right to employ all the processes of the company in Paris, and then proceeded to improve its sulphuric acid manufacture so as to bring it up to the standard of that at la Folie.[265] The high reputation which Holker long enjoyed is shown by the fact that in 1830 we find Clément making approaches for the possible purchase of his process,[266] perhaps for the Saint-Gobain company. It is regrettable therefore that we have no further information on the particular methods Holker used. The detailed description sent by Holker's son to Girardin in 1851 has not been traced.[267]

We do, however, have some details of a plant used in the 1820s by Anselme Payen and Cartier, and we can cite this as an illustration of the degree of sophistication which production methods had then reached among leading manufacturers.[268] Payen and Cartier employed a series of three chambers, after which the effluent gas passed through a gently sloping channel to a chimney. Steam was injected constantly into the terminal channel, and from time to time into each of the chambers. Manufacture was so regulated as to maintain the acid on the floor of the first chamber at a strength of 48–50° Baumé (60–3 per cent), that in the second chamber at 46–8 per cent, and in the third chamber at 16–20 per cent. (With chambers in series a weaker acid is naturally formed in later chambers, as the sulphur dioxide becomes used up, but Payen and Cartier seem also to have deliberately maintained it at a lower strength on the principle that acid production was thereby favoured.) The chambers were built at successively higher levels, so that with the regular withdrawal for use of a part of the acid from the first chamber, it could be replaced by running in weaker acid from the second, while similarly the second chamber received acid from the third. This passage of acid through the set in countercurrent to the gases was to be a standard feature of multi-chamber systems. Payen and Cartier claimed to obtain at least 300 parts of acid from 100 of sulphur, a yield equal to that obtainable by the intermittent method. The production rate for a given chamber space was claimed to be one-third higher than for intermittent operation, labour costs were halved, saltpetre was used in the low proportion of only 8 per cent, and there was a saving of nine-tenths of the fuel employed by the intermittent method in the combustion furnace. Comparison of the plant here described with the simple chamber and carriage portrayed in Pl. 2 well illustrates the rapid strides made since the end of the eighteenth century.

It seems to have been with the general adoption by continuous manufacturers of such elaborations as multiple chambers and steam injection that the continuous method finally established its supremacy, and it appears largely to have displaced intermittent working in the 1830s. Quoting Grouvelle again:

This system [Holker's], applied everywhere from 1830 to 1840, tripled the production of acid in Europe for an equal chamber space, and so brought down selling prices that it was necessary to seek everywhere large and new uses for a great part of the acid produced.

The nitrous gases

The old hopes of saving saltpetre costs by using air currents, oxidizing agents, or heat instead were shown by Clément and Desormes's theory to have been misguided. At the same time their theory suggested directions in which economies on this head might more properly be sought. In the first place, since the saltpetre was now revealed to act by virtue of the nitrous gases it furnished, one might look to obtain those gases from some cheaper source. In the second place, since the nitrous gases were not consumed but remained intact among the waste gases released, one might try to collect them and use them anew. Though Clément and Desormes did not spell out these implications, they did remark that a major benefit to be expected from the theory was an almost total saving of the saltpetre employed, evidently having recycling of the gases in mind. In the event the response proved to be rather limited, but both implications did receive some attention.

In September 1810 Clément wrote that

This theory, demonstrated by experiment and calculation, has received the approval of the Institute; many manufacturers have doubted it; a single one has profited from our indications, but he has not taken the consequences very far; however, he has saved saltpetre in his manufacture.

The manufacturer of whom Clément here spoke had dispensed with saltpetre altogether, by utilizing instead the nitrous gases evolved in the preparation of oxalic acid, from nitric acid and sugar. This provided him with an effectively free supply, and at the same time enabled him to obtain a higher yield of sulphuric acid—up to 280 parts from 100 of sulphur—by employing the gases in larger quantity. Clément did not say who the manufacturer in question was. It may have been Cartier, whom we find listed as making both sulphuric and oxalic acids in 1812 (see p. 51), but it was perhaps more probably Chaptal *fils*, whose works at les Ternes was certainly using the method early in 1811. This is clear from a letter written to him in February by his father, who was then temporarily keeping an eye on the works.[269] Chaptal *père* wrote that the method was still being used since the market for the oxalic acid was not yet saturated; but significantly he added that it would be wise to develop as an alternative the production of corrosive sublimate (mercuric chloride, made via the nitrate), since this would find a larger sale. The applicability of such methods, indeed, was unfortunately limited by the market for their products.

Clément rightly saw that the only kind of economy likely to be generally applicable was not the adoption of such alternative sources of nitrous gases, but the recovery of the gases after use:

There are few practicable means known for economically obtaining the nitrous gas, but since it is not destroyed after the service it renders . . ., one can recover it, something which has not yet been carried out, so that our theory has not really been utilised up to the present.

It was a process for recovering 'this precious gas . . . which is employed only as an instrument, a tool', which was one of the principal innovations described in a patent taken out by Clément on 4 November 1810.[270] (The other chief feature, to which we shall return later, was the use of pyrites in place of sulphur.) Clément gave few details, but his intention was to follow the chamber with 'a kind of artificial saltpetre bed', in which the nitrous gases would be absorbed in alkaline materials to form nitrate; from the nitrate he then planned to regenerate nitrous gases by reaction with sulphuric acid and a reducing agent. The chamber was evidently to operate by continuous combustion, using an external furnace. By his method, Clément expected to save three-quarters of normal saltpetre costs, which at 1810 price levels meant a saving of about 12 per cent in his estimate of total production costs, for concentrated acid. It is clear that he was describing a proposal, rather than a process he had actually tried, and it is doubtful whether he ever did put it into practice. He seems later to have lost interest in the patent, for he failed to pay the second part of the fee due, which led in due course to the patent's being revoked, in 1824.[271] After its publication, Dumas drew attention to it in 1828, in his textbook, and suggested that one might perhaps pass the chamber exit gases through a channel containing lumps of lime or moistened chalk. The general idea of using an alkaline absorbent continued to receive some sporadic attention as the century wore on, but never found more than rare and isolated application.

Clément was not the only person to be giving thought by 1810 to the collection of the nitrous gases. Chaptal in October wrote to his son of the need to avoid the loss of these gases, presumably with their collection in mind (see p. 78), but of his interest in the matter we regrettably know no more. We are rather better informed about some developments in Rouen, however. There, at the end of 1809, manufacturers found themselves obliged to collect their nitrous gases by order of the authorities,[272] an anti-pollution rather than an economy measure, but a requirement which we shall see did lead one manufacturer to devise a system for recycling as well as absorbing the gases. Intervention in the chemical industry by the Rouen authorities had been provoked by the rapid growth of soda manufacture there during 1809. The torrents of fumes from the new soda furnaces brought unprecedented pollution problems, and in October the Prefect consequently introduced measures to control the chemical industry's development, including the appointment of a small commission to act as an advisory and supervisory body (see p. 285). When, immediately afterwards, the manufacturer Lefrançois applied to bring two new lead chambers into operation, this commission judged that the chambers were likely to inconvenience neighbours, but helpfully suggested that the objection might be overcome by absorbing the nitrous gas from the chambers in a solution of iron sulphate. Their suggestion was based on the reaction between nitric oxide and ferrous sulphate, which had been discovered some thirty years earlier by Priestley and had since come into familiar use in laboratory analysis. The method was tried on one of Lefrançois's chambers (with leaden pipes to carry the effluent gas from its roof vents to vats of sulphate), and upon hearing that it had perfectly succeeded, that the expense was negligible, and that Messrs. Lefrançois were very pleased, the Prefect acted speedily, on 6 December, by ordering all manufacturers in the vicinity of Rouen to fit similar devices. This they duly did, though to what extent they actually used them is probably another matter. They were an additional complication, which did entail some extra expense and which slowed down the ventilation of the chambers. It is likely, moreover, that the pollution problem posed by the nitrous gases had been overestimated. Any nuisance caused by sulphuric acid manufacture probably owed more to the occasional release of sulphur dioxide from 'hard' chambers than to the regular evacuation of nitrous gases from 'sweet' chambers (the commissioners indeed recognized 'hard' chambers to be the more particular culprits, but mistakenly ascribed the cause of these to a larger than usual release of nitrous gases). The use of an iron sulphate absorbent was not to prove of any lasting value and within two years it seems for the most part to have been quietly forgotten.

One might have expected that the Rouen manufacturers would have sought to elaborate this absorption arrangement into a recycling system, for it so happens that the reaction involved is a reversible one, and the gas could readily have been re-expelled from solution simply by heating. It is disappointing, therefore, to find no record of attention being given to this possibility. In 1810, however, the manufacturer Pelletan did devise a recycling process using a different absorbent, and this he patented on 12 November.[273] His method is of particular interest in being a precursor of that later perfected by Gay-Lussac, and in due course adopted throughout the industry. Pelletan was a leading pioneer of the Leblanc soda industry in France, and his career will call for more extended discussion in Chapter 4. For the present we need only note that in 1809 he had become an associate in the Rouen works of Jean Holker. When the official absorption system was announced, he and Holker showed some lack of enthusiasm for the idea. Pelletan complained to the Prefect that it would be an unnecessary expense, since the supposed nuisance did not exist, arguing by calculation that the quantity of nitrous gases released was very small, and remarking that it was made all the smaller by their partial retention in the acid of the chamber.[274] It was the absorption of the gases by sulphuric acid itself which was to become the basis of his recycling process. That they dissolved in the acid to some extent had been obvious since the seventeenth century from the contamination of the product, and was not in itself remarkable: weak sulphuric acid dissolves the gases in a similar fashion to plain water, forming (ultimately) nitric acid. When of sufficient strength, however, sulphuric acid absorbs the gases in a different manner, forming nitrosylsulphuric acid, and it is in this property that its particular value lies.

$$NO + NO_2 + 2H_2SO_4 \rightleftharpoons 2NO.HSO_4 + H_2O$$
$$4NO + O_2 + 4H_2SO_4 \rightleftharpoons 4NO.HSO_4 + 2H_2O$$

These reactions begin to occur at an acid strength of about 58 per cent and proceed to a rapidly increasing extent above this.[275] Though the chemical interaction of sulphuric acid with oxides of nitrogen was virtually unexplored at this time, Pelletan evidently had some practical inkling of the acid's particular value as an absorbent: he was perhaps aware from manufacturing experience (as Payen showed himself to be a decade later)[276] that when the acid on the chamber floor rose above about 60 per cent it began to retain the gases in a more marked fashion, giving a more strongly contaminated product. At all events, Pelletan chose as his absorbent chamber acid of strength 45–50° Baumé (55–63 per cent). The gases were absorbed by passing them through a continuous 'rain' of this acid in a long leaden channel (not explained further) terminating in a chimney. The arrangement no doubt derived from his recent experience at Holker's works in combating pollution by the soda furnaces, for his strenuous efforts in that direction had included absorbing the fumes in a rain of water (see p. 288). The lead chamber to which the absorption channel was attached was equipped with an external furnace for the combustion, and was connected by means of glass tubes with a number of retorts for the supply of nitrous gases.

Everything being thus disposed, one lights the sulphur, which burns without interruption, with the help of a steady draught; into one of the retorts one introduces a certain quantity of nitric acid, with one-tenth of pure sulphur; when, by this method, one has obtained a certain quantity of sulphuric acid of forty-five to fifty degrees, one fills all the retorts with it, and successively boils them, introducing into each a little sulphur. At this strength and at this temperature, all the nitrous gas is disengaged from the sulphuric acid, and goes off to form more; but cold sulphuric acid at fifty degrees dissolving it readily, it is again retained in the long channel, and thus serves continually, save for a small loss.

The method described has clear deficiencies. Apart from the obvious inconvenience of glass apparatus, and the doubts one might have about mechanical provisions for the 'rain' of acid, the sulphuric acid Pelletan employed was in any case too weak to exploit very fully its special absorbent properties, and the composition of the chamber exit gas—later recognized to have an appreciable bearing on the absorption—must have been very much a matter of chance, and presumably less than ideal; with only a single chamber the gas would be likely, for example, to contain residual sulphur dioxide, which would chemically hinder the desired absorption. Whether Pelletan actually tried the method is uncertain, though it is quite possible that he did. After the end of the Napoleonic Wars it is interesting to encounter him in Manchester in 1815–16, applying for English patents for a 'new method or methods of making sulphuric acid', but whether these methods resembled those described above we do not know, since no specifications were enrolled.[277] Pelletan says nothing about recycling in his *Dictionnaire de chimie*, of 1822–4, there giving only a brief and unremarkable sketch of manufacture by the familiar intermittent method

The development of a viable recycling process was to prove much more difficult than Clément and Pelletan suspected, and was to come only three decades after their own early thoughts on the matter. During the intervening years evidence of interest is scanty. The incentive was reduced, of course, when the return of peace in 1815 brought prices down from their wartime values. Manufacturers now imported their saltpetre from England and in this way were soon able to have it at prices lower than they had ever paid before.[278] They could often, moreover, partly offset their saltpetre costs by turning to profit the sulphate of potash in the combustion residues. Before the Revolution these residues had found little use, but their application to alum manufacture in the early nineteenth century gave them appreciable value: the Saint-Gobain company, for example, in 1817 reckoned on recouping from its residues a third of its saltpetre costs, and in the south of France such residues sold in the early 1820s at 22–4 fr. per 100 kg.[279] After March 1819, the raising of the import duty from 14 fr. 30 to 79 fr. 75 per 100 kg (including a standard 10 per cent surcharge) made saltpetre a good deal dearer in France than in England, where the duty was equivalent to less than a franc,[280] but even so in the mid-1820s it still sold in Paris a little more cheaply than before the Revolution, at 160–70 fr. per 100 kg.[281] In the chemical literature the idea of recycling the nitrous gases seems to have gone unremarked until its mention in Dumas's textbook in 1828. Clément himself, surprisingly, does not appear to have discussed the subject in his course at the *Conservatoire* in the mid-1820s (although he did speak of an arrangement he had devised for feeding back into the chamber the steam evolved when the acid was concentrated, noting as a subsidiary advantage that this also returned the small quantity of nitrous gases which it had retained).[282] Attempts at recovering the gases were made at the la Folie works in 1832 by Jean Holker's son, under the direction of Darcet, but these were unsuccessful and we do not know the method employed.[283]

It was in 1842 that a viable process was finally developed, by Gay-Lussac at the Chauny works of the Saint-Gobain company. After proving itself in large-scale use the process was covered by French and English patents in the same year.[284] Gay-Lussac's work on the subject included trial of alkaline solutions, which he found to be effective absorbents, but it was to sulphuric acid that he gave his final preference since although more difficult to handle, being more corrosive, it allowed the nitrous gases to be more readily disengaged again. It is interesting to note that it had been Gay-Lussac (along with Thenard) who as a member of the *Bureau consultatif des arts et manufactures* had examined and approved the patent application of Pelletan in 1810.[285] Gay-Lussac's method differed from Pelletan's, however, in using stronger acid (82 per cent), which was far more effective. There is a world of difference, moreover, between the rather primitive arrangements outlined by Pelletan and the elaborate plant developed at Chauny. The plant's central feature, of course, was the absorption tower which came to bear Gay-Lussac's name. In this, the exhaust gases from the chamber rose through a descending stream of acid, dispersed over the surfaces of the tower's coke packing. The tower was undoubtedly inspired by the essentially similar device invented over twenty years earlier by Clément, as a general-purpose system for absorbing any soluble gas on laboratory or industrial scale. Clément had perceived the outstanding efficiency of a counterflow arrangement in a suitably packed column, combining as it did a very large area of contact between gas and liquid with a very small pressure against the gas's entry. After being patented in 1821[286] his column had been widely publicized in the chemical literature under the name 'cascade absorbante', the name by which Gay-Lussac referred to his own tower.

Even after Gay-Lussac's success, it was to be a generation before his method began to come into very widespread use.[287] To many manufacturers it looked disturbingly complicated and of doubtful profitability: they tended to underestimate the savings it could bring, and failed at first to appreciate what was to prove a further significant advantage, that by allowing the nitrous gases to be employed in larger quantity it enabled an increase in output to be achieved. The early process, moreover, did still leave something to be desired in its provisions for returning the gases to the chambers after their collection in the Gay-Lussac tower. Early methods of disengaging the gases, besides being troublesome, involved diluting the acid so as to weaken its hold, and the acid had therefore to be concentrated again before it could be reused in the tower. The problem found its definitive solution in 1859, with the invention by John Glover, in England, of his denitrating tower. Glover's method still began by diluting the Gay-Lussac acid, but the acid was then trickled down a second packed tower against a rising stream of hot sulphurous gases, on their way from the combustion furnace to the first chamber; and these gases, at the same time as they carried off the oxides of nitrogen, also concentrated the descending acid so that it emerged at a strength of some 80 per cent, ready for reuse. The method also presented the important secondary advantage that by using chamber acid for the dilution of that from the Gay-Lussac tower, it was possible to concentrate a factory's entire output to a strength of 80 per cent at no cost. It was from about 1870 that the Gay-Lussac recycling process passed into general use, as its advantages came to be more fully realized, and as the Glover tower set the final seal on its success.

Attempts to use pyrites in place of sulphur

The roasting of pyritic ores—as a preliminary to copper smelting, for example—was well known to evolve sulphurous fumes, and so it must have been fairly obvious to think of applying such materials to sulphuric acid manufacture. Chaptal in 1781 made a passing, naive suggestion, that one might collect sulphuric acid from heaps of roasting ore by simply suspending over them large leaden hoods: in effect a gigantic bell process. During the Revolutionary and Napoleonic periods a number of attempts were made to apply pyrites to acid manufacture, when wartime conditions heightened the incentive to replace imported sulphur by an indigenous substitute.

The earliest endeavour known was that which Sorel tells us to have been made in 1793 by d'Artigues in response to the sulphur shortage of that time.[288] D'Artigues was then a young man of nineteen, assisting his adoptive father, Jourdan, in the direction of the large Saint-Louis glassworks at Munzthal (near Bitche, in Lorraine). His chemical interests had already shown themselves two years earlier in experiments on the manufacture of minium, a material employed in glass-making. It is possible that his interest in sulphuric acid was with a view to use of the acid in the manufacture of soda (see p. 230). From Sorel's account it appears that he tried to burn the pyrites in some kind of furnace—presumably communicating with a lead chamber—but that his efforts were unsuccessful:

... sometimes the sulphides, charged into a furnace that was too cool, and ill-ignited, refused to burn; sometimes, on the contrary, too brisk a combustion caused the distillation of a part of the sulphur from the recently charged mass, or at certain points in the mass raised the temperature near to fusion; the softened fragments welded together and stopped the draught; in short, it was impossible to obtain a continuous and steady disengagement of sulphurous gas, an indispensable condition for normal and economical manufacture.

D'Artigues was later to establish a works in Belgium for the extraction of sulphur from pyrites,[289] but he is not known to have pursued any further his attempts to make sulphuric acid. His distinguished industrial career was to be essentially as a glass manufacturer: he was a leading figure in the development of crystal glass in France, and creator of the great works at Baccarat which today remains among the most highly reputed in Europe.

Another to concern himself with pyrites was Chamberlain at Honfleur. Since one of Chamberlain's chief activities was the extraction of iron sulphate from local pyritous deposits, we can readily understand his interest in employing these same materials in his lead-chamber plant. He included the use of pyrites among the various alternative modes of manufacture described in his sulphuric acid patent of 1801. The pyrites was to be burned in the heated furnace—supplied with oxygen as a supposed substitute for saltpetre—whose use with sulphur we have described earlier (p. 62). From the rather subsidiary place of pyrites in the patent we may doubt whether Chamberlain was actually employing it in regular manufacture at that time. From a pamphlet he published in 1822, however, it would appear that his acid was now made from pyrites, though perhaps only indirectly:

it is established that pyrites and other sulphides, which are inexhaustible in France, contain at least 50 per cent of sulphur which one can very easily obtain by simple and very economical processes. The sulphur, in measure as it is disengaged from the sulphides, is decomposed, and one obtains perfectly pure sulphuric acid of 56 or 58 degrees [72–75%] without the needless and great expense of saltpetre. The production and combustion of sulphur are without interruption; the acid obtained weighs nearly four times more than the sulphur employed; the acid contains neither lead, nor potash, nor nitrous nor muriatic acid, as are contained in all other acid obtained by processes generally used in France. The processes for this interesting production are keenly sought by England.

The prime difficulty facing early attempts to use pyrites was that of securing its satisfactory combustion. When Clément made its use a main feature of his patent of 1810 (p. 83), he proposed to promote the combustion by an admixture of charcoal, or alternatively by employing a pyrites which occurred naturally mixed with fuel, such as the deposits of the Aisne and the

Oise. The pyritous mixture was to be made into briquettes with a little clay and water, and then burned in a furnace. With Sicilian sulphur imports at this time severely restricted, and the price consequently at its wartime peak, the savings promised by pyrites were quite spectacular, as can be seen from Clément's estimates: whereas 100 kg of sulphur then cost 140 fr., an equivalent quantity of pyrites (250 kg) was reckoned at a mere 4 fr., plus transport. Apart from private profit, Clément also had in mind the potential national importance of this substitution.

Not a single kilogramme of sulphuric acid is now made without using foreign sulphur. I hope that before long not a single kilogramme of this sulphur will be used for this purpose; importation will no doubt diminish by more than half, we shall no longer each year bear millions of francs to the Sicilians and consequently to the English, we shall find its value in the soil of France, in enormous masses of ore hitherto without use, and in peace as in war we shall no longer have to fear foreign competition, since sulphur for the manufacture of sulphuric acid will cost less in France than in Sicily itself.

Like Chamberlain, Clément had particular reason for interest in pyrites, as an associate in the important works at Verberie for the extraction of alum from the pyritous deposits of the Oise. He does not appear to have worked his patent, however. Any plans he might have had for adding sulphuric acid to his manufacturing activities were no doubt undermined by the depression of the Empire's later years. Lunge considered that his method would in any case have been unsatisfactory, since the passage into the chamber of carbon dioxide from the fuel would have been injurious to the reactions. Clément himself had supposed that its only effect would be to slow the reactions down a little, by diluting the gases, and he had proposed to compensate for this by an improvement in chamber design (see p. 94).

At about the same time as Clément was giving thought to the matter, the use of pyrites was also receiving attention from the Swiss inventor Isaac de Rivaz,[290] a figure remembered by historians of technology as one of the early would-be pioneers of steam transport. Towards 1808 de Rivaz had invented a plant of a general-purpose kind for the conduct of chemical distillations, and following trials in his saltpetre works at Martigny (Switzerland), he installed a small works for the exploitation of this invention at Thonon, by Lake Geneva, under the direction of a pharmacist colleague called Biehly. The plant employed cylindrical iron retorts mounted horizontally in a furnace, with a series of Woulfe bottles as condensers, the essential novelty being the use of a suction pump to draw through the plant a constant stream of air; this arrangement was supposed, among other things, to speed production by allowing recharging of the retorts without cooling. The plant was at first applied largely to the distillation of nitric acid, but in 1809 de Rivaz sought to extend its use to the production of sulphuric acid by burning pyrites in the retorts. The fact that he continued to employ a series of Woulfe bottles, rather than a lead chamber, no doubt contributed to the very limited success he was to achieve. After protracted development difficulties, the Thonon works does seem to have made sulphuric acid in this way to some extent during 1811–12, but the process was inefficient and production of the acid was abandoned in 1813 when falling prices on the French market made it no longer economical. De Rivaz's faith in his methods was nevertheless unshakeable, and he subsequently introduced them at Lyons, where his eye had been caught by the rich pyrites deposits of Sain-Bel. Having failed in 1812 to sell his methods to the firm of Oblin, Bouvier, & Cie, he at length succeeded in 1815–16 in persuading an unfortunate goldsmith by the name of Antoine Michel to set up a small works for their exploitation in the Saint-Clair district. De Rivaz's processes, now supposedly 'improved', were in reality more complicated and inefficient than ever, and with manufacturing costs exceeding the value of the products the Lyons venture was presumably short-lived.

Among the early experimenters with pyrites we can perhaps also include Curaudau, a chemist and alum manufacturer in Paris, who in April 1811 presented to the First Class of the Institute some 'observations on the conversion of sulphide of iron to sulphuric acid',[291] though we can only guess that it was to acid manufacture that he here referred. His remarks prompted Vauquelin to speak of similar experiments made by the mining engineer Lefroy. Of these we again know nothing further, but we may note that in 1809 Lefroy had been concerned at the high cost of sulphuric acid for soda manufacture, and had devised, and put to use in the Aisne, a soda process which in place of sulphuric acid employed the lixivium of weathered pyrites (i.e. iron sulphate; see p. 265). A number of early soda manufacturers, in fact, evaded the use of sulphuric acid by this kind of expedient, and it perhaps diverted attention to some extent from the application of pyrites to the manufacture of acid, allowing as it did its use in the production of soda in a more direct fashion.

With the possible exception of Chamberlain, none of the early experimenters is known to have enjoyed any appreciable success in his efforts. Major success was not to come until the 1830s, when Claude Perret's works at Lyons mastered the technical problems and profitably took up the use of pyrites on a large scale, setting the example which others were then to follow.[292] Techniques at Perret's works were developed over the course of several years by his two young sons, Michel and Jean Baptiste, together with Jules Olivier. Their first method of burning the pyrites resembled in principle those of Chamberlain and de Rivaz in using a kind of muffle furnace (the combustion being supported, in other words, by external heating). It was on this basis that the works switched to the regular use of pyrites in 1833. This initial method had its drawbacks, however, in particular the expense of the fuel required, and the definitive solution came with the development in 1835–6 of a means of burning the ore alone, without external heat, in a burner somewhat resembling a lime kiln. The task proved less easy than one might imagine, for it was found in early trials that even when the pyrites could be made to burn properly, the resulting sulphurous gases did not react in a satisfactory way in the chamber; but in due course the secret of success was discovered to lie in suitable regulation of the draught to the burners. It was on the basis partly of their success with pyrites that the Perret family firm saw a remarkable growth towards the middle of the century, to become one of France's largest chemical enterprises. Other manufacturers were rather slow to emulate their example, but by the later 1860s pyrites had largely replaced sulphur in French acid plants.[293] In Britain it found widespread use rather more quickly. As in France, its major adoption was prefaced by sporadic minor experimentation over a number of years, with the patent of Hills and Haddock in 1818 the evidence commonly cited of early interest in the matter (the patent is said not to have been successfully worked). The first extensive endeavours to employ pyrites in Britain came in response to the notorious Sicilian sulphur monopoly of 1838–40, which set in train a gradual movement away from sulphur over the following two decades.

The chambers

From the moment lead chambers were introduced into France the sizes that were built rapidly grew. The primitive chamber which De la Follie described in 1774 had a capacity in the region of 1300 cu. ft. (French measure), and that in his 1777 account was of about 3000 cu. ft. Before the end of the century chambers of 20 000 cu. ft. were being erected, and this then became quite typical in the early nineteenth century, though there was wide variation. The 27 chambers at Rouen in 1809 ranged from about 6000 to over 30 000 cu. ft., with an average of a little under 20 000 (see Table 2.1). The 74 chambers in the region of Marseilles in the later 1820s ranged from about 8000 up to some 60 000 cu. ft., and perhaps even higher in the case of some exceptional chambers owned by Vidal, which were 192 ft. long; the average was 20 000 cu. ft.[294] By way of comparison we may note that in late-nineteenth-century practice a range of 8000–116 000 cu. ft. was indicated, with principal chambers generally 21–58 000 cu. ft.[295] There were, of course, good economic reasons for the growth in size, since the cost of a chamber depended to a large extent on its surface area, and this increased at a slower rate than its capacity. A typical 20 000 cu. ft. chamber, for example, measuring $50 \times 20 \times 20$ ft., has eight times the capacity of a chamber measuring $25 \times 10 \times 10$, but only four times the surface area. The limited data available suggests that sizes grew rather more rapidly in the early French industry than in Britain. Certainly, no imitation is encountered in France of the extraordinarily laborious and extravagant practice found in some British works, of employing very small chambers in large numbers: at Burntisland in Scotland, for example, in 1805, there were 360 chambers each a mere $8 \times 4 \times 6$ ft. (192 cu. ft. English measure), while the Prestonpans works in 1813 was still operating 108 chambers of only $14 \times 10 \times 4\frac{1}{2}$ ft. (630 cu. ft., English measure, equivalent to 521 cu. ft. in French measure).[296]

With regard to the construction of the chambers, Arvers in 1817 made a curious allegation to the effect that until the turn of the century it had been believed by Rouen manufacturers that the joints between the leaden sheets should be left unsoldered, so as to allow the outside air some access to the interior. It was on this faulty construction that he blamed the great loss of sulphur dioxide. It is true that in the pioneering English works of Samuel Garbett the sheets had not been soldered but simply riveted together, and we have seen that in 1765, at an early stage in John Holker's projects at Rouen, there had been a no doubt abortive plan to bring over from England a workman able to join lead sheets without solder. This, though, was probably simply through exaggerated early fears for the corrosion of any solder employed. It is doubtful whether any such practice as Arvers describes can have been more than an isolated eccentricity, and we can be sure that by the beginning of the nineteenth century it was always by soldering that the sheets were joined, at Rouen as elsewhere.[297] Chambers were generally built at least two or three feet above the ground, to allow access underneath for inspection and repair. Sometimes they were raised to a height of six feet or more, so that the space beneath could be employed for storage or manufacture: Jean Holker, for example, built half a dozen soda furnaces under one of his chambers at Rouen. As protection against the weather it was evidently common in the south simply to roof the chamber over (see Fig. 2.10), but in the north some more substantial covering was commonly provided, with the chambers often housed in buildings: this was the case at Chaptal's Ternes works, for instance, and was the rule in the early nineteenth century at Rouen, where each chamber generally had its own building, sometimes sharing the space with furnaces for other manufactures. At Charles-Fontaine, the Saint-Gobain company's five chambers were all housed under a single hangar, a hundred yards long.

Lead chambers were naturally items of heavy capital expense, chiefly on account of the quantity of lead involved. A typical 20 000 cu. ft. chamber built near Rouen in 1809 was said by its owner, Lefrançois, to have cost over 40 000 fr., and we can take this figure as fairly representative, for Clément in 1827 wrote that it was rarely for less than 40 000 fr. that a chamber of this size could be built (apparently including the cost of the sheltering building).[298] A lead chamber was thus comparable in cost with a medium sized steam engine, this being the sum one would have had to pay in the 1780s for the purchase, from the Périer brothers' Chaillot works, of a single-acting Watt-type engine with 30-inch diameter cylinder.[299] A chamber did not, of course, last forever. Both the tin solder and the lead sheets were gradually corroded, so that regular repair was needed to stop up leaks, and eventually there came a point at which it was necessary to scrap the chamber and build again.

Clément in 1827 spoke of a life expectancy of ten years, and reckoned that to cover running repairs, interest on capital, and eventual replacement, one should allow 15 per cent of a chamber's capital value as an annual expense in estimating manufacturing costs. By his calculations this amounted to an expense of 10 fr. for each 100 kg of sulphur burned (about half the price of the sulphur).[300]

The high cost and impermanence of lead caused some thought to be given to possible alternative materials for chamber construction. Glass sheets, glazed bricks, vitrifiable stones, terracotta, sulphur, and mastics, are all described as having been tried or proposed before 1810. A very ambitious essay in this direction was made by Chaptal in the 1780s, as the culmination of a series of systematic researches to find an acid-resistant mastic. He began by building a wooden test chamber, measuring $30 \times 12 \times 6$ ft., whose walls he coated successively with the various mastics which small-scale experiments had suggested might be suitable. Each was tested in use for a period of one or two months, after which the coating was generally found to have cracked, softened, loosened, or dissolved. After 12–15 months of such attempts, however, he lighted on a mastic of resin, wax, and turpentine, which was suitably resistant, and in 1789 he reported that the chamber with this lining had now been in constant use for three years; its only drawback was a danger of melting the mastic, but this he avoided by performing the combustion in an external furnace. Encouraged by this success, he went on to erect an enormous chamber measuring some 84 feet long, 44 feet wide, and 29 feet high, with a brick floor, walls of plastered masonry, and wooden ceiling, the whole lined with two or three coats of mastic. This must have been considerably larger than any chamber hitherto built, in France or Britain. Chaptal reckoned the mastic coating to have cost only about 6–7000*l.*, as compared with the 80–90 000*l.* which it would have cost to line a chamber of that size with lead. Alas, after eighteen months' use the roof fell in, and the construction of the chamber had been such an immense labour, and had so injured Chaptal's health, that he did not rebuild it.[301] Later, a chamber built of masonry with a coating of a sulphur mortar was patented by Clément in 1827, and was tried by him in the soda works at Dieuze,[302] but without success. No satisfactory alternative to lead was ever in fact to be found.

A minor development which we may note finally was the idea of fitting chambers with internal partitions. Clément was an early figure to think of this. Having calculated—he does not say how—that existing chambers were ten times larger than theoretically necessary, he proposed in his 1810 patent to increase their efficiency by dividing up the space to give the form of a folded pipe; in this way the gases would be better mixed in passing through, and the reactions would thereby be promoted. Dividing curtains of lead were to find some use during the second quarter of the century, but while sound in principle they proved unsatisfactory in practice since such partitions, exposed on both sides to the hot chamber gases, were very rapidly corroded.

Platinum vessels for concentration

The concentration of the chamber acid was at first carried out in a rather laborious fashion, in glass (or sometimes stoneware) retorts, generally heated in rows on galley furnaces. One step towards greater economy came with the growing use at the beginning of the nineteenth century of large leaden pans instead, but unfortunately the acid could not be taken beyond about 78 per cent in these, for the lead would then have been threatened, and so it remained necessary to complete the concentration in glass. It was with the development of vessels fabricated from platinum[303] that it became possible to dispense with glassware altogether, with significant gains in economy and safety.

As a new and precious metal, platinum had begun to attract scientific and technical interest in the middle of the eighteenth century, when the first samples found their way to Europe from the gold mines of Spanish South America. The refining and working of the metal proved to present special difficulties, particularly since its high melting point made it virtually infusible, but there were gradually evolved a variety of techniques by which it could be sufficiently purified to become malleable and therefore workable under the hammer or press. By the end of the *ancien régime* it was beginning to find some limited use for jewellery and ornamental items, and was also applied to the production of small crucibles, highly prized by chemists for their resistance to attack and their high melting point. The earliest production of vessels of a size for industrial use occurred in Britain, thanks to the work of the chemist Wollaston. In the opening years of the nineteenth century, Wollaston conducted fundamental researches into the impurities present in the native metal, and developed a superior refining technique which enabled him to market a consistent and malleable product in considerable quantity and at a reasonable price. With this platinum he in 1805 supplied an acid manufacturer in London with the first platinum boiler to be made, weighing about 12½ kg and holding 30 wine gallons (114 litres); between 1809 and 1818, he then furnished acid manufacturers in various parts of the United Kingdom with a further fifteen such boilers, ranging between 91 and 178 litres.

While platinum production in Britain rested in Wollaston's hands, in France it was monopolized by the Janety family, whose involvement with the metal went back to the 1780s. Janety *père* had then been Europe's leading pioneer in this field, and he is well remembered for his production in the 1790s of the reference standards for the new metric system. By 1812 the concern had passed into the hands of his son, who in that year exhibited to the *Société d'encouragement* two platinum boilers which were the first sizeable vessels to be made in France, though with capacities of 22 and 16 litres they were clearly still much smaller than those made by Wollaston. They were intended to serve for the parting of gold and silver, or the concentration of sulphuric acid, and it was presumably a vessel similar to these which Janety supplied in about 1812–13 to the sulphuric acid manufacturer Alban, the first in France to adopt platinum for the concentration of his acid.[304]

Alban's use of platinum for some years seems to have remained very much an exception, essentially no doubt on account of the especially high price of the metal in France. In 1813 the German chemist Gehlen complained that, ounce for ounce, Janety charged over three times the price of Wollaston for his laboratory crucibles.[305] The general adoption of platinum vessels in France began only after the Empire, when in rivalry with Janety there appeared an important new platinum-producing concern—that of Bréant, Cuoq, & Couturier—which soon succeeded both in bringing down prices (from 30 fr. to 15–18 fr. an ounce), and in producing better ware.[306] The new concern originated when Cuoq and Couturier, who were merchants in Lyons, came into possession of a considerable quantity of native platinum and approached Bréant, assayer at the Paris Mint, to see if he could refine it for them. This was presumably some time after September 1814, when Bréant moved to his Paris post from la Rochelle. Though Bréant had no previous experience of platinum he did now develop very effective refining methods, superior even to the still secret processes of Wollaston. He was no doubt encouraged in the work by Darcet, who combined his chemical manufacturing interests with the post of *vérificateur général des essais* at the Mint, and who therefore had a dual reason to follow closely work in this field. The first item made with Bréant's platinum, in 1816, was a large boiler for the concentration of sulphuric acid which Darcet passed to Chaptal for trial, presumably at les Ternes. It weighed 20 kg and was made from four sheets of platinum soldered together with gold. In its use of soldered joints it resembled all the boilers of Wollaston and also vessels made by Janety, but Bréant recognized this mode of construction to be undesirable, since gold seams after a time began to leak, requiring expensive repair. He therefore made it his aim to avoid joints altogether by producing ingots of such a size and purity as to be capable of being worked into large vessels of a single piece, and his success in this constituted a notable advance on Wollaston's products. By February 1817 the journal of the *Société d'encouragement* could report the production of a single-piece boiler of 162 litres capacity, weighing 15 kg, and others were said to be being made, all for sulphuric acid concentration. At the industrial exhibition of 1819 a single-piece boiler of 200 litres capacity was displayed. The vessels were sometimes perhaps sold simply as open boilers, but they were soon also offered complete with a platinum still-head, to be connected to a leaden coil to condense the vapours (Fig. 2.11), and there were also available by the early 1820s water-cooled platinum syphons for speedy recharging.[307]

From 1817 platinum vessels progressively passed into general use. One early purchaser, for example, was Dubuc at Rouen, who in December 1817 happily declared that he had just bought an alembic weighing 28 kg, for

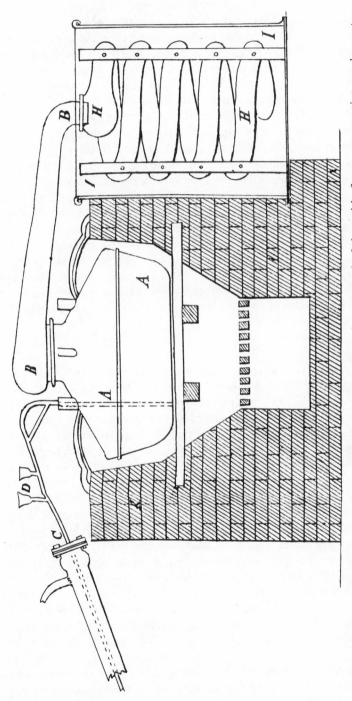

Fig. 2.11. Platinum still, 1832. Fitted with a platinum syphon for removal of the acid after concentration, and a water-cooled coil to condense the vapours. (A composite illustration based on the plate in Payen and Cartier (1832).)

which he had evidently paid 24 000 fr. but from which he was expecting great savings.[308] The adoption of platinum was rather gradual, however: an observer in Lyons in 1821–2 spoke of platinum as being employed in only one of the four works there, while in the region of Marseilles the first platinum vessel may well have been that introduced in 1823 at the soda works of Chaptal *fils*.[309] The hindrance, clearly, was the price, for even after Bréant's work they necessarily remained expensive items. In the mid-1820s they were said by Clément to cost 20–30 000 fr., the lower figure, according to Dumas in 1828, buying a retort able to distil 150 kg of acid at a time. They were thus comparable in cost with a smallish lead chamber. The advantages, on the other hand, were considerable. In terms simply of size their superiority was less than might be imagined (a glass retort typically held 50 kg according to Dumas), but they allowed for much speedier operation, with the possibility of commonly three or four, and in some cases six or seven boilings a day, as compared with a time of perhaps three days for a single boiling in glass.[310] They brought savings in labour and fuel, and also in breakages, for glass retorts frequently shattered in use, each surviving on average only five boilings. By the mid-century platinum had for the most part replaced glass, save in certain regions such as Montpellier, where the exceptionally low price of glassware made this still the more economical apparatus to employ.[311]

Conclusion

While some of the technical developments considered above were to reach fruition only after 1820, the basic improvements in the conduct of the chambers seem to have been rapidly adopted, and by the end of the Empire any remaining use of the old carriages was probably rare. A circumstance contributing to the speedy implantation of the improved practices was the rise of the soda industry soon after they were devised, for the advances were incorporated in the many new chambers built to meet that industry's demand. By the Restoration the yield generally obtained was probably about 250 parts of acid for each 100 of sulphur burned, with some manufacturers doing significantly better still. The progress was reflected in the price of the acid, and particularly in its price relative to that of sulphur, as can be seen from Fig. 2.12. In Paris and Rouen the acid was now perhaps cheaper than in Britain, despite the probably higher cost of materials and fuel. At the beginning of 1816, for example, the St. Rollox works in Scotland was selling acid at $3\frac{1}{4}$–$3\frac{1}{2}d$ a pound[312] (72–77 fr. per 100 kg), while the price in Paris was about 65 fr. and was still falling with the peace. In the 1820s the scanty data available for Britain shows the acid to be selling on Tyneside at £18.10s. a ton[313] (44 fr. per 100 kg), while prices in Paris and Rouen ran at about 30 fr.

With the yield commonly obtained now over 80 per cent of the theoretical, there were only relatively small gains still to be made in this respect. As the century progressed, the chief directions for further gains in efficiency were to be in reducing the quantity of nitrate employed (by recycling the nitrous gases), and in raising the production rate for a given chamber space. In the first decade of the century the typical burning rate was about 6 kg of sulphur a day for each 1000 cu. ft. of chamber space. Clément in 1810 said that this was the principle on which chambers were erected. By the 1820s there are signs that the rate was already increasing, for Clément taught in 1825 that

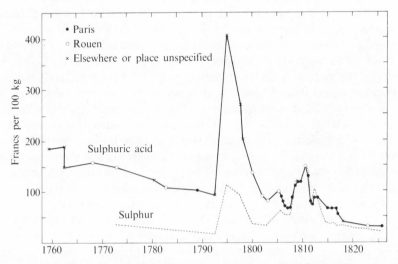

Fig. 2.12. Prices of sulphuric acid and sulphur in northern France, 1759–1825. Prices are for concentrated acid and crude sulphur, and are mostly for Paris and Rouen. The 1759 price is for Tournemine-lès-Angers, and the 1792 and 1797 prices are for Honfleur. In the few cases where no place was specified the reference was usually probably to Paris. Prices in Paris and Rouen were closely similar, but elsewhere there was some regional variation at the end of the period, e.g. 55 fr. per 100 kg at Couternon in 1819, 50 fr. at Montpellier in 1822, and 39 fr. at Mulhouse in 1830 (*Ann. ind.*, 1st series, **3** (1820), 188–9; Creuzé de Lesser (1824), 538; Soc. ind. Mulhouse (1902), vol. 2, 627–32). (Sources (in roughly chronological order): Chassagne (1971), 233; AN, $F^{12}1506$, [8]; Demachy (1773), 40–1; AN, $F^{12}1506$, [1]; $F^{12}650$, [1]; *J. de Normandie*, 7 Jan. 1789; AN, $F^{12}1966^L$, [4]; $F^{12}1508$, [20]; $F^{12}2234$, [1]; Arvers (1817), 52; Dieudonné (1804), vol. 2, 157; *J. du commerce*, 1802–18, 1825 (periodical prices-current); AN, $F^{12}1554$; F^73041; $F^{12}1966^L$, [6]; $F^{12}2245$, [21]; *Annuaire* (1823), vol. 1, 219; *Recherches*, vol. 3 (1826), Tables 114, 115.)

about 10 kg a day could be burned, while statistics from Marseilles in the later 1820s suggest a rate there possibly approaching 15 kg (with a yield of 260 parts of acid per 100 of sulphur). In the 1830s, Clément was said by Ure to have achieved a rate of nearly 25 kg at the Saint-Gobain works and at Dieuze, a figure which Ure described as 'immense'. Half a century later, the rates encountered generally ranged from a poor 12 kg up to about 35 kg, and approached 50 kg in some exceptional plants pursuing particularly intensive methods.[314]

The sulphuric acid industry was regarded with some pride in France after the Empire as one of the country's most efficiently and intelligently conducted manufactures. 'This branch of industry is today one of the most perfect that France possesses,' wrote Chaptal in 1819; 'one can even say that it has reached its perfection, since by known analysis of the acid obtained, it is proven that there is not an atom of sulphur lost in the operation. Other countries are still far from similar results.'[315] One of the French innovations—the use of combustion furnaces—is first encountered in Britain at the end of 1807, when Tennant fitted a furnace to one of his chambers at St. Rollox; he extended the method to all his chambers in 1811.[316] It seems likely that the inspiration came from Chaptal's advocacy of the furnace method in his *Chimie appliquée aux arts*, which appeared in 1807 in both French and English editions; in the same year the method was also indicated in Aikin's *Dictionary of chemistry*, which drew on Chaptal's book. After St. Rollox, the next known use of furnaces was in the Tyneside works of Doubleday and Easterby, whose first chamber was erected in 1809–10 using plans supplied by St. Rollox.[317] As for the practice of blowing steam into the chambers, our earliest definite knowledge of this in Britain is again at St. Rollox, where it is said to have been introduced in about 1813 or 1814. That these improvements were far from having general currency we can judge from Parkes's failure to make any reference to them in his *Chemical essays* of 1815, though he was a close observer of the industry and discussed it at length. In adopting furnaces, the St. Rollox and Tyneside works still operated in an intermittent fashion. The continous mode of working, with a steady draught through the chambers, arrived in Britain only after the fall of the Empire, when it was known as 'the French plan'. Gamble, in Dublin, claimed to have been the first in the United Kingdom to introduce it.[318] Exactly when is not known, but he established his Dublin works only in or about 1814, and operated it for a time by other methods before adopting continuous combustion, with great secrecy.

Apart from France's technical lead, we have already noted the lead she had acquired in the volume of her production. We should not close this chapter, however, without remarking on one significant respect in which the British industry was the stronger. This was in the field of exports, where almost certainly Britain had a considerable dominance. Though Darcet in 1817 spoke of French acid competing with British in the markets of Belgium, Germany, Switzerland, Spain, and Italy,[319] the quantities involved were not large. Over the six years 1820–5 (the earliest for which the records inspire confidence), French exports averaged 423 453 kg a year, well under 5 per cent of her production; figures for earlier years were probably lower.[320] We have no comparable set of figures for Britain, but Hardie has spoken of exports amounting to 2000 tons a year (2 million kg) by the end of the eighteenth century, perhaps echoing Clapham, who gave this figure for 1790.[321]

NOTES

1. De la Follie (1774), 335. He perhaps referred to the project of Brown (discussed below), though that was only ten years earlier.
2. Since 1664 oil of vitriol had paid duties totalling 20 *livres* per 100 pounds. On 26 June 1762 this was reduced to 3*l*.15*s*., bringing oil of vitriol into line with 'aigre de vitriol' which had paid only that amount since 1664, although it was simply another form of sulphuric acid. In 1762 the price of oil of vitriol before duty was 65–70*l*. per 100 pounds. (AN, $F^{12}1506$, [8].)
3. De la Follie (1774), 335–6.
4. Dardel (1966), 123
5. This has earlier been treated (with a few inaccuracies) by Baud (1933*c*), (1934). On Holker's career in general see Henderson (1972), 14–24, and Rémond (1946).
6. AN, $F^{12}1506$, [4].
7. Full account in André-Félix (1971), 33–72.
8. It is presumably the same Brown, however, whom one finds associated with Murry in 1770 (André-Félix, 69, 71).
9. The Garveys were an expatriate Irish family prominent in Rouen's trade and industry (Dardel (1966), 213).
10. The material on Holker's new project is mostly in AN, $F^{12}1506$, [8].
11. Rémond, 83, 95; AN, $F^{12}740$, [1], [2], [3].
12. AN, $F^{12}740$, [3]; $F^{12}1506$, [8].
13. AN, $F^{12}1506$, [2].
14. AN, $F^{12}1506$, [8].
15. AN, $F^{12}1506$, [7].
16. AN, $F^{12}1506$, [6].
17. AN, $F^{12}1506$, [2].
18. Magnien (1786), vol. 1, p. lvi; vol. 2, pp. 248–9.
19. AN, $F^{12}1506$, [8]; Demachy (1773), 41.
20. AN, $F^{12}822^{C}$, [1].
21. AN, $F^{12}1506$, [2], [10]; $F^{12}879$, [1].
22. AN, $F^{12}740$, [3].
23. *Annuaire* (1823), vol. 1, 219; Matagrin (1925), 29.
24. AN, $F^{12}1506$, [10].
25. Ballot (1923), 543; AN, $F^{12}1508$, [17].
26. Rémond, 119–21; *Dict. Nat. Biog.*
27. Dardel (1966), 200; AN, $F^{12*}172$, f° 363r.
28. Louis and Jean Baptiste (see Scagliola, 58–9).
29. De la Follie (1774); Demachy (1775), vii–viii.
30. De la Follie's contributions to the development of local industry were to bring him an appointment as inspector of manufactures shortly before a chemical accident led to his premature death in 1780 (Lebreton (1865)).
31. Chaptal (1782), 152.
32. AN, $F^{12}1506$, [11].
33. Dardel (1966), 199.

34. Ballot (1923), 543.
35. AN, $F^{12}1506$, [11], [9]; Dardel, 199.
36. ADSM, C 80, [1].
37. Arvers (1817), 48–9. I have presumed in what follows that the 'Messrs Cartame and Strouble' referred to by Arvers were in fact Carstens (Dardel, 200) and Stourme. Arvers is frequently inaccurate.
38. Ballot (1923), 543.
39. Wallon (1897), 128–33; Young (1931), vol. 1, 258.
40. Ballot (1915), 48.
41. AN, $F^{12}1506$, [11]; $F^{12}1507$, [4]; $F^{12*}30$, p. 16.
42. AN, $F^{12}650$, [1].
43. AN, $F^{12*}106$, pp. 648–50.
44. AN, $F^{12}1506$, [3].
45. Procès-verbal, 1781, p. 353; AN, $F^{12}1506$, [13], [12].
46. Procès-verbal, 1789, pp. 512–13; Thomas [1936], 195.
47. ADH, C 2739.
48. ADH, C 2741; AN, $F^{12}1507$, [5].
49. Ballot (1923). 543.
50. AN, $F^{12}879$, [1]. There was a Malvesin listed as a manufacturer of copper sulphate in 1804 (Herbin de Halle, Atlas, 106).
51. See the informative article by Rolants (1922); also Leclair (1901).
52. CNAMA, [3].
53. Pingret and Brayer (1821), Pl. 34; AN, $F^{12}1552-1553$.
54. Demachy (1773), 37; Biog. univ. Supp. (valuable notice on Athénas); AN, $F^{12}1508$, [12], [23].
55. Arvers (1817), 49; Dardel (1966), 200; AN, $F^{12}1508$, [17].
56. AN, $F^{12}2234$, [2], [3]. Chamberlain was full of projects (here abundantly documented) and in the course of the 1790s persistently but unsuccessfully sought Government backing for the extension of his enterprise.
57. The standard biographical dictionaries (e.g. Dict. biog. fr.) confuse him with the political figure Aimé Marie Alban, to whom he might have been related. Both appear to have originated from Pont-de-Veyle.
58. AN, $F^{12}1506$, [5].
59. MC, Étude XLVIII, [1].
60. AN, $F^{12}1506$, [5], [3].
61. BML, [1].
62. AS, D^3B^673, 30 Oct. 1772; D^3B^679, 20 Nov. 1778. On the Buffault family see Szramkiewicz (1974), 37–45.
63. Boysse (1887); Guerlac (1961), 207; J. Payen (1969), 252.
64. Correspondance [1780], 58–9.
65. Demachy (1780), 529–30.
66. Correspondance, 41.
67. Thiéry (1787), vol. 2, 642–4; Correspondance [1780]; J. de Normandie, 7 Jan. 1789; BML, [1].
68. Alban and Vallet (1785).
69. Faujas de Saint-Fond (1783–4), vol. 2 (Première suite), 232–8.
70. Vallet had shared the technical direction at Javel with Alban since 1780 or earlier. Like Alban he was said to hail from Pont-de-Veyle (BML, [1]).

71. On Chaptal see primarily Chaptal (1893), Pigeire (1931); and on the Montpellier scientific background, Dulieu (1958).
72. Thomas [1936], 184–5.
73. Dulieu (1950–2).
74. Béchamp (1866), 56.
75. Gerber (1925–7), Pl. XIX–XXI.
76. Gerber, 192.
77. The total value of mineral acid sales in 1788 was 87 217*l*. (*Procés-verbal*, 1789, p. 511).
78. Béchamp, 56–7; BUM, [1].
79. CCM, [3]; AN, $F^{12}1507$, [1], and $F^{12*}107$, pp. 241, 818.
80. AN, $F^{12}724$, [2].
81. ADR, 1 Q 328*, f° 15; for other purchases see 96 Q 3*.
82. The following account is based on: AN, $F^{12}2234$, [7]; ADR, 1 L 524.
83. Notes of departmental administration, *circa* 1796 (ADR, 1 L 524).
84. Laferrère (1972), 395.
85. AN, $F^{12}1506$, [1].
86. AN, $F^{12}1506$, [9], [11]. A petition of about June 1778 was signed by: Anfrye; Auvray, Payenneville, & Cie; Stourme; Fleury *frères* & Cie; Bunel, Blamany.
87. AN, $F^{12}1506$, [3], [1].
88. AN, $F^{12*}166$, p. 288.
89. AN, $F^{12}724$, [1]. The provision lasted until 1 Jan. 1788.
90. AN, H 1438.
91. Dossie (1758), 158; Dickinson (1939), 49; Clow (1952), 138–9; Guttmann (1901), 6–7; *Memoirs Manchester Literary and Philosophical Society*, 1 (1785), 241.
92. AN, $F^{12}650$, [1]; *J. de Normandie*, 7 Jan. 1789; AN, $F^{12}1966^L$, [4].
93. Gerber (1925–7); Lavoisier (1862–93), vol. 6, 69–73; *Procès-verbal*, 1788, pp. 379–80, and 1789, pp. 510–13; AN, H 940, [1]. After the failure of the 1787 petition to Paris, the benevolent *États de Languedoc* offered Chaptal instead a compensatory premium of 5 per cent on his sales, but he renounced this on receiving the apparent concession in 1789.
94. AN, $F^{12}1506$, [3]; $F^{12*}167$, p. 9; $F^{12}1508$, [6]; $F^{12}652$; Gerbaux and Schmidt (1906–37), vol. 1, 430, 591, and vol. 2, 328, 364, 652, 659, 680.
95. *Arch. parl.*, 1st series, vol. 31, 379, and vol. 44, 393; *Tarif* (1853–4), vol. 2, 820–1.
96. *Moniteur*, 6 Sept. 1797.
97. Braudel and Labrousse (1970–7), vol. 2, 256–7.
98. According to the *Bureau de la balance du commerce*, in the twelve years between 1776 and 1790 for which figures were available, imports totalled 106 800*l*. in value and exports 114 700*l*. Exports first exceeded imports in 1781, and the missing years were 1783, 1785, and 1786 (AN, $F^{12}2234$, [1]). In 1787, according to the records of the same *Bureau*, imports amounted to 16 100 pounds (value 9847*l*.) and exports to 44 500 pounds (value 24 590*l*.) ('Aperçu' (1794), table).
99. On the Revolutionary manufacture of arms and munitions see particularly the standard study by Richard (1922).
100. *Arch. parl.*, 1st series, vol. 74, 590; Richard, 560; AN, $F^{12}1508$, [22].
101. AN, $F^{12}1508$, [22].
102. AN, $F^{12}1508$, [22].

103. AN, $F^{12}1508$, [15].
104. AN, $F^{12}1508$, [17].
105. CNAMA, [3].
106. *Biog. univ. Supp.*
107. AN, $F^{12}1508$, [17], [28].
108. Pigeire (1931), 176–96; Chaptal (1893), 36–52; AN, $F^{12}1508$, [3].
109. AN, $F^{12}2234$, [7]; ADR, 1 L 524, 1 Q 724 (piece no. 78), and 2 Q 261* (f° 16–17); Tissier (1822).
110. Calvet (1933), Ch. 1.
111. AN, $F^{12}2234$, [1].
112. Aulard, vol. 19, 63, and vol. 20, 365; AN, $F^{12}1508$, [20].
113. Debidour (1910–17), vol. 2, 72.
114. Law of 30 Aug. (*Moniteur*, 6 Sept. 1797).
115. AN, $F^{12}1508$, [27].
116. AN, $F^{12}2234$, [1]. Exports, however, were also high at 170 000 fr., largely perhaps owing to Chaptal's sales to Spain (see p. 42).
117. AN, $F^{12}1966^L$, [4].
118. Gerbaux and Schmidt, vol. 4, 668; AN, $F^{12}2234$, [7].
119. Chaptal (1807), vol. 4, 106; AN, $F^{12}652$.
120. Fourcroy (1800), vol. 2, 70.
121. Dumas (1828–46), vol. 1, 172.
122. AN, $F^{12}1966^L$, [1].
123. *Bull. lois*, 3rd series, no. 7, piece 52.
124. Magnien and Deu (1809), vol. 2, part 2, pp. 473–4.
125. AN, $F^{12}1966^L$, [4]; $F^{12}2234$, [1].
126. AN, $F^{12}1966^L$, [2], [3].
127. Imports in the years 1806–13 were always below 25 000 fr., while exports ranged between 143 905 fr. and 251 259 fr. (AN, $F^{12}2512$, [1]).
128. Aulard, vol. 16, 634.
129. Herbin de Halle (1803–4), *Atlas*, 29; Haber (1969), 44.
130. AN, $F^{12}1508$, [5].
131. Lefebure's associate was perhaps related to a later professor of pharmacy at the Strasbourg school, Charles Frédéric Oppermann, a student of Liebig. (For these personal details on Lefebure and Oppermann I am indebted to Mr John Perkins.)
132. *Notices* (1806), 226; AN, $F^{12}937$, and $F^{12}1569$, [1].
133. On Kestner see particularly Soc. ind. Mulhouse (1902), vol. 2, 576–9. Additional details from AN, $F^{12}937$, and $F^{12}1569$, [1]; AMM, 10 F 1.
134. It does not (recognizably) figure in the statistical returns of the mayor in the later years of the Empire (ADBR, $M^{14}6$ and $M^{14}47$). The Kestner brothers do, however, appear in the Marseilles *Almanach* in 1817 as manufacturers of oil of vitriol at 121 rue Paradis.
135. AN, $F^{12}937$.
136. This paragraph derives mainly from: AML, 'Il Guillotière', [2]; ADR, 'Ateliers incommodes', [1]; Bertin *et al.* [1848]; Coignet (1894), 97; Tissier (1822).
137. AN, $F^{12}2234$, [6].
138. Bertin *et al.*; Coignet, 97.
139. Parisel (1836).

140. Dubois, 80.
141. On Perret's enterprise see Laferrère (1960), (1972). On that of Estienne and Jalabert see more particularly Bertin *et al.*; Coignet; 'La Vitriolerie' (1972).
142. AN, $F^{12}937$.
143. BMM, [1]; *Notices* (1806), 44; ADBR, $M^{14}47$; AMM, 22 F 1.
144. ADBR, $M^{14}47$; AMM, 22 F 1; *Notices* (1806), 44; Masson (1919), 271–2, 343; id. (1926), 51–2. F. A. Porry was continuing to refine sulphur and to manufacture a variety of chemicals in 1819 (*Ann. ind.*, 1st series, **3** (1820), 191).
145. On these various works see ADH, 109 M 5, and more particularly 109 M 139.
146. Chaptal (1893), 52.
147. Pigeire (1931), 224–9; Béchamp (1866), 56–7; Dulieu (1950–2); MC, Étude XXVI, [1], [2].
148. AN, $F^{12}937$.
149. Thomas [1936], 228, 250, 265, 291.
150. 'Prix courant' in *J. du commerce*, 18 Feb. 1802; Herbin de Halle (1803–4), *Atlas*, 29; Huet (1803–4), Table 15.
151. AN, $F^{12}937$; $F^{12}2234$, [2]; Chamberlain (1822).
152. INPI, [6].
153. ADSM, Series M (unclass.), [1]; *Almanach comm.*, 1815. Two works were reported to be operating at Ingouville in the 1820s (Dupin (1827), vol. 2, 22).
154. AN, $F^{12}937$.
155. Mentioned in ADSM, 5 MP 1253, [4].
156. Dieudonné (1804), vol. 2, 157–8.
157. Kuhlmann (1926), 4.
158. AN, $F^{12}1552–1553$.
159. Brayer (1824–5), vol. 2, 265.
160. Kuhlmann (1926), 12, 49.
161. ADSM, 5 MP 1253, [1].
162. ADSM, 5 MP 1335, [3]; Arvers (1817), 49, 57.
163. AN, $F^{12}2234$, [6]; ADSM, 5 MP 1335, [3], and 5 MP 1252, [1]; *Tableau* (1808), 28.
164. ADSM, 5 MP 1253, [4].
165. AN, $F^{12}937$; ADSM, 5 MP 1048, [1], and Series M (unclass.), [1]. Lefrançois was described in 1810 as having been a merchant and subsequently a banker.
166. AN, $F^{12}1508$, [17]; ADSM, 5 MP 1335, [3], and 5 MP1252, [1]; *Tableau* (1808), 28.
167. ADSM, 5 MP 1194, [6], [2].
168. ADSM, 5 MP 1335, [3]; 5 MP 1194, [5].
169. ADSM, 5 MP 1253, [4].
170. ADSM, 5 MP 1335, [3]; 5 MP 1252, [2]; 5 MP 1194, [3].
171. Projects of Pierre Bougon and Jean Baptiste Bougon *fils* (producers of nitric acid and other chemicals in the Sotteville district); Jacques André Germain ('entrepreneur de bâtimens' in Rouen); and Vallée (a merchant in Rouen). ADSM, 5 MP 1194, [1], [3], [4].
172. ADSM, 5 MP 1253, [4].
173. AN, $F^{12}871^A$, [1].
174. ADSM, 5 MP 1252, [1].

175. ADSM, 5 MP 1252, [3], [5]; AN, F^{12}2244, [10]; *Tableau* (1817), 142; *Annuaire* (1823), vol. 1, 220.

176. Duchemin (1890), 189, 275–6; Baud (1932*a*), 270, 383; Haber (1971), 305.

177. Pigeire (1931), 224–9; Kersaint (1961); MC, Étude XXVI, [7].

178. MC, Étude XXVI, [3]. The partnership was also to exploit Chaptal's 'maison de commerce et de commission' in Paris.

179. MC, Étude XXVI, [4]. Chaptal *fils* had 80 shares and Berthollet *fils* the remaining 20. From the profits Chaptal *fils* was to receive 5000 fr. a year and Berthollet *fils* 3000 fr. Surplus profits were to remain in the business, to be paid out only on expiry of the partnership (30 July 1817).

180. MC, Étude XXVI, [5], [7].

181. MC, Étude XXVI, [8].

182. PP, [3], no. 170; Chaptal (1807), vol. 3, 42.

183. AN, F^{12}1551, doss. 'Seine'.

184. Szramkiewicz (1974), 43.

185. MC, Étude LV, [1]. The agreement recognized the property as belonging equally to the three of them; Alban received an annual salary of 2400*l*. as director. The Buffault brothers (Philippe Jean Baptiste and Alphonse Jean) were both then embarking on distinguished careers in public administration (see Szramkiewicz, 37–45). According to Szramkiewicz the Javel works was owned by the Buffault family until 1846, when Philippe sold it for 250 000 fr.

186. MC, Étude CIII, [2].

187. Gille (1963), 61.

188. AS, DM^{12}24.

189. *Exhibition* (1852), 38–9.

190. Lambeau (1912), 73–4.

191. MC, Étude CIII, [1]. The works still figures in the *Almanach du commerce* in 1831, by then under the name Ador, Bonnaire, & Monod.

192. AN, F^{12}2245, [1].

193. AS, D^{31}U^3, année 1816, no. 130. The other associates were F. J. M. Marc *fils* (successor to F. L. S. Marc), J. F. Tochon (*négociant* in Paris), and G. Homberg and J. M. Michel (both *négociants* at le Havre).

194. Oudiette (1812), 423, 453; *Alm. comm.*, 1816, 156.

195. Mactear (1877), 17, giving a list which, however, appears incomplete; Clow (1952), Ch. 6.

196. Typically, 100 parts of salt were reacted with the equivalent of 75 parts of concentrated acid. For salt figures see p. 277.

197. INPI, [4]; Chaptal (1819), vol. 2, 175.

198. CNAMB, [1], vol. 2, cahier 1, p. 131. Clément allowed 1 million kg of sulphur for purposes other than acid manufacture, and from the rest evidently assumed a yield of 2·45–2·75 kg of acid per kg of sulphur. If we apply a similar calculation (assuming a yield of 2·5) to import records for 1818–21, the following estimates result:

1818 9·79 million kg acid (net sulphur imports 4·92 million kg)
1819 14·28 (6·71)
1820 13·86 (6·54)
1821 21·33 (9·53)

Import figures for 1815–17 lead to impossibly low results, perhaps because affected by trading disturbances connected with the end of the Empire. (Data from AN, $F^{12*}251$ and $F^{12}7591$).

199. Parkes (1815), vol. 2, 441; Hardie and Pratt (1966), 18. The estimate of the latter is unexplained but is roughly in accord with Britain's sulphur imports of 4600 tons in 1820 (Haber (1969), 13; Ure (1839), 1218).

200. Mellor (1922–37), vol. 10, 89–90, 332.

201. One author claimed that by using his funnel-shaped condenser, $\frac{1}{2}$ oz. of oil of sulphur was obtained from 16 oz. of sulphur, other methods allegedly giving scarcely $\frac{1}{2}$ oz. from 16 lb. ('Basil Valentine' (1678), 128). The chemist Homberg claimed to obtain $1–1\frac{1}{2}$ oz. with his own condenser, this being much more than with an ordinary bell (Henkel (1757), 159). The theoretical yield from 16 oz. sulphur would be about 50 oz. of concentrated acid.

202. Stahl (1766), 159–63; Macquer (1777), articles 'Acid, vitriolic', 'Sulphur', 'Volatile sulphureous acid', 'Spirit of sulphur'.

203. For Lavoisier's fragmentary comments see e.g. Guerlac (1961), 226, and Lavoisier (1862–93), vol. 2, 271, 549, 647, 716–17. We might note, as an additional complication, that in the eighteenth and early nineteenth centuries the modern distinction between acids and their anhydrides was not made, so that the term 'sulphuric acid', for example, was applied both to SO_3 and to H_2SO_4. For convenience we shall similarly blur the distinction in our discussion; it does not affect the basic arguments.

204. De la Follie (1777), 142; Lavoisier (1789), vol. 1, 240.

205. Based on a set of pre-war prices given by the local authority at Honfleur in 1797 and probably reasonably representative: 7–8$l.$ per 100 pounds for sulphur, 17$s.$ per pound for saltpetre, 8–10$s.$ per pound for oil of vitriol (AN, $F^{12}1966^{L}$, [4]). I have assumed a yield of one part of acid from one of sulphur. Two decades earlier, Demachy gave data pointing to corresponding cost figures of about 32 per cent for saltpetre and 28 per cent for sulphur (Demachy (1773), 40–1).

206. AN, $F^{12}1506$, [1].

207. Demachy (1780), 54–5; *Enc. méth. Arts*, vol. 2 (1783), 293; *Enc. méth. Chymie*, vol. 1 (1786), 357–8.

208. AN, $F^{12}2242$, [1].

209. Berthollet (1789b), 181.

210. Pajot Descharmes (1798), 44–6; Du Porteau (1790); Acad. sci. (1910–22), vol. 1, 256, 281; CNAMA, [4].

211. Chaptal (1782).

212. Cadet de Gassicourt (1802), (1804), (1805); *Bull. Soc. enc.* **1** (1803), 62.

213. Chaptal (1807), vol. 3, 34–9.

214. Cadet de Gassicourt (1804).

215. Derrien (1804).

216. INPI, [6]. Alternatively the oxygen was passed direct to the chamber, to combine with the sulphureous fumes there. It was generated by heating manganese dioxide with dilute sulphuric acid, a method perhaps likely to yield more steam than oxygen.

217. Mellor (1922–37), vol. 10, 89, 332.

218. Fourcroy and Vauquelin (1797), 243. The same experiment but using a porcelain

tube was indicated by Bouillon-Lagrange [1798–9], vol. 1, 114–15. Cf. however Berthollet (1803), vol. 2, 132.

219. *Dict. sci. biog.*; *Dict. biog. fr.*; Dunoyer (1842). Clément is sometimes confused with Desormes, since after marrying Desormes's daughter he adopted the name Clément-Desormes.

220. Desormes and Clément (1806). According to Anselme Payen, the same theory had been privately proposed some years earlier by Pluvinet, in a letter to Chaptal on the properties of sulphur (*Dict. tech.,* vol. 1 (1822), 133). On the Pluvinet brothers, manufacturers of sal ammoniac and business associates of Payen's father, see below.

221. Chaptal (1782), 152; Berthollet (1804), vol. 1, 353. For an interesting English anticipation, which used such a calculation in 1799 to support a heat interpretation, see Mactear (1877), 14–15.

222. Chaptal (1789b), 91.

223. Partington, vol. 4, 601–2; Duecker and West (1959), 103–16.

224. The mechanism now generally accepted is that of Lunge and Berl:

(1) $2NO + O_2 \rightarrow 2NO_2$
(2) $H_2O + SO_2 \rightarrow H_2SO_3$
(3) $H_2SO_3 + NO_2 \rightarrow SO_5NH_2$ (sulphonitronic acid)
(4) $2SO_5NH_2 + NO_2 \rightarrow 2NO.HSO_4 + H_2O + NO$
(5) $2NO.HSO_4 + SO_2 + 2H_2O \rightleftharpoons 2SO_5NH_2 + H_2SO_4$
(4a) $SO_5NH_2 \rightleftharpoons H_2SO_4 + NO$
(5a) $4NO.HSO_4 + 2H_2O \rightleftharpoons 4H_2SO_4 + 4NO + O_2$
(5b) $NO.HSO_4 + HNO_3 \rightleftharpoons H_2SO_4 + 2NO_2$
(Imperial Chemical Industries (1955), 22–3).

225. Lemay (1949); Kuznetsov (1966); Collins (1976). Clément was one of the first to recognize the common catalytic character of the later discoveries.

226. Acad. sci. (1910–22), vol. 3, 417–19.

227. Gay-Lussac (1807), 241–9.

228. Berthollet (1789a), 61–2; id. (1803), vol. 2, 132.

229. Rougier (1812), 64–5; Chaptal (1819), vol. 2, 65.

230. Arvers ((1817), 52) described the yield at Rouen as having gradually risen to 100 parts of acid by about 1800, from an initial 80. De la Follie's early description of a Rouen plant ((1777), 142) implied a yield of about 100, and Chatel's works near Rouen was apparently obtaining a yield of that order in 1794 (AN, $F^{12}1508$, [17]: annual capacity 3000 bottles of chamber acid of 150 pounds each, consumption 300 000 pounds of sulphur). Elsewhere, a yield of 100 is indicated for the Lille works in 1800–1 (Dieudonné (1804), vol. 2, 158), while at Javel figures given by Alban in 1794 show a yield of only 80, though a low yield would have favoured his argument (AN, $F^{12}1508$, [20]). Demachy ((1773), 40) gave a yield of 88 for production in glass globes.

231. Bouillon-Lagrange [1798–9], vol. 1, 107–8.

232. Payen and Chevallier (1824) found commercial grades of sulphur at Rouen in 1819 to contain between $1\frac{1}{2}$ and 11 per cent of solid impurity, and suggested an additional allowance of 4 per cent for water. Payen later gave a figure of 3–10 per cent for the (solid?) impurity ((1855), 99). Thenard ((1813–16), vol. 1, 164) indicated $8\frac{1}{2}$ per cent, and Girardin ((1846), vol. 1, 200) 3–4 per cent.

233. *Dict. tech.*, vol. 1 (1822), 125.

234. AN, F¹²1508, [17]. After extracting the sulphur the remaining ash contained about 3/8 of potassium sulphate. In Scotland in 1807 a relatively efficient manufacturer like Tennant bought up residues containing 25–50 per cent sulphur from other producers (Mactear (1877), 16).

235. Arvers (1817), 52–5.

236. The Javel manufacturers had at first tried a chamber fitted with a bench around its internal walls, on which a workman (by entering?) arranged dishes of material; these were then ignited from outside using a long pole (Demachy (1780), 529).

237. Chaptal (1807), vol. 3, 44. And on British methods: Mactear (1877), 14–16; Guttmann (1901); Parkes (1815), vol. 2, Ch. 7.

238. Fourcroy (1800), vol. 2, 59; Bouillon-Lagrange (1812), vol. 1, 289.

239. De la Follie (1774); Demachy (1775), vii–viii.

240. Dossie (1758), 158–62; Demachy (1773), 38–9; id. (1775), vi–vii.

241. ADSM, 5 MP 1335, [3]. Holker's was the works omitted.

242. Lemery (1697), 418 (himself denying its value).

243. Mactear (1880–2), 417–18.

244. *Encyclopédie*, 'Soufre'.

245. De la Follie (1777), (1819). In the same year Keir in England seems to speak of filling chambers with steam before combustion (Macquer (1777), 'Spirit of sulphur'), but this reference to steam was probably in fact made with glass globes in mind (see Macquer (1771), vol. 2, 742).

246. When, for example, the above-mentioned report of 30 Oct. described each chamber enumerated as being with furnaces (plural), it must have referred to a boiler as well as a combustion furnace.

247. Parnell (1844), vol. 2, 378.

248. Quoted by Mactear (1877), 15.

249. Chaptal (1789*b*), 91; id. (1807), vol. 3, 46.

250. ADSM, 5 MP 1252, [1].

251. Theoretically, one pound of sulphur requires 76·3 cu. ft. of air (at 20°C. and normal pressure); in late-nineteenth-century practice a working excess of 25 per cent was allowed in the air supply, giving 95·1 cu. ft. per pound (Lunge (1891–6), vol. 1, 303–10). De la Follie's 1774 data point to a chamber space in the region of 150–260 cu. ft. per pound, and his 1777 data to one of 230–390 cu. ft. Bouillon-Lagrange ([1798–9], vol. 1, 107) described the combustion of 6 kg of sulphur per 1000 cu. ft. a day, which would correspond to about 260–270 cu. ft. per pound if we assume (say) 3 charges. Pelletan in 1809–10 indicated a chamber space of about 185–95 cu. ft. per pound (ADSM, 5 MP 1335, [2]), and A. Payen in 1822 one of about 205–15 cu. ft. (*Dict. tech.*, vol. 1, 127–8).

252. *Description*, vol. 12 (1826), 94–5.

253. *Description*, vol. 8 (1824), 367.

254. *Enc. méth. Chymie,* vol. 3 (1796), 586.

255. Creuzé de Lesser (1824), 538; Béchamp (1866), 56–7 (quoting Bérard). See also Dumas (1828–46), vol. 1, 200–1, 207.

256. De Peyre (1959), 70.

257. For early accounts of the continuous process, and comparison with the intermittent method, see: Rougier (1812), 65–74; Tassaert in *Enc. méth. Chymie,*

vol. 6 (1815), 157–60; A. Payen in *Dict. tech.*, vol. 1 (1822), 126–30; Péclet (1823), 131–4; Kuhlmann (1827); Dumas (1828–46), vol. 1, 200–11; Villeneuve (1821–34), vol. 4, 781–4.

258. Demachy (1780), 529.

259. *Enc. méth. Chymie*, vol. 6 (1815), 157–60; AN, 26 AQ 1, [1].

260. Lunge (1891–6), vol. 1, 369, 375, 380–1, 471–82.

261. AN, 26 AQ 1, [2].

262. Girardin (1846), vol. 1, 200–1; id. in *Précis*, année 1851–2, pp. 22–4.

263. In about Aug.–Oct. 1810 Pelletan referred to the factory's five lead chambers 'of which some are of a new construction' ([Pelletan], 4).

264. Laboulaye (1867–70), *Complément*, article 'Insalubres (Établissements et opérations)'.

265. AN, F^{12}6728, [1].

266. BMR, [1].

267. It is not in BMR, Collection Girardin, which does, however, have two letters from the younger Holker referring briefly to his father's work.

268. Dumas (1828–46), vol. 1, 210–11. Payen later spoke of having used some such plant from 1818 ((1855), 126), but this may be an error since at that time he was occupied with an intermittent method (*Dict. tech.*, vol. 1 (1822), 126–9).

269. De Peyre (1959), 71.

270. Clément's application of 8 September, from which all the quotations here come, is in INPI, [4]. The published patent description, in *Description*, vol. 12 (1826), 92–6, is an edited version of this.

271. *Bull. lois*, No. 10, 18 Dec. 1824, piece no. 175.

272. The following account is chiefly from: ADSM, 5 MP 1048, [1]; 5 MP 1252, [1]; 5 MP 1335, [2].

273. *Description*, vol. 8 (1824), 367–8. The brief published description is substantially the same as that furnished by Pelletan in his application of 1 October (AN, F^{12}1017^A; INPI, [5]).

274. ADSM, 5 MP 1335, [2].

275. Duecker and West (1959), 105.

276. *Dict. tech.*, vol. 1 (1822), 129.

277. Patents no. 3946 (3 Aug. 1815) and 3998 (18 March 1816). His French patent was for 'un nouveau procédé propre à fabriquer de l'acide sulfurique'.

278. A report of 2 Jan. 1817 spoke of manufacturers obtaining saltpetre at less than 1 fr. 80 per kg by importing it from England (AN, F^{12*}194); in 1817–18 it probably fell considerably below this. In 1817 saltpetre imports for sulphuric acid manufacture and other private use were 791 785 kg, as against 26 000 kg now bought from the State (*Arch. parl.*, 2nd series, vol. 22, 715).

279. *Dict. tech.*, vol. 1 (1822), 125; J. P. J. Darcet (1817); AN, 26 AQ 1, [2]; Creuzé de Lesser (1824), 538. They fetched 12 fr. 50 at Marseilles in the later 1820s (Villeneuve (1821–34), vol. 4, 783).

280. *Tarif* (1853–4), vol. 2, 820–1; Tooke (1838–57), vol. 2, 409. French manufacturers were paid a compensatory premium on any sulphuric or nitric acid they exported (*Arch. parl.*, 2nd series, vol. 23, 81).

281. *Recherches*, vol. 3 (1826), Tables 105, 114.

282. CNAMB, [1], vol. 2, cahier 1, pp. 130–44.

283. BMR, [2].

284. *Description*, vol. 90 (1859), 463–9; English patent no. 9558 (15 Dec. 1842), in name of Sautter. On Gay-Lussac see Crosland (1978).

285. INPI, [5].

286. Patent of 23 Aug. 1821, in *Description*, vol. 21 (1831), 282–6.

287. Lunge (1891–6), vol. 1, 509–11.

288. Sorel (1902–4), first volume, 78–9 (giving no source). On d'Artigues's career in general see *Dict. biog. fr.*

289. Briavoinne (1839), vol. 1, 388–9; Dumas (1828–46), vol. 1, 132–3. It seems possible that it was sulphur rather than sulphuric acid which was the object of his early experiments, too, and that Sorel's account is an embroidered error (see A. Payen (1855), 103).

290. For a detailed account see Michelet (1965).

291. Acad. sci. (1910–22), vol. 4, 475.

292. Laferrère (1960), 475–85; Perret (1867); *Description*, vol. 75 (1851), 139–41.

293. Lunge (1891–6), vol. 1, 30.

294. Villeneuve (1821–34), vol. 4, 781–3.

295. Lunge (1891–6), vol. 1, 362 (converted from English to French measure).

296. Mactear (1877), 15–16.

297. For details of construction methods see: *Enc. méth. Chymie*, vol. 6 (1815), 156–9; *Dict. tech.*, vol. 1 (1822), 131–2; Dumas (1828–46), vol. 1, 195–7.

298. ADSM, 5 MP 1048, [1]; patent of 7 August 1827 by Clément, in *Description*, vol. 48, 435–7. At Charles-Fontaine, the five chambers built in the years 1811–17 employed between 13 500 and 24 000 kg of lead each, at a cost of 103 fr. per 100 kg in 1811–13, falling to 72 fr. 40 in 1817. The lead, at 94 745 fr., formed about two-thirds of the chambers' total cost. (AN, 26 AQ 1, [2]).

299. J. Payen (1969), 150.

300. Cf. figures for Marseilles in Villeneuve (1821–34), vol. 4, 781–4, where a life expectancy of 20 years is spoken of.

301. Berthollet and de Dietrich (1789), 49–51; Chaptal (1791a), 771–3; id. (1807), vol. 3, 43–4. The large chamber was used mostly for making alum rather than acid, but the process was similar.

302. Patent cited above, and Ancelon (1879), 178.

303. For a general account see McDonald (1960).

304. Mérimée (1817), 35; Héricart de Thury (1819), 97. Probably Pierre Alban, of Vaugirard.

305. McDonald, 118, 134.

306. Besides sources previously cited see: *Dict. biog. fr.* (on Bréant and Cuoq); L. Costaz (1819), 168, 364; Péligot (1888); *Ann. ind.*, 1st series, **2** (1820), 280–2.

307. *Dict. tech.*, vol. 1 (1822), 130–1.

308. ADSM, 5 MP 1335; *Ann. ind.*, 1st series, **3** (1820), 186; *Annuaire* (1823), vol. 1, 220.

309. Tissier (1822), 245–6; Péclet (1823), 134–5; AN, F¹²6728, [1].

310. ADSM, 5 MP 1335; Mactear (1877), 16.

311. A. Payen (1855), 142–3.

312. Gilbert (1952), 325.

313. Clapham (1868–71), 38–9, 42.

314. On burning rates: Bouillon-Lagrange [1798–9], vol. 1, 107; INPI, [4]; CNAMB,

[1], vol. 2, cahier 1, p. 140; *Description*, vol. 48, 435; Villeneuve (1821–34), vol. 4, 782–4; Ure (1839), 1221; Lunge (1891–6), vol. 1, 371–4.

315. Chaptal (1819), vol. 2, 65.
316. Mactear (1877), 16.
317. Clapham (1868–71), 38; Dickinson (1939), 53.
318. Allen (1907), 46–7.
319. J.P.J. Darcet (1817).
320. AN, $F^{12*}251$. Annual export *values* are available for the years 1806–19 (AN, $F^{12}2512$, [1]), but it would be rash to draw conclusions about weights from these, since in the accompanying import tables, where both weights and values are given, the relationship between the two is mysteriously erratic. For the years 1816 and 1817, export weights of 500 000 kg and 550 000 kg are given in $F^{12*}251$, but these seem to include the saline vitriols as well as sulphuric acid; deduction of the export figures for iron, copper, and zinc sulphates ('Tableaux' (1818)) would leave acid exports of 326 367 kg and 307 620 kg. Imports were negligible.
321. Hardie (1950), 9; Clapham (1868–71), 38. The figure is unsupported and is given for 'England'. See also Landes (1969), 109, who gives the same figure for 'Britain'.

3

THE DEVELOPMENT OF CHLORINE BLEACHING

ONE of the outstanding technological innovations of the late eighteenth century was the application of chlorine to the bleaching of textiles, a development which sprang from the recent discoveries of chemical science, and which was grounded in the supply of cheap sulphuric acid from the new chemical industry. The gas chlorine was discovered in 1773, during that remarkable heyday of pneumatic chemistry when so many of the common gases were coming to light in rapid succession. Its potential as a bleaching agent was first recognized in 1785 by the French chemist Berthollet,[1] whose work quickly aroused such widespread interest that by the later 1780s large-scale application was already beginning both in France and in Britain, promising important improvements in the bleaching of linen, hempen, and especially cotton goods (the bleaching of the animal fibres—wool and silk— was a different matter, and was unaffected by the developments with which we are here concerned). Although its general adoption was rather slower than has sometimes been imagined, and its impact less simple, the introduction of chlorine can nevertheless fairly be described as a revolutionary advance in the bleaching treatment, establishing the bleaching industry on its modern basis, and it is rightly famed as one of chemistry's chief contributions to the Industrial Revolution. We shall begin our account of its development in France by first sketching something of the character and methods of the bleaching trade before chlorine appeared on the scene.

Traditional bleaching

With the growth of the linen and hempen manufactures in France during the seventeenth and eighteenth centuries, bleaching too had developed on a commercial scale. It is true that there was a long continuance of bleaching as a more or less primitive domestic activity. In the non-industrial Cantal, for instance, it was still the general practice about 1800 for individuals to bleach cloth themselves, simply by spreading it on the banks of streams and there watering it daily until its greyness had sufficiently lightened;[2] among the poor, similar practices seem to have persisted even in areas that were industrially more advanced, as in Flanders where peasants bleached their cloth by spreading it on the grass verges of the roads. Bleaching on a commercial scale, however, was well established in the textile regions by the second half of the eighteenth century, probably employing some hundreds of workers in the major centres. Bleaching districts of note included Lille, Valenciennes, Beauvais, Rouen, Louviers, Laval, Troyes, Reims, Senlis, and Saint-Quentin, the latter being one of the leading centres where bleaching had been established for nearly two centuries. Bleaching was largely conducted as a separate and specialized trade, the bleacher processing goods for merchants and producers at advertised or agreed rates, though towards the end of the *ancien régime* one also finds some early examples of its integration with the manufacture or printing of textiles. Most bleaching was of cloth but a good deal of yarn was bleached, too, for lace, embroidery, sewing thread, and the weaving of certain types of fabric. The method employed was basically similar for both.

The function of the bleaching process was twofold: not only to whiten the cloth (or yarn), but also, more generally, to cleanse it of materials which would detract from its finish or interfere in subsequent dyeing or printing. In essence, the process used dated back to ancient times, and consisted in alternately steeping the cloth in an alkaline lye and exposing it on bleach-fields to the action of the sun and air.[3] The former operation, known in British terms as 'bucking', was generally carried out by the method which the French called 'coulage': the goods were packed into a bucking tub and an alkaline lye was ladled over them from an adjacent copper, in which it was gradually heated to boiling and into which it drained back after filtering down through the material. In this way the hot lye was repeatedly circulated through the goods, generally for between twelve and twenty-four hours. The lye was made from the various vegetable ashes which constituted commercial potash and soda, sometimes with the addition of lime. The second operation, known in British terminology as 'grassing' or 'crofting', involved spreading the cloth out on fields and there exposing it for anything from half a day to a fortnight. To prevent the wind furling the pieces and blowing them across one another, which would interfere with their proper exposure, the most careful bleachers pegged them down or held them down by cords. In many places (at Rouen, for instance, and at Caen), once the cloth had been laid out it was simply left, apart perhaps from being turned so as to expose both sides. In the best bleach-fields, however, such as those at Saint-Quentin and at Beauvais, it was considered important for efficient bleaching that the goods be periodically watered to keep them moist; here the fields were cut by channels from which water was drawn and then hurled over the cloth by workers using long-handled scoops. Bleachers sometimes went to considerable lengths to supply their fields with water: in the 1760s a leading textiles printer in Brittany—Danton of Tournemine-lès-Angers—built an aqueduct some 40 feet high and 600 yards long, hewed a reservoir 200 feet long and 30 wide out of the rock, and constructed a system of stone-lined channels totalling perhaps several kilometres, to supply the bleach-field associated with his works.[4] These bucking and grassing treatments were the principal operations of bleaching and they were repeated alternately, usually between ten and twenty times. A number of subsidiary operations, however, also entered into the process. Thus, before the first bucking the cloth was given a preliminary steep in water for several days, to remove the size applied in the course of weaving; old lyes, lime, or cow dung were sometimes added to assist in this. Also, towards the end of bleaching, when the repeated buckings and grassings were beginning to bring the cloth to a white condition, it was customary to introduce into the cycle of operations a number of soapings (in which the cloth was washed with soap), and in some places several acid baths or 'sours'. These were composed of whey, buttermilk, or sour milk, or, failing that, of fermented bran water, or rye, and the goods were steeped for a time which might vary between a day and a fortnight but was commonly two or three days. Two or three such soapings and sours seem to have been common, though sometimes as many as four or five were given. Finally, it should be added that interspersed throughout this whole sequence of operations there were rinsings in clear water, performed by hand or increasingly by various machines. As can well be imagined, within this general pattern there was a good deal of detailed variation, according to type of cloth, local tradition, and the personal preference of individual bleachers.

Bleaching was clearly a laborious and lengthy business and for this the slowness of the grassing stage was chiefly responsible. At its fastest, as at Saint-Quentin (and also Beauvais), it might accomplish the bleaching of light, fine linens in three to four weeks, granted favourable weather. Bleachers there, urged on by impatient merchants, followed a somewhat hectic routine involving bucking and grassing to a daily cycle: the goods were bucked through the evening and night, rinsed and laid out on the fields at dawn, and there exposed and watered through the day. In this way, by intensive lyeing, the time spent in grassing was reduced to a remarkable minimum, though only at the expense of a greater weakening of the cloth, and perhaps of a higher cost. Elsewhere, with the cloth spending several days on the field between buckings, the whole bleaching process commonly took between two and five months, depending on the type of cloth and the particular practices followed. A time of as much as ten months is mentioned for some heavy goods[5] but that is exceptional. Besides its slowness the grassing of goods had other inconveniences, too, for it rendered bleaching dependent on weather and hence on season. It could not be practised very satisfactorily during the winter months and to prevent the spoiling of cloth by imprudent attempts at winter bleaching a government regulation of 20 December 1740 forbade the exposing of goods on the field between 1 December and 1 March each year. This ban was reiterated in a regulation of 6 February 1781 and in the 1780s was enforced against some attempted contraventions, though enforcement may have been patchy.[6]

Bleaching remained an art still heavily dominated by the routine practice of traditional skills. In Britain, however, the industry was beginning to receive some scientific attention from the mid-eighteenth century. In Scotland particularly, the Board of Trustees for Manufactures, Fisheries, and Improvements was actively promoting the development and improvement of bleaching, encouraging the experiments of such individuals as the chemist William Cullen and the physician Francis Home.[7] The work of the latter was of especial importance, his book *Experiments on bleaching*, published in 1756, being generally regarded as the foundation work in the scientific study of bleaching. It described researches which Home had first reported in lectures, attended by bleachers among others, and now published in response to a petition from his hearers. In his researches Home sought to bring chemical enlightenment to the empirical processes of the bleachers. Thus, he analysed the various alkaline ashes employed for lyes in order to understand the reason for their widely varying qualities, known to bleachers only through experience; he examined the effect of adding lime to the lye and was able to affirm its advantages; he investigated the causes and cures of the hardness of water; he experimented on the acid baths, replacing the traditional organic sours with mineral acids; and he endeavoured to understand the chemical nature of the various operations and so to develop a theory of bleaching. Home's work was complemented by the more academic researches of his famous contemporary Joseph Black, whose classic study of carbonates provided a firm theoretical basis for some of Home's findings. In particular, the effect of lime in enhancing the efficacy of alkaline lyes was now explained by its removal of carbonic acid, so rendering the alkali caustic. In modern terms:

$$K_2CO_3 + Ca(OH)_2 = 2KOH + CaCO_3.$$

The old prejudice against addition of lime, based on its known corrosive properties, was thus shown to be unfounded—for the lime was precipitated from the lye as carbonate—and this contributed to the spread among bleachers of the practice of causticizing lyes with lime, one of the significant improvements of the eighteenth century. Scotland was not alone in seeing developments of this kind. In Ireland, where bleaching came under the eye of the Irish Linen Board, the industry received chemical attention from James Ferguson, while in England, too, there is evidence of interest on the part of chemists.[8]

In France the improvement of bleaching does not seem to have attracted interest to the same degree. One does find some inspectors of manufactures endeavouring to spread the knowledge and practice of superior methods: Roland in Normandy and Picardy, for example, and Brisson in the Lyons region. But though Home's book was published in French translation in 1762,[9] the hope of its translator that it would draw French scientific attention to the art did not bear fruit. When Brisson in 1766 wrote a small booklet of instruction for the bleachers of his region, he was not even aware of Home's work. Having later made its acquaintance he deemed it far beyond the intelligence of bleachers, and for their benefit reduced Home's four hundred pages to a little pamphlet 'of a quarter-hour's reading', which he published in 1780.[10] While the thousand copies of this which appear to have been distributed presumably made some small contribution to the education of bleachers, it hardly constituted a notable addition to the literature. When towards 1785 Roland wrote the article on bleaching for the *Encyclopédie méthodique*, he searched hard for earlier writings on the subject, but apart from Home's book could find scarcely anything published and thought little of what else he did find (Brisson's brochure, for example). Of necessity, Roland wrote his own account chiefly on the basis of his experience as inspector of manufactures, and it remains the fullest, indeed almost the only, account of French eighteenth-century bleaching practices to have been published. Roland complained of the slowness of bleachers to adopt improved methods: 'How few French bleachers apply reason to their art!' One example which might be cited as illustrating an apparently greater inertia in France than in Britain is the delayed adoption in France of sulphuric acid sours. In investigating the souring stage, Home had established that it was not the fermentation in the acid bath which was important, as had sometimes been thought, but simply its acidity; and that consequently mineral acids such as sulphuric acid could profitably replace organic acids like buttermilk, a particular advantage being that they acted more quickly. Sulphuric acid in Home's view produced the same effect in five hours as buttermilk in five days. Following Home's work the use of sulphuric acid sours seems quite quickly to have become common in Britain. In France, however, Roland in 1785 remarked that despite the greater convenience, lower cost, and quicker action of mineral acids, only a very small number of bleachers had yet adopted them. Bonjour in 1790 spoke of a prejudice against sulphuric acid among merchants,[11] and in the early 1790s its use by the bleachers of Mayenne was evidently still considered exceptional.[12] It was only with the growth of chlorine bleaching that sulphuric acid sours came into general use in France, following their advocacy by Berthollet as part of his new bleaching method.

This difference in character between the traditional bleaching industries of France and Britain is one factor relevant for an understanding of why the new chlorine bleaching, though beginning as a discovery of French chemistry, was nevertheless to be rather slower in establishing itself in its country of origin than in Britain.

Berthollet's early researches on chlorine

Berthollet's first researches on chlorine were reported to the Academy of Sciences in April 1785,[13] a dozen years after the discovery of the gas by the Swedish chemist Scheele. By this time Berthollet was a leading figure in the scientific circles of the capital. A physician by profession, he had since 1776 been contributing actively to chemical research, and his growing stature as a chemist had found recognition in 1780 with his election to the Academy. This was the period, of course, when phlogiston theory in chemistry was being challenged by the new oxygen doctrines of Lavoisier, and Berthollet's chief motivation in taking up the study of chlorine was connected with this controversy. Lavoisier's researches of 1783–5 demonstrating the composition of water had notably strengthened his case. Swayed by this from his former phlogistic position, Berthollet undertook his study of the properties of chlorine to test the merits of the rival theories in their interpretation. The gas had been regarded by Scheele and other phlogistic chemists as dephlogisti-cated muriatic acid (i.e. hydrochloric acid deprived of phlogiston). Berthollet's researches convinced him that Lavoisier's view of the gas as oxygenated muriatic acid was superior in explanatory value, the notion of phlogiston being redundant. His paper of April 1785 announced his qualified conversion to Lavoisier's doctrines and has some significance in the history of chemistry in that Berthollet was the first prominent chemist to align himself with Lavoisier against the phlogiston school.

If Berthollet's prime concern was theoretical, he also had good reason for interest in the potential practical applications of chlorine, for in February 1784 he had secured the post of chemical adviser to the *Bureau du commerce*, a post vacated by Macquer's death.[14] This was said by Fourcroy (one of the contenders) to be the best paid scientific post in the country,[15] and its salary of 6600 *livres* a year[16] enabled Berthollet now largely to abandon his medical practice and devote his full attention to chemistry. The post also, of course, brought obligations and these Berthollet took very seriously, conscientiously devoting particular attention henceforth to the applications of chemistry. His specified responsibilities were twofold. Basically, as general chemical consultant, he had a duty to examine and report on chemical inventions referred to him by the *Bureau*, this carrying a salary of 2000 *livres*. In addition to this Macquer had in 1781 been given a more specific task, with a salary of 4600 *livres*: to engage in researches on dyeing and on their basis draw up a general treatise on the art. Since Macquer's declining powers had prevented him from accomplishing this second task, it too was taken over by Berthollet, and was to result in 1791 in his *Éléments de l'art de la teinture*. Although not specifically mentioned in his letter of appointment, it seems that Berthollet's functions also included technical oversight of the state dyeworks at the Gobelins (though the designation 'director of the Gobelins' sometimes encountered in the secondary literature would appear a little misleading). In view of these duties, it is understandable that chlorine's property of destroying vegetable colours, already noticed by Scheele, should have attracted Berthollet's particular interest. He immediately saw that this bleaching action might find valuable application in his researches on dyeing, as a tool for investigating the relative fastness of dyestuffs; and while his initial paper did not explicitly suggest its application to large-scale bleaching, he may already have been beginning to think of the great simplification and abbreviation which chlorine might be expected to effect in that process, for he did mention incidentally that after several hours in a chlorine liquor cloth acquired a white comparable to that resulting from exposure on bleach-fields. By the end of 1785 he was engaged in experiments directed towards the use of chlorine for large-scale bleaching.

An important feature of Berthollet's investigations of chlorine was his use of the gas in aqueous solution. Scheele had found it to have only a very small solubility and had concluded that its properties were therefore best studied in the gaseous state. In this he had been followed by Bergman. Berthollet, however, found that although the solubility was small it was nevertheless significant, and by using the recently introduced Woulfe's apparatus he was able to obtain fairly saturated solutions conveniently and without discomfort (the gas, of course, being a poison and an irritant). His use of such solutions must have contributed considerably to the ease and success of his studies. It was also in the form of its aqueous solution that he sought to apply chlorine to bleaching, and we shall see that in this he was to be followed by most of the other early workers on the subject. Bleaching by chlorine was to prove much less straightforward than Berthollet at first imagined and the difficulties involved will call for consideration in due course. For the moment suffice it to say that it was not until 1789 that techniques had developed sufficiently for him to feel able to publish a first description of the method, a description which then became the chief guide of other early workers. Prior to this publication, nevertheless, Berthollet made no secret of his work. From an early stage he made it clear that he had no wish to exploit the process for personal gain, and he communicated it freely. This resulted in experimentation with the new bleaching spreading with extraordinary speed, not only in France but in Britain, too. In Britain James Watt was one of the first to introduce it, following a visit to Paris in November–December 1786 during which the process was shown to him by Berthollet. The French chemist's remarkable openness was demonstrated with particular force when Watt soon afterwards offered to secure on his behalf a patent and parliamentary privilege for the exploitation of the new process in Britain: Berthollet declined the offer.[17]

The early interest of Lavoisier and the agriculture administration

Berthollet's bleaching experiments aroused early interest in government circles in a body called the *Comité d'administration de l'agriculture*,[18] and in particular in one of its leading members, Lavoisier. This body had been formed in 1785, in response to the agricultural crisis, and consisted of a small group of experts from the Academy of Sciences and the *Société d'agriculture de Paris*; it met under the presidency of Gravier de Vergennes, head of the *département des impositions et de l'agriculture* of the *Contrôle général*. Although intended only as a consultative body, the committee was in practice soon seeking to take a more active role in the formulation and execution of policy, and among the early projects which it conceived was one to promote the cultivation of flax and hemp in France, with the promotion of the spinning and weaving of these products as a necessary corollary.[19] The committee was approached with proposals towards the end of 1785 by a certain Diot, and as a result a scheme was elaborated to establish in Paris a works which would serve as a school of spinning, training spinstresses who would then go out to provide instruction in the provinces. This was to be centred on the family of Lefebvre (Diot's nephew), skilled textile workers from the Saint-Quentin area. Lack of government funds prevented the immediate implementation of the scheme but as a preliminary step a small works was set up in Paris in 1786, financed privately by the members of the committee themselves along with other well-wishers; it was said to be on the new boulevard, near the rue Montparnasse. There the Lefebvre family began to occupy themselves with the production of various textiles, in particular of linens imitating kinds of cloth previously made in cotton or silk. Despite a continuing lack of government finance this venture proceeded to develop, and a prospectus was issued in the autumn of 1787 aimed at raising further funds by sale of shares. From this it is learned that a spinning school was now to be set up, under the supervision of the Church of Saint-Sulpice; and the enterprise was evidently now widening its scope to embrace cotton manufacture, too, for the spinning machine of de Barneville was to be adopted for the production of fine cotton yarn, with which it was proposed to introduce into France the manufacture of 'superfine' muslins. By the beginning of 1788 it appears that the chief output of the enterprise had become cotton goods: English-style muslins and dimity.

When these schemes were still nascent, Lavoisier on 9 December 1785 drew the attention of the committee to the contribution Berthollet's new bleaching method might make to the promotion of the linen and hempen manufactures. At his instigation, samples of fine linens submitted by Diot were sent to Berthollet for bleaching. The results were approved in the meeting of the following week and it was thereupon decided that Vergennes should write to Berthollet for more information, and invite him to collaborate with the committee for the development of his valuable discovery. In reply, Berthollet announced in January that he was engaged in large-scale experiments for the bleaching of cloth, and volunteered to inform them of the results he obtained.[20] His experiments no doubt met with more difficulties than anticipated for in fact no more is heard until July 1786, when Lavoisier again presented to the committee a piece of cloth bleached by Berthollet's process (in three days). Although the result of only a small trial on a single piece, Lavoisier felt it could be concluded that the cost would be less than that of ordinary bleaching. Berthollet was congratulated on his readiness to communicate his discovery freely to bleachers and it was resolved to recommend him to the Controller-General as meriting government reward. The active interest which Lavoisier was taking in the subject is shown by his announcement in mid-September that he was then engaged in mounting a workshop for chlorine bleaching in Lefebvre's works. He proposed, too, that a demonstration and trial of the new bleaching be conducted at the Arsenal on 20 September, though this does not seem to have materialized.[21] A chlorine-bleaching plant did subsequently come into use in Lefebvre's works; exactly when and with what success is not clear, but it was apparently in operation around the beginning of 1788 as a subsidiary to the manufacture of cloth there. It was not to last long. By April 1788 Lefebvre's whole enterprise had closed, for reasons which were said to have been quite other than any lack of success, and which we may conjecture to have been financial. A contributing factor was no doubt the dissolution of the *Comité d'administration de l'agriculture* itself in September 1787, for political reasons.

Although short-lived and probably very limited in scale, this venture nevertheless perhaps has the distinction of having been the first in France to take up the new bleaching commercially. And two men associated with it— Bonjour and Bodeau de Grandcour—were subsequently to play a notable part in the growth of chlorine bleaching elsewhere.

The spread of interest

In 1787, and more particularly in 1788, experimentation with the new bleaching began to spread. One of the earliest instances of which we have knowledge is that of a Nantes firm, Évesque & Cie, who launched a venture under the title 'manufacture du blanc serein'.[22] In October 1787 they bleached samples of cloth and yarn, both cotton and linen, from half a dozen prospective clients. Then, armed with certificates of satisfaction and with the supporting signatures of some thirty-six local merchants, they petitioned the Controller-General for the award of an exclusive privilege for their process, claiming to have discovered it only after many years' work. Soon afterwards they admitted that the process they were using was Berthollet's, which naturally nullified their request for a privilege. The subsequent fortunes of the concern are not known. Chlorine bleaching was later said to have been in use in one works in Nantes before the Revolution (see p. 136), but that of course may have been a different venture.

The obstacles to success confronting the first experimenters were formidable. Quite apart from the intrinsic technical difficulties in the process, and the fact that the details of its large-scale application were still only in course of being worked out, there must also in the provinces have been considerable ignorance of such as had already been learned. Prior to the appearance of Berthollet's first description in mid-1789 there was no significant publication on the subject, merely the scantiest of indications in odd press reports. Thus, although Berthollet was free in communicating information, knowledge of the technical developments being made would tend to be confined to that circle enjoying contact with him, and others wishing to try the new method must often have been working rather in the dark. One such, for example, was the inspector of manufactures Roland.[23] Roland was naturally interested to try the new bleaching method which had come upon the scene so soon after he had written his article on bleaching for the *Encyclopédie méthodique*. The only details he had, however, were from a newspaper item in the *Courier de l'Europe*, which prescribed simply steeping the fabric in a solution of salt of tartar (potassium carbonate) and then plunging it into a solution of chlorine. This Roland tried, with various modifications, but he understandably found that he could not get satisfactory results, and he was left rather sceptical about the large-scale practicality of the process. Similarly discouraging results were obtained by experimenters in the bleaching region around Laval and Mayenne, in the west.[24] A local official there reported in March 1789 that 'several have been ensnared by it', among them the bleacher de la Jubertières, who had had much cloth spoiled in unsuccessful attempts. Those who had tried the process had concluded that it was not viable and few bleachers in fact had thought it worthwhile to attempt it. They raised many objections against the new method, their deepest feelings being revealed when they asked: what would become of the 1000 or 1200 men and women then employed in the bleach-fields of Laval and Mayenne, if the new method did become established?

How widespread such experiments were we do not know, but we can be certain that success proved elusive. Amid the early failures, the successful development and establishment of the process was spearheaded by a handful of pioneers, and it is to these that we now turn.

Bonjour at Valenciennes

The pioneer venture to which Berthollet himself attached especial importance was that of his sometime assistant F. J. Bonjour. Since it was on this that the question of government encouragement for the new bleaching particularly focused, it will be worth considering the affair in a little detail.

Like Berthollet, Bonjour had come to chemistry from a medical background.[25] He had received his doctorate in Paris in 1781 but is said to have been too sensitive to enjoy the life of a physician and so practised little, devoting his attention rather to botany and chemistry. By 1783 he had become *démonstrateur* at the Queen's botanical garden, and perhaps also by about this time he had a junior teaching position as *répétiteur* at the *École des mines*.[26] The turning point in his career came with Berthollet's appointment in 1784 as chemical adviser to the *Bureau du commerce*, for this enabled Berthollet to engage an assistant and Bonjour was the man he chose. Bonjour was clearly no mere bottle-washer in this post but an able chemist in his own right, devoting his spare time to preparing the French translation of Bergman's *Dissertation on elective attractions*, which appeared in 1788, with anti-phlogistic notes also by Bonjour.

As Berthollet's assistant, Bonjour collaborated in the researches on chlorine bleaching from the outset, and then, probably in 1787, he took the bold step of leaving Berthollet's laboratory to attempt a career in its commercial exploitation. The details of his transition from laboratory to industry are obscure. He played a part in the Paris venture of Lefebvre, discussed above, but his main activities were to be in Flanders. As early as 30 July 1787, Berthollet, after mentioning public trials of the new bleaching, remarked that it was now sufficiently proven to be taken up by two large establishments in Flanders,[27] perhaps a reference to early projects involving Bonjour. Certainly, by the beginning of April 1788 Bonjour was engaged in launching chlorine bleaching there,[28] and soon afterwards information on his activities begins to become available, for by early June the difficulties he was encountering had led him to approach the government for support, in the form of an exclusive privilege.[29] From his petition to Tolozan (who as *Intendant du commerce* was head of the trade and industry administration), it emerges that Bonjour had gone to Valenciennes at the invitation of a bleacher there named Cuvellier. Cuvellier had professed a desire to establish the new process in his works, but when Bonjour had arrived in Valenciennes he had found his efforts opposed not only by the other bleachers of the town but even by Cuvellier himself. Bonjour had nevertheless carried out a successful public demonstration, bleaching fifty-three pieces of linen (lawns) in four days, and he had also succeeded in finding new prospective backers for a venture. When he had begun to look for a suitable location, however, the established bleachers had conspired to frustrate his plans; they took possession of all the likely premises, so that the only place he was able to find suffered the dual disadvantage of being over a league ($2\frac{1}{2}$ miles) from the town and of requiring some expenditure for the provision of a water supply and for generally rendering it suitable. This was presumably the location which Berthollet later described as having been provided by the *comte* de Bellaing. Bonjour's company of prospective backers were understandably hesitant, and were prepared to finance him only if granted the protection of an exclusive privilege. Bonjour therefore requested exclusive rights to exploit chlorine bleaching within a radius of 10 leagues (25 miles) of Valenciennes, for a period of fifteen years. This was to guard against the possibility that after spending large sums on research and development the company might find their concern undermined by the appearance of a more favourably situated rival. But when on 1 July this request was taken by Tolozan to the *Bureau du commerce* for a decision, it was rejected. Not so easily deterred, Bonjour immediately presented a revised petition. He now requested a privilege for a radius of only 2 leagues (5 miles) around Valenciennes, and a similar area around Cambrai; and he made the interesting suggestion that his works should serve as a school for the new bleaching, offering to receive there, free of charge to the government, as many pupils as the administration might care to send. As we shall see, this request ultimately proved no more successful than the first.

Bonjour's failure to obtain a privilege is the more surprising in that his petitions had the whole-hearted support of Berthollet, both as inventor of the process and as technical consultant to the *Bureau*, for it was of course to Berthollet that the matter was referred for advice. Reporting on Bonjour's first petition, Berthollet spoke approvingly of the young man's talents, and of the desirability of seeing a successful large-scale establishment formed, to demonstrate the value of the process. He recommended that it would be appropriate to promote the development of the new bleaching by the award of limited privileges not only to Bonjour, but also subsequently to such others as might have the necessary competence. Following the perhaps unexpected failure of the first petition, Berthollet proceeded to argue at greater length and more vigorously in favour of the second. He enumerated the advantages of the new bleaching, but also the difficulties involved, both the technical problems and the obstacles arising from prejudice; he emphasized the high costs which would certainly be entailed in the first large-scale establishments; he pointed out that an exclusive privilege was one of the least onerous forms of encouragement the government could give, since after all it cost nothing; and he argued against the growing trend towards laissez-faire policies by contrasting the examples of Poland and Italy, where such freedom prevailed, with that of England, whose government was in the habit of readily granting privileges far more extensive than that sought by Bonjour. The comparison with England was given added force when the inspector of manufactures Brown returned from a visit across the Channel, probably in October. He reported that Bourboulon de Boneuil, formerly of the Javel firm, was establishing a chlorine bleaching works in England and had obtained an exclusive privilege from the British government: France risked seeing the British reap the rewards of a French discovery. That Bourboulon had obtained a privilege was in fact untrue but it made a good argument!

The reasons for the negative attitude of the *Bureau du commerce* are not known, no explanation being recorded for their decision of 1 July. A number of observations might be made, however. In the first place, it is true that the drift of government policy was away from the award of privileges. That the *Bureau* was thinking hard about such awards may be seen from its suspension on 5 June 1788 of an earlier decision in favour of a manufactory of white lead, pending reconsideration of the best means of encouraging that branch of industry.[30] On the other hand, this did not mean that the practice had ended, and in fact in August 1788 an exclusive privilege was granted for soda manufacture which was far wider in geographical scope than the request of Bonjour (see Chapter 4). It seems likely that a major consideration affecting the *Bureau*'s attitude was continuing doubt about the merit of the new bleaching process. In a memoir of 6 October, Bonjour spoke of a prejudice which had arisen against it because others who had attempted its use, employing imperfect methods, had obtained products which although white initially had yellowed on subsequent laundering; he asked the administration to arrange rigorous trials of his own method, to prove the fastness of the white obtained. A subsidiary factor may have been the vociferous hostility of the traditional bleachers, for in the same month Berthollet warned Tolozan: 'I have reason to believe that you will receive from Valenciennes some further memoir against the new bleaching'. It is possible, too, that Berthollet's position as the *Bureau*'s consultant did not give him so strong a hand for the promotion of his invention as one might have expected, since there might have seemed cause for doubt about the objectivity of his advice on this matter. Certainly, in arguing the case for Bonjour's second petition, Berthollet took care to make clear that he had no hidden interest in the enterprise, adding that if he had wished to make his fortune from the discovery he would have taken up Watt's offers regarding an English privilege.

It was following conversations between Berthollet and Tolozan at the beginning of October that the proposal emerged for rigorous trials of the quality of Bonjour's white, and it seems likely that Tolozan informally gave Berthollet to hope that if such trials proved satisfactory a privilege might be forthcoming; for in urging the need to proceed with them at once, Berthollet explained that it would be important for Bonjour to establish his works before the winter. It may be, too, that such an understanding enabled Bonjour to make some progress with his venture. At all events, by 10 November a bleaching works described as 'newly established' had been set up by Bonjour in association with Philippe Constant, a merchant and cloth finisher of Valenciennes. It was in the village of Aulnois, about a league from Valenciennes, presumably the location mentioned earlier. There, between 10 and 13 November, the trials were conducted. Under the eye of Crommelin, the local inspector of manufactures, Bonjour bleached samples of a variety of linen and cotton goods, some of them sent specially by Tolozan, others procured locally. The plan was that with seals attached for identification these would then be sent to Paris for testing by government commissioners. Berthollet had specifically asked not to be included among these, for fear of charges that he had influenced the findings. Alas, before reaching that stage the trials were largely vitiated through the negligence of a workman, who accidentally prepared too concentrated a sulphuric acid bath (or perhaps too strong a chlorine liquor—the accounts vary), as a result of which over half the cloth samples were so degraded that some of them tore apart on subsequent rinsing. It may well have been this unfortunate accident which finally lost Bonjour his privilege, for the trial could only be judged inconclusive.

When Berthollet published his first account of chlorine bleaching, in mid-1789, hope of a privilege for Bonjour had apparently been abandoned and Berthollet spoke of it regretfully as an opportunity missed. While the *Bureau* might conceivably have believed that the new bleaching would best develop with experimentation proliferating freely, Berthollet, for his part, was convinced that the technical difficulties would result rather in a proliferation of failures, and he was anxious at the discredit these tended to throw on his process. If Bonjour had received the support he needed, Berthollet reflected, he might have formed a strong establishment, whence trained practitioners could have carried the process successfully to other provinces, so avoiding the fruitless attempts of many individuals. As it was, Bonjour's backers had so lacked confidence in the Valenciennes venture that they had launched another at Courtrai, which Berthollet evidently considered an unwise dissipation of effort.

Bonjour's subsequent fortunes in Valenciennes are not entirely clear. After the embarrassment of November 1788 he asked to try again, and Tolozan immediately set about procuring another batch of samples. The new trials did not take place until May 1790, however, by which time Bonjour's affiliations in the town had apparently changed. There is no longer any mention of Constant or of the Aulnois works; instead Bonjour now seems to have effected a reconciliation with his former adversary, for he was engaged at the bleaching works of Jacques Cuvelier and P. J. Cuvelier *fils*, in the faubourg Notre-Dame. Chlorine bleaching was established there in regular use, it being reported that '150 pièces sont en cours de blanchissage journalier'.[31] The official samples were this time bleached without mishap and were then referred by the *Bureau* to a group of inspectors of manufactures for examination. The report—by Lazowski, Brown and Lepage—spoke approvingly of the results and was enthusiastic about the process: 'the inspectors are deeply persuaded', they concluded, that the new process 'could not be too much encouraged, and that one cannot too strongly promote the discovery of M. Berthollet ... they applaud the course the administration has taken in having it verified, so as to show in its full light an invention essential to the prosperity of commerce'.

Another inspector, Brisson, in a separate report spoke with equal enthusiasm. The *Bureau* itself, however, remained cool. Its secretary, Abeille, remarked that only time and experience would tell whether the white would withstand storage and washing (the inspectors by this time having apparently forgotten quite what it was they were supposed to be testing for). And after hearing the results, on 22 August, the *Bureau* merely came to the limp conclusion that:

experience alone being able to teach all the advantages which should result from the process of bleaching invented by M. Bertholet, one cannot but conceive the wish that there be formed several establishments putting this process into use.

By this time the Revolution had in any case largely deprived the *Bureau* of its former powers.

Welter at Lille and Armentières

Berthollet's second protégé, J. J. Welter,[32] seems to have enjoyed a more ready success. Son of a forgemaster in the mining area of Rédange in Lorraine, Welter had attended the *Collège royal* in Luxembourg and had then moved to Paris, where he studied at the *École des mines* for three years from 1783. There he met Bonjour, who was a *répétiteur* at the school, and the two struck up a close friendship. In 1787 Welter joined Berthollet as laboratory assistant, presumably as successor to Bonjour, and then in the following year, when still only twenty-five, he followed Bonjour to the bleaching centres of the north. His subsequent career shows him to have been a very capable technical chemist, gifted with an inventive turn of mind, and already during his time with Berthollet he was beginning to make notable contributions to bleaching technique. Whereas Bonjour had concentrated on the bleaching of cloth, Welter concerned himself particularly with yarn, whose bleaching by traditional methods was rendered even more troublesome than that of cloth by the difficulty of ensuring satisfactory exposure to air and sun, and for whose bleaching chlorine therefore offered perhaps even greater advantages. With two associates he established a works at Lille, which Berthollet in mid-1789 described as enjoying much success.[33] Berthollet added that Welter had already begun several others, one of them evidently at Armentières. In February 1790, outlining the progress the new bleaching had made in France, Berthollet spoke with particular approval of the Lille and Armentières works, remarking that chlorine bleaching had met with 'a complete success in some establishments, such as those of Lille and Armentières, which I cite for preference among all those I know'.[34] Welter was to be one of his chief sources for the successive accounts of bleaching technique he published.

Descroizilles at Rouen

In Rouen, the local academy's offer of a prize in 1786 for the winter bleaching of cotton can perhaps be taken as an indication of early interest in Berthollet's discovery there, although in fact no mention of Berthollet's work was made in the prize programme.[35] The process to be used was left open, the prize being offered to whoever should carry out satisfactorily the speediest and cheapest bleaching of a one-pound sample of cotton yarn during the months of January, February, and March. The prize attracted four entrants in each of the years 1787 and 1788 but we know the method employed in only one instance, that of an unnamed 1788 entrant who had used chlorine. His results were judged unsatisfactory and the use of chlorine expensive. The prize of 300 *livres* was not awarded and was thereafter withdrawn.

The development of chlorine bleaching in Rouen really dates from the arrival in the town towards April 1788 of a certain Bodeau de Grandcour.[36] Grandcour had worked with Bonjour in Paris, as one of the directors of the

Lefebvre venture, and on a later occasion he described himself as 'élève de Berthollet pour le nouveau blanchiment'.[37] He had come to Rouen, he said, with the encouragement of the *Intendant* of the province, and of the town's chamber of commerce, to try the application of the new bleaching to the local textiles. Local interest is shown by his subsequent receipt of proposals from several manufacturers and bleachers. A trial of the process was arranged before members of the chamber of commerce, and this duly took place, according to Arvers at the premises of the local mechanician and physics teacher Scanegatty.[38] Unfortunately, the trial proved rather less than successful: Arvers relates that the samples could bear no comparison with traditionally bleached goods, and even Grandcour himself could only claim that the experiments 'have allowed one to foresee the possibility of a complete success'. He realized that on both technical and economic grounds the application of the process there was going to be far from easy. Showing a somewhat modest opinion of his own chemical capabilities, he decided to seek a chemist as an associate and on local recommendation he lighted upon the industrial chemist F. A. H. Descroizilles. The latter was destined to become France's most renowned practitioner of the new bleaching.

Descroizilles came from a long line of Dieppe apothecaries.[39] He had studied chemistry in Paris and had then in 1777 settled in Rouen, where he began teaching chemistry with the title 'démonstrateur royal de chimie'. In 1778 he was received a master apothecary and he then practised that profession for the next ten years. During this time he was also interesting himself in industrial matters. He travelled a good deal, formed a friendship with the inspector of manufactures Roland, and was showing that enthusiasm for invention which a later report by the local authority was to describe as a 'mania'. ('He exhausts himself in projects and in discoveries. This mania will always be harmful to the prosperity of every establishment he might direct' (1810).)[40] One of his early inventions was that of flashing lighthouses, in which the time period of the flashes served to identify the location, and a later notable invention was the filtration technique for brewing coffee. By the end of 1787 he had left his apothecary's shop and set up as an industrial chemist, after having already earlier been busy retailing mordants, acting as a chemical consultant, and (since 1785) engaging in the manufacture of various salts employed by local industry.

By April 1788 Descroizilles was working on chlorine bleaching with Grandcour and had already made some important advances. Satisfactory large-scale development, however, was to absorb much further effort and considerable expense. That Descroizilles was able to persevere with it he owed primarily to the support he received from the brothers Alexandre and Pierre Nicolas de Fontenay,[41] who were leading merchants in Rouen, with textile manufacturing interests, notable for having been among the earliest in France (about 1787) to introduce large-scale mechanical spinning by English machines. They had sufficient faith to finance Descroizilles through his early difficulties, and their faith was eventually rewarded, for in February 1790 Berthollet was able to report: 'I learn that Mr. Décroisille who encountered many obstacles at the beginning, has finally overcome them, and that his bleaching is already in great demand at Rouen'.[42] Berthollet added a few months later that Descroizilles's works was in full activity, bleaching coarse cottons, fine linens, stockings, and caps, at almost the same price as the traditional bleachers, the results being highly esteemed.[43] Descroizilles was to become one of Berthollet's most ardent admirers and devised a whole new vocabulary enshrining the name of his hero: in his 'bertholleries' (workshops) at Rouen, the 'bertholleurs' (workmen) 'bertholleyed' (bleached) the cloth, using 'berthollet' liquor whose strength they regulated with a 'berthollimeter'.

As for Grandcour, he soon left Descroizilles and turned up in Lyons in the summer of 1789 to play a similar role in introducing chlorine bleaching there.[44] A public demonstration took place on 27 August at which in the presence of some fifty leading figures from local scientific and commercial circles, a bleaching liquor provided by Grandcour was successfully tested by Macors, a pharmacist and teacher of chemistry in the town. Within three weeks the liquor (of unknown composition) was on sale at 5 *sous* a pound from the shop of Macors, pending the establishment of a larger outlet. Macors was said to have already increased the strength of the liquor by a mechanical improvement in its preparation, and a prospectus was published giving directions for its use. Macors was evidently a man of some repute and became a correspondent of Berthollet when the latter, in gathering information on dyeing for his book, established relations with a number of 'correspondents of known merit' in the provinces.[45] He was to be one of the principal figures in the development of the new bleaching in Lyons.

The interest of chemical manufacturers: the Javel company

Among those taking an early interest in the new bleaching were chemical manufacturers. Thus, Chaptal in Montpellier was experimenting with the bleaching properties of chlorine before 1789.[46] His experiments do not seem to have led him to any industrial venture but he did indicate some new applications which were subsequently to be taken up, namely its use for the cleaning and removal of disfiguring marks from books and prints (a process employed at the *Bibliothèque nationale* about 1800),[47] and for the bleaching of rags in papermaking (see below). Another, slightly later, instance is that of Jean Holker, who by 1794 had introduced a chlorine-bleaching plant in his chemical works at Rouen.[48] There is some indication, too, that by 1793–4 the Nantes manufacturer Athénas had an interest in chlorine bleaching; an apparatus devised by him for the preparation of chlorine liquors was published in 1801.[49] But the earliest and most important instance of contribution by a chemical concern to the growth of the new bleaching is that of the Javel company.

Berthollet relates how at an early stage in his experiments he was invited by the Javel manufacturers to demonstrate to them the preparation of chlorine and its use in bleaching.[50] He happily obliged, travelling out to Javel on two occasions, taking his apparatus with him. This was perhaps as early as 1786 and must certainly have been before May 1787, for by then Berthollet was no longer using quite the method he had demonstrated. As we have noted earlier, the Javel manufacturers were quick to seize the chance of a new outlet for their products. A travellers' guide to Paris published in 1787 mentioned that they had very recently reduced to practice the bleaching method proposed by Berthollet, and were offering hydrochloric acid for sale complete with instructions for the preparation of chlorine.[51] Moreover, appreciating the difficulty for bleachers of having to prepare chlorine themselves, the company soon developed a commercial bleaching liquor for direct use, and it is here that their major contribution lies. They began to market this liquor extensively at the turn of 1788–9, when they advertised it by circular letters to likely customers and by insertions in the Paris and provincial press.[52] By then they had obviously been supplying it in a smaller way for some time: although their claim to have been producing it for 'some years' was clearly an exaggeration, their product had been known to Berthollet in March 1788.[53]

Berthollet denounced the firm, feeling that the bleaching liquor to which they attached their name was a mere plagiarization of that which he had demonstrated to them.[54] In this he was less than fair, however, for the change they had made, though simple, was important. At the time of his demonstration Berthollet's practice had been to add a little potash to the chlorine water, since this alleviated its offensive smell. The Javel manufacturers, instead of adding potash to the chlorine solution after its preparation, adopted the practice of receiving the chlorine gas *into* a solution of potash. This made it possible to prepare much more concentrated solutions, owing to the reaction of chlorine with potash to form—as we now know—potassium hypochlorite ($KOCl$); the latter is more soluble than chlorine itself but like chlorine has bleaching action. The Javel manufacturers were thus able to produce a bleaching liquor which was sufficiently concentrated to be transported with reasonable economy, and so commercially marketed; for use, the liquor was to be diluted with eight to twelve parts of water. It may also be that the firm were the first to recognize, albeit vaguely, the value of employing alkali in such liquors in excess[55] (see below). With sound commercial instinct the firm called their product 'lessive de Javel', deliberately avoiding reference to the oxymuriatic acid (chlorine) which entered into its composition, since any name mentioning acids might be expected to arouse apprehension in a public acquainted with their corrosive nature. The product was essentially the same as modern hypochlorite bleaches, familiar domestically under such brand names as 'Domestos', and such liquid bleaches are still universally known in the French-speaking world as *eau de Javel*. The chief chemical difference is that soda is now used instead of potash, since it is cheaper.

The Javel company advertised their product in 100-pound bottles at 7 *sous* a pound, available from their depots in Paris, Rouen, Lyons, and Orleans. They gave detailed directions for its use, although Berthollet said the method they prescribed would be adequate only for cotton goods. Their hope, clearly, was to open up a potentially immense industrial market, but in this they were to be disappointed. In the event, *eau de Javel* was not to become important as a commercial bleach for industrial use. It was too expensive, and though much more concentrated and more stable than a simple aqueous solution of chlorine, it remained a rather bulky item for large-scale transport and did gradually lose its bleaching properties when stored. Instead, industrial users came to prepare their own hypochlorite liquors while as a commercial product *eau de Javel* found its niche as a convenient small-scale bleach for domestic and laundry use. Its manufacture for such purposes subsequently spread.

Chlorine bleaching was also the focal concern of a venture launched in Liverpool as an offshoot of the Javel works, and some brief account of this is here appropriate.[56] The English venture arose as a by-product of the bankruptcy of Bourboulon de Boneuil (a member of the Javel company), in his position as treasurer-general to the Count of Artois. The bankruptcy was accompanied by charges of embezzlement, and within about three weeks of its declaration on 5 March 1787, Bourboulon had fled to England to evade arrest.[57] Making the best of his enforced exile, Bourboulon must have seen in the textile industry of Lancashire tempting prospects for the exploitation of chlorine bleaching, still little known in Britain. He was accompanied or followed to England by Matthew Vallet, one of Javel's technical directors, and by June or July the two were in Liverpool, soon stirring up considerable commotion with their schemes. The prominent part they played in the growth of chlorine bleaching in England has been shown by Musson and Robinson. By late August they had given demonstrations of the new bleaching and had also prepared a certain quantity of soda from salt, their intention clearly being to apply residues from chlorine production to the manufacture of soda. In February 1788 they petitioned the Commons for a Parliamentary monopoly, with more extensive rights than an ordinary patent, but opposition was organized by those like Watt who felt their own interests threatened, and the move was defeated. Bourboulon did, however, secure some limited patent coverage in March 1789. In June 1789 he and Vallet began advertising their bleaching liquor (*eau de Javel*), having by then established a works outside Liverpool. Three years later the concern had failed. French chemists from the works are said, nevertheless, to have continued to contribute to the growth of the new bleaching in Britain. In particular, Vallet now joined the firm of Ainsworth, near Bolton, the largest bleacher in Lancashire, and probably set up bleaching plant for other firms too. He is commemorated in the name of the hamlet of Vallets, near Bolton. There may in fact have been more than one Vallet active, for Chapman speaks of a certain Victor Vallet.[58]

It is not known what relations existed between the Javel works and the Liverpool venture. There does not seem to have been any financial link, but in 1788 those at Javel were still in correspondence with 'our friends at Liverpool'.

Printed-textile manufacturers: Widmer and Haussmann

Manufacturers of printed textiles also figured among the early experimenters and this can readily be understood. Theirs was an industry which had enjoyed a dramatic growth and which included large concerns; and since it employed techniques of some chemical sophistication, one might expect to encounter there the skills requisite for the new bleaching. Moreover, apart from the possible application of chlorine to the bleaching of raw cloth, there was a more particular reason why it should attract the attention of textile printers: as a means of removing unwanted dye in their printing operations. The commonest of the industry's basic techniques consisted in printing the design onto the cloth using a paste containing a mordant, and then dipping the cloth into a dye bath (usually madder), whereupon the dye adhered firmly to the mordanted areas and the pattern thus appeared. Some dye also remained more weakly attached to the unmordanted parts of the cloth, and this unwanted dye then had to be removed so as to leave a white background. In technical parlance, the 'grounds' had to be 'cleared'. Textile printers did this by a process resembling the traditional bleaching of raw textiles, with treatments in baths of cow dung and of bran alternating with exposure on bleach-fields. Reliance on this procedure naturally introduced into textile printing inconveniences similar to those suffered by the bleaching trade—an element of seasonal dependence, for instance—although the clearing of grounds was a somewhat less lengthy business. The application of chlorine to the clearing of grounds was early attempted by Berthollet himself, but he found that the chlorine bleached the pattern as well as the unwanted residual dye. The problem was solved in 1788, when Thomas Henry in Manchester and Descroizilles in Rouen found that the bleaching action could be limited to the unwanted dye by using a solution of chlorine in potash or soda, instead of plain chlorine water,[59] and this discovery was then soon taken up by others. The Javel manufacturers, with their typically quick eye for a new opportunity, were advertising their *eau de Javel* for this purpose by the beginning of 1789, after experiments of their own. Berthollet early in 1790 spoke of trials of chlorine in textile printing being made in Lyons, and of an establishment which was to be formed for the purpose at Saint-Quentin.[60] The most notable early instances of the adoption of chlorine by textile printers, however, are those of Haussmann at Colmar, and more particularly Widmer at Jouy.

The use of chlorine in alkali for clearing grounds was communicated by Berthollet to Oberkampf, the founder-director of France's most celebrated printed-cotton works, at Jouy near Paris.[61] Oberkampf quickly began experiments on the subject and in mid-1789 was reported to be continuing them with a certain Royer.[62] Before long the development of chlorine bleaching at Jouy was being particularly pursued by Oberkampf's young nephew, Samuel Widmer, who was then just beginning to play that role in the works which was subsequently to become so important. Widmer had been brought to Jouy from his native Switzerland as a boy, and had been carefully educated for his intended role as Oberkampf's assistant. A thorough grounding in the skills of the factory had been followed by the study of physics with Charles and of chemistry with Berthollet, and then from the early 1790s Widmer began to take a growing hand in the technical direction of the enterprise. It was both a reflection of his scientific education and an indication of the character the works was increasingly to assume that one of his first acts was to install there a chemical laboratory. And having studied with Berthollet, it is no surprise that his earliest notable technical concern was with chlorine bleaching. By 1791, under Widmer's direction, chlorine was in regular use at Jouy for clearing grounds, though it still presented some problems which limited this use.[63] By now, too, Widmer was also concerning himself with the bleaching of raw cloth. His investigations in the summer of 1791 into the comparative printing characteristics of cloths bleached by chlorine and by the traditional method pointed to the superiority of the former, and the bleaching of raw cloth in this way subsequently became a regular part of the business: a step towards Oberkampf's ultimate ambition of embracing within his enterprise the entire manufacturing sequence, from the purchase of unspun cotton to the sale of the final printed fabric. The establishment of chlorine bleaching at Jouy was just the first of a succession of important technical innovations for which Widmer was to be responsible, and which secured the firm's position at the very forefront of its field and brought it major honours in the Napoleonic period. Both Widmer and Oberkampf enjoyed the friendship and esteem of the leading scientists of the capital. In particular, Berthollet retained close relations with them over the years: in 1791 he referred to investigations which he and Widmer had undertaken together into a particular bleaching problem; in 1792 he is said to have visited Jouy for the installation there of a chlorine bleaching plant;[64] subsequently it was on the experience of Widmer, along with that of Welter, that Berthollet was especially to draw in writing his later accounts of bleaching practice; and at the beginning of the nineteenth century Berthollet was to send his own son to study applied chemistry for several years at the Jouy works, so reciprocating the compliment he had received earlier from Oberkampf.

The other manufacturer to be mentioned is Jean Michel Haussmann, notable as the earliest known adopter of chlorine bleaching in the east of France. In the 1770s, Haussmann had established a printed-textile works at Logelbach, near Colmar in Alsace, after having earlier been a chemist with the leading German firm of Schüle.[65] He was evidently a competent chemist, having studied the subject when initially destined for a career in pharmacy; he published papers on chemical subjects and like Macors was chosen as a correspondent by Berthollet when the latter was gathering information on dyeing for his book. His factory enjoyed the status of a *manufacture privilégiée*, and prospered so as to be employing some 1200–1400 workers by the end of the *ancien régime*. From 1786 he had also begun to undertake the weaving of cloth (by the putting out system) and so had a dual reason for interest in chlorine bleaching. He began experiments with the new bleaching perhaps as early as 1788, and it was in large-scale use at his works by the winter of 1790–1.[66] He informed Berthollet that during that winter his use of chlorine to clear the grounds of some 3–4000 pieces had spared him great embarrassment, and that he was counting on using the process for a larger quantity in the winter to come. He had also used chlorine for bleaching some 300 pieces of raw cloth prior to printing. The subsequent history of chlorine bleaching there is not known.

Chlorine bleaching and the Revolution

During the early years of the Revolution, the publication of Berthollet's practical guide in 1789, and the abolition in 1790 of the tax on salt (an essential raw material), helped promote the growth of interest in the new bleaching. To the ventures already indicated, others might now be added. Thus, in February 1790 Berthollet reported that the new bleaching was beginning to be used in Orleans.[67] He mentioned, too, that its use was beginning at Laval, which perhaps suggests that the discouraging early experiments there were now faring better (Jubertières in particular seems to have eventually succeeded).[68] In 1791 the pharmacist Nicolas Schemel established a works near Metz which coupled chlorine bleaching with the manufacture of mineral acids and salts.[69] And by about that time chlorine seems to have been in use in a bleaching works at la Glacière near Paris, by Ribaucourt (something of a chemist and later a director of alum works in the Oise and the Aveyron).[70] Another works near Paris was that established at Bercy in 1791 by Pajot Descharmes, of whom more will be said later; this was unusual in that its aim was not the bleaching of raw goods but the decolourization of dyed and printed cloth for re-use, to help make up for the shortage of raw cotton from which France was then suffering.[71] There was also a works near Lille established by Bernard (p. 238), and one at Troyes formed by Bosc (p. 241), and there were no doubt a good many other ventures too. With the worsening of the war between 1792 and 1794, however, and the consequent commercial and industrial disruptions, the progress of chlorine bleaching was interrupted.

Bonjour and Welter were now particularly unfortunate in the location of their works close by France's imperilled northern frontier, within the battle zone of the Flanders campaigns of 1792–4. Both abandoned their operations when it became impossible to procure supplies.[72] Valenciennes, the seat of Bonjour's enterprise, was bombarded by the Allies for forty-three days in the summer of 1793. During the siege, Bonjour himself served as a gunner until injured in a blast, after which he put his chemical skills to use as a pharmacist in the service of the health officers. By the disinfectant application of chlorine (or perhaps more probably of hydrogen chloride?—see below, p. 182), he is said to have contributed greatly to the overcoming of contagion in the town's hospitals. He was apparently back in Paris by mid-1794, for he must surely have been the Bonjour—described as previously employed in the offices of the naval department—who was named a member of the newly reorganized central gunpowder administration on 8 August, alongside Chaptal and Champy.[73] Bonjour was not to resume his career in bleaching. In Paris he had a junior teaching post for a time under Guyton de Morveau, at the newly formed *École polytechnique*; then in 1797 he became government commissioner in the salt works of eastern France, a post he held until his death. His friend Welter also returned to Paris at the height of the Revolution, obtaining a post in 1794 at the *École polytechnique* under Berthollet, and at the same time working as a mineralogist in the military research establishment at Meudon. Welter did go back to bleaching, however, in 1797, and his later activities will be referred to in due course.

The works of the third outstanding pioneer, Descroizilles of Rouen, is not known to have suffered any remarkable interruption. Presumably, though, like all works not already afflicted by other ills, it must at least have felt the severe sulphuric acid shortage in 1794–5. Descroizilles himself was temporarily drawn into other activities at the height of the Revolution. From the middle of 1792 he was increasingly active politically in Rouen and Dieppe, and was imprisoned for a time in 1793. In January 1794 the Committee of Public Safety named him inspector in charge of the Revolutionary gunpowder and saltpetre programme in a group of ten northern departments, and then in March he was called to Paris to join the central administration heading that programme, along with Chaptal, Carny, and three others.[74] This position presumably occupied him until the reorganization of the administration in July.

How many ventures came to an end as a result of the Revolutionary disturbances it is impossible to say, but there were certainly some. A clear casualty was the works of Schemel at Metz, totally destroyed in 1792 during the siege there. Less dramatically, the bleaching plant of Holker in Rouen had ground to a halt by late 1794[75] and is not known to have been revived. It was perhaps also the events of the Revolution which put paid to chlorine bleaching in Nantes, where towards 1804 the works which had operated before the Revolution was said no longer to exist; all local cloth was now again being bleached by the traditional method, in the thirteen large bleaching works along the river Erdre.[76] An instructive example comes from Tarare, near Lyons.[77] There, a certain Simmonet, mayor of the town, later related how his brother-in-law had begun experimenting with chlorine bleaching in about 1792 ('He retained a chemist here for several months'). He succeeded in a small way, but was still encountering difficulties in its large-scale execution when the overturning of his workshop by Revolutionary events came as a last straw, causing him finally to abandon the enterprise. One wonders how many faltering experimenters, still struggling with the technical difficulties, similarly had the final seal put on their discouragement by the additional problems the Revolution brought. All told, the Revolution must have set back the growth of chlorine bleaching in France by several years. It was only from the later 1790s, with the restoration of relative peace and the general resurrection of industry, that the new bleaching could resume its progress and enter upon a course of continuous development which was to bring it into common use in the early 1800s.

An important factor assisting its development then was the growth of a literature on the subject.

The growth of a literature

Berthollet's own researches were essentially confined to the laboratory, and the large-scale application of chlorine bleaching was chiefly worked out by such figures as Bonjour, Welter, Descroizilles, and Widmer. Nevertheless, it will have become obvious from the account so far given that Berthollet himself remained very much at the centre of things. Through his personal contacts and correspondence he kept closely in touch with developments not only in France but also (so far as British reticence permitted) in Britain. He must certainly have been the most completely informed man in Europe and was ideally placed to write the first detailed description of the new bleaching. This appeared in mid-1789 in the second volume of the important new journal Berthollet had helped found, the *Annales de chimie*, and this account immediately won international recognition as the standard source on the subject. It received wider circulation in France in 1792, when it was reprinted by Fourcroy in the *Encyclopédie méthodique*, with the accompaniment of a number of subsidiary papers which had by then also appeared.[78] In 1795 it was also issued separately as a brochure, in a pirated reprint.

By this time Berthollet dismissed his first account as being now outdated, and he published a new description early in 1795. These were the months when the government was turning its attention to the revival of trade and industry, now that the Revolution had passed its peak, and as it happens Berthollet himself had a position of special responsibility in this regard.

Following the abolition of the old government ministries, the nation's industrial affairs were now in the hands of the *Commission d'agriculture et des arts*, to which Berthollet was appointed in September 1794, apparently as its head: his position must thus have approximated to that of a minister of agriculture and industry.[79] Under Berthollet's leadership the commission saw its task not only as the re-activation of France's industrial life but also as the promotion of major improvements in her industry, to enable it to sustain foreign competition, and to this end weekly meetings were instituted bringing together leading figures from the fields of technology, science, and the administration. In the event the commission had time to achieve little before in turn being abolished when the government of the Convention gave way to that of the Directory in November 1795, industrial affairs then passing to a re-created Ministry of the Interior. One measure it did take, however, was to begin publication of a new government journal, the *Journal des arts et manufactures*, aimed at the promotion of technological advance and intended as a companion to the two official technical journals which already existed, the highly reputed *Journal des mines* and *Feuille du cultivateur*. It was in the second number of this journal that Berthollet's new account of chlorine bleaching appeared, probably in about April 1795. In the same year the account was also put out separately as a brochure, by the same publisher, and there may in fact have been several issues in brochure form, for in 1804 Berthollet wrote that 'a number of separate editions were produced'.[80]

Berthollet's writings on bleaching reached their definitive form in 1804, in the newly updated account he wrote for the second edition of his *Éléments de l'art de la teinture*. As one of the standard books in its field during the early nineteenth century this presumably found a wide circulation. Moreover, its account of bleaching also came to be included in a main rival dyeing text of the period, that by Homassel, first published in Year VII (1798–9): a book which is now forgotten but which in its day was probably more appealing to practical men than Berthollet's own work. Homassel had been chief dyer at the Gobelins ('chef des teintures . . . des Gobelins') for several years towards the end of the *ancien régime*, and his book can be seen as his response as an artisan to Berthollet's scientific tomes, exemplifying that artisanal hostility towards academic science which formed one of the undercurrents of the Revolution. Despite his general scorn for the scientists Homassel could not deny the value of chlorine bleaching, and the first edition of his book included a poor account by himself of bleaching by 'acide marin phlogistiqué' (*sic*). When a second edition was called for in 1807 the book was revised by Bouillon-Lagrange, who amended the chemistry and expunged the diatribes, and who, on coming to the subject of bleaching, was so impressed by Berthollet's new description that he confessedly abandoned his own draft and simply copied Berthollet's account more or less verbatim. The continuing popularity of the book is shown by the appearance of an unchanged third edition in 1818, and a fourth edition in 1857.

The successive accounts by Berthollet deservedly dominated the field but by the beginning of the nineteenth century there were also two other extended treatments available. The more substantial of these was the book by Pajot Descharmes: *L'Art du blanchiment*, written towards 1791 when he was sub-inspector of manufactures at Abbeville. Descharmes had a particular interest in chemistry (he will figure again in our account of the soda industry), and was inspired by Berthollet's 1789 publication to take up experiments himself, so as to be able to promote the new bleaching in his area. He soon became convinced that if the process was to establish itself it would need to be made easier, less dangerous, and more economical, and this was the avowed purpose of his book. In fact, although Descharmes identified the problems well enough it is clear that he had no very notable improvements to offer. Moreover, through having been delayed in publication by the Revolution, the book was already beginning to be outdated in some respects by the time it appeared. It was originally to have been published by the 'administration générale du commerce' (the *Bureau du commerce*?) in 1791, but publication was prevented by that body's suppression the same year, and so the work finally appeared only in 1798, with one or two minor additions but no apparent revision of the basic text. The work was nevertheless of some merit and must have been a useful complement to the writings of Berthollet. It was a good deal longer than any of Berthollet's accounts, running to nearly three hundred pages, this partly reflecting Descharmes's comparative verbosity but also the greater detail into which he entered on some aspects, explaining carefully, for instance, the important matter of how to prepare and use lutes for sealing joints. The book was sufficiently valued for an English translation to be published in 1799.

Rather shorter was the book by O'Reilly, published in 1801 under the title *Essai sur le blanchiment*. O'Reilly was editor of the *Annales des arts et manufactures*, a journal he had founded in 1800. In the first volumes he had published several short articles concerning bleaching developments in Britain, and it was no doubt partly on the strength of this that he was invited by Chaptal, then Minister of the Interior, to write the *Essai*. The work had the merit of indicating recent developments but suffered from O'Reilly's approach to the subject more as a journalist than as an experienced practitioner or original investigator, and from the haste with which it was composed (O'Reilly announced proudly that he had put it together in less than a month). Though it covered the use of chlorine quite extensively, the main focus of interest was in another development, that of steam bleaching, a particular concern of Chaptal himself, and the circumstances of the book's publication will be considered more fully in that connection later.

These various general accounts of bleaching practice, together with the periodical literature which also began to appear, provide a basis for tracing the technical evolution of the new bleaching. It is to this aspect that we now turn.

The development of the bleaching treatment

The first chemist to give serious attention to bleaching theory was Francis Home, in his researches towards 1756 on the traditional bleaching process. On the basis of the considerable loss in weight suffered by linen goods during bleaching (up to some 30 per cent), Home adopted the view that the colouration of the raw textile was due to coloured impurity, and that bleaching consisted in the removal of this so as to uncover the native white of the basic fibre. This he believed to be effected partly by the bucking operation, in which colouring matter was washed out by the alkaline lye; and partly by the exposure on bleach-fields, whose purpose he conceived to be to allow evaporation from the cloth of a further portion of oily colouring matter which he supposed to have combined with alkali to form a volatile soapy material. Watering of the cloth on the field assisted in this, since the soapy material was carried off with the evaporating water, while the sun served to promote evaporation by its warmth. Home seems to have been broadly followed in his interpretation by the first notable French writer, Roland, and some such view was perhaps the general one, although there was a suspicion that sunlight also played some more particular role in the grassing stage. Traditional bleaching was thus seen to consist essentially in the laborious progressive removal of colouring matter from the fabric.

When the bleaching effect of chlorine was discovered, the gas was regarded as acting by decolourizing the colouring matter rather than by removing it, and this was interpreted by Scheele and other phlogistic chemists in terms of the phlogiston theory. Viewing chlorine as dephlogisticated muriatic acid, they supposed that its action on vegetable colours was to extract phlogiston from them, thereby itself reverting to ordinary muriatic acid (hydrochloric acid), while the colouring matter, in losing its phlogiston, at the same time lost its colour. This view fitted the theory of such phlogistic chemists as Bergman and Kirwan that colour was directly caused by the presence of phlogiston in a body, a theory which found some empirical basis in the colour changes observed when iron salts, for example, passed from one degree of phlogistication (or as we would say, oxidation state) to another. Broadly speaking, when Berthollet began his researches on chlorine, he simply stood the phlogistic interpretation on its head, interpreting gains and losses of phlogiston in terms of losses and gains of oxygen.[81] In the preparation of chlorine, he envisaged the reaction as consisting in the transference of oxygen from 'manganese' (manganese dioxide) to muriatic acid, so regarding the chlorine produced as oxygenated muriatic acid, or oxymuriatic acid. This supposed compound retained its oxygen with only a very weak affinity and readily abandoned it to other materials, being what we would term an oxidizing agent. It was to this oxidizing action that Berthollet attributed the decolourizing power of chlorine. Whereas phlogistonists had viewed the loss of colour as resulting from loss of phlogiston, Berthollet saw it as a consequence of the combination of the colouring matter with oxygen, from the oxymuriatic acid.

In taking up the application of chlorine to practical bleaching Berthollet seems initially to have hoped that this decolourizing action would in itself be a sufficient bleaching treatment. Of his earliest experiments we know only the main points, as he later recounted them.[82] Thus, at first he tried to bleach by simply immersing the fabric in a chlorine solution, and of course if left there a sufficient number of hours it would indeed be whitened. The results were not satisfactory from a practical point of view, however, for he found that if a strong chlorine solution was used then the cloth itself was attacked and weakened, while if a dilute solution was used this problem was avoided only to be replaced by another: though appearing well bleached initially, the cloth tended to yellow on subsequent storage or washing. He was to find his way out of this problem, he tells us, by looking to the practices of the traditional bleacher. Already in his 1785 paper he had seen an analogy between the action of chlorine and the traditional operation of grassing, evidently regarding grassing as contributing to the bleaching process by effecting decolourization through slow atmospheric oxidation. The analogy was not quite so obvious as might appear to the modern eye, for Home, as we have seen, had regarded the grassing stage in quite a different way, and Berthollet was probably the first to interpret it as involving oxidation. At all events, he came to reflect that since in the traditional treatment grassing alone was insufficient, requiring to be alternated with buckings in alkali, perhaps exposure to chlorine should similarly be alternated with alkaline lyes. This practice he began probably in the latter half of 1786, and he found that satisfactory results could indeed be obtained in this way, that the length of time the fabric had to spend in the chlorine could now be reduced, and that weak chlorine solutions could now be employed without the problem of later yellowing.

Berthollet now saw the role of oxidation in bleaching—whether by chlorine or by exposure on bleach-fields—as being not only to decolourize the colouring matter but, even more important, to render it soluble in alkali, its subsequent extraction with an alkaline lye being essential for a good, permanent white.[83] It was this insight, he implies, which guided him to the use of lyes. Detailed researches into the precise mode of action of chlorine, reported to the Academy in May 1790, confirmed him in this general theory. His experiments indicated that raw textiles contained a certain proportion of colouring matter which was immediately soluble in alkali, and so could be removed simply by lyes, but also further colouring matter which resisted such treatment and became soluble only after oxidation. There was a good deal of truth in the basic understanding of chlorine's bleaching action at which Berthollet had now arrived. The modern chemist would broadly agree in seeing bleaching as a question of eliminating colouring matter (and other impurities) so as to reveal the basic white of the pure fibre (cellulose), the treatment being based partly on immediate solution in alkali and partly on decolourization and solubilization with a bleaching agent. Chlorine bleaches are still considered to act largely by oxidation, though it is now appreciated that the reactions occurring can vary with conditions (chlorination, for instance, also playing a significant part in some circumstances); the precise reactions are in fact highly complex and are still not fully understood. Of course, Berthollet was wrong in regarding chlorine as oxymuriatic acid: following Davy's researches in 1810 it was recognized to be an element and not a compound at all. On the other hand, in his studies of chlorine Berthollet was usually concerned with its behaviour in aqueous solution, where as we now know it reacts to form hypochlorous acid (HOCl) to a considerable degree:

$$Cl_2 + H_2O = HOCl + HCl$$

It is this hypochlorous acid, an oxidizing agent, which is of central importance in the bleaching action of chlorine preparations, and it can be seen that Berthollet's concept of oxymuriatic acid fits the situation rather well.

The problem of yellowing or browning, after having been solved once over by Berthollet's adoption of alkaline lyes, recurred in 1788.[84] This was no doubt a consequence of the progression to more difficult cloths or simply of the scaling up of operations. A certain Caillou was among those who encountered it, for example, when after successful small-scale experiments in Paris, mostly on cotton, he went to Saint-Quentin to apply the new bleaching to the linen cloths of that region. The problem was also met with by Descroizilles in Rouen, and by Berthollet himself in his laboratory experiments, though it was successfully avoided by Bonjour and by Welter. We have already seen how the doubts which it cast on the merit of the new bleaching led the administration, towards the end of 1788, to arrange trials of the permanence of Bonjour's white. By 1789 Berthollet had traced the trouble to inadequate lyeing, establishing that it was important for lyes to be thorough and sufficiently hot. He explained this phenomenon of colour reversion by an ingenious elaboration of his bleaching theory.[85] In the colourless compound formed by the oxidation of the colouring matter he supposed that the oxygen subsequently tended to combine with part of the hydrogen present, to be then released as water; the increased proportion of carbon in the now dehydrogenized colouring matter gave it a brown colour. This, of course, occurred on the fabric if the colouring matter had not been properly removed, and proceeded slowly on simple keeping and more rapidly if the temperature was raised, by laundering for instance. The theory was related to ideas Berthollet was developing at the same time on the colour changes produced in dyestuffs by the air. Though somewhat naïve, it did correctly stress the importance of effective extraction of the solubilized impurity, if colour reversion was to be avoided.

Having profited from the example of traditional bleaching in his introduction of alkaline lyes, Berthollet soon tried the effect of the other basic traditional operations—soaping and acid baths—and by mid-1787 was finding these, too, to be of value.[86] As a result, when he came to publish his first description of the new bleaching, in 1789, it had become a fairly elaborate business. After preliminary removal of the weaver's size, the treatment for linen cloth involved between four and eight buckings with alkali, a similar number of immersions in chlorine water, a soap wash, a sulphuric acid bath and a number of rinsings. Soon afterwards Berthollet even re-admitted exposure on bleach-fields into his treatment. In 1790, in a supplement to his description, he explained that Welter had found a tendency for linens to retain a certain yellowish tint, which could be better removed by grassing for three or four days than by further application of chlorine. By Berthollet's second account, of 1795, the procedure prescribed had become even more involved: the number of alkaline lyes and chlorine immersions remained about the same (typically about eight lyes and six immersions), but the treatment now also included between two and four soap washes and a similar number of acid baths and exposures on the bleach-field, the whole sequence interspersed with numerous rinsings. The procedure indicated in his final 1804 account remained essentially unchanged. These details are for linen and hempen goods. Cottons were much more readily bleached: in 1789, two or three lyes and immersions were prescribed; in 1795, a soaping and an exposure on the field as well; in 1804, five lyes, three or four immersions, an acid bath, and a soaping, but no exposure on the field, this now being declared unnecessary for cotton goods.

Thus, chlorine did not after all provide a simple replacement for the old bleaching procedures. The complex cycle of traditional operations was retained basically intact, with the major exception that the lengthy exposure of cloth on bleach-fields was largely replaced (and for cottons entirely replaced) by short immersions in a chlorine liquor. In one respect this fact no doubt eased the growth of chlorine bleaching, since the established bleachers did not in the event find their plant and their skills suddenly made redundant; rather, they were invited to adopt the use of chlorine as a detailed modification to their normal mode of working. This point was recognized by the Javel manufacturers when they advertised that their bleaching liquor could be used

without upsetting the operations employed by ordinary bleaching works; . . . it is destined to go absolutely hand in hand with the customary treatments, whose effect it will simply accelerate.[87]

In another respect, however, the necessary retention of so much of the traditional procedure added to the difficulties of the early pioneers, for there was consequently incorporated into the new bleaching a dependence on much of the empirical artistry of the old. Success called not only for a basic competence in chemistry but also for a sufficient familiarity with traditional bleaching skills. Thus, while at first Berthollet particularly stressed the need for chemical expertise—writing in July 1788, for instance, that even when the most precise details of the new bleaching had been given 'I affirm that it will be capable of execution only by people instructed in chemistry'[88]—in 1795 he balanced this with the observation that

to attain perfection in a large establishment, to profit from all the processes which have been introduced into bleaching works by long experience, it is indispensable to have followed with great care all the details of an ordinary bleachery, and even to gather together some workers accustomed to the principal manipulations which are there carried out.[89]

Although Berthollet and his co-workers did introduce some measure of chemical enlightenment into bleaching, it was to remain in high degree an art in which experience was of paramount importance.

Bleaching liquors

One of the major difficulties which experimenters found in the use of chlorine concerned its effect not on the cloth but on themselves, for it is, of course, a highly toxic gas. Though the general practice was to employ it in aqueous solution, such solutions exhaled the gas to a disagreeable extent. A concentration of chlorine in the air as low as 5 parts per million by volume is sufficient to cause discomfort, and 50 parts are described in modern texts as producing dangerous effects within an hour; as is well known, chlorine was used as a poison gas during the First World War. A particularly graphic description of its fairly advanced effects was penned by Pajot Descharmes, recounting the miseries he himself experienced when he began experimenting on the basis of Berthollet's 1789 description:

Running of the nose, asthmatic affection of the breast, headache, tears and smarting of the eyes, bleeding at the nose, the sensation known by the name of the teeth set on edge, pains in the small of the back, and even spitting of blood, are the ordinary inconveniences to be expected, when the pure oxygenated muriatic acid is used as is prescribed in the *Annales de Chimie*. . . . The strong expectoration to which I was exposed, agitated the system so much, that I found it impossible to retain any food on my stomach, and was for forty-eight hours, without intermission, not only deprived of sleep, but continually emitting saliva, with acid and corrosive humours from the eyes and nose in such abundance, particularly from the eyes, that it was sometimes five or six hours before I could open them to support the light.[90]

Descharmes tried working with a kind of gas-mask he devised, or with a handkerchief moistened with alkali tied around his face; as a palliative he recommended chewing liquorice, and with a hint of desperation he suggested that the operations should be conducted in a shed through which a strong draught blew. Sufferings to quite this degree were no doubt exceptional but all the early experimenters must have experienced the discomforts of the gas to some extent.

A remedy to the problem had in fact been discovered incidentally by Berthollet in his earliest scientific studies of chlorine.[91] He had found that chlorine (oxymuriatic acid) was a very much weaker acid than hydrochloric acid (muriatic acid), a point of great theoretical interest since it made chlorine a striking anomaly to Lavoisier's theory of acidity, which expected acidity to increase with oxygen content. Since a chief characteristic of acids is their reaction with alkalis to form neutral salts, Berthollet had naturally been interested to study the action of chlorine on alkalis, and he had found that some kind of weak combination occurred, the addition of alkali to chlorine water having the effect of removing its colour and odour but not its decolourizing action. This property he put to use when he began to develop the application of chlorine to bleaching, adopting the practice of adding potash to his bleaching solution so as to alleviate the odour. The use of alkali was subsequently recognized to have other advantages too. As the Javel manufacturers were perhaps the first to see, if chlorine was passed into a potash solution much stronger liquors could be obtained than by just dissolving it in water. This was obviously important if the liquor was to be transported as a commercial product, and could be expected also to be of advantage to a bleacher preparing his own liquor since it would facilitate handling within the works. The readier solution of chlorine in alkali than in water also meant that the preparation of such liquors was easier, and that simpler apparatus would suffice. Moreover, it must presumably have been fairly early appreciated that an alkaline solution of chlorine would retain its bleaching action better on standing than plain chlorine water, an advantage if it was not to be used immediately. Such advantages in the use of alkali proved to be counterbalanced by drawbacks, however, both technical and economic. The simple economic disadvantage was the important one, but we shall consider first the technical drawback since it seems to have been this which at first particularly impressed Berthollet, the scientist.

In the spring of 1787, Berthollet made a discovery[92] which for the next thirty years was to cast scientific doubt on the wisdom of employing additives, such as the alkalis, in bleaching liquors. In experiments conducted in the further pursuit of his theoretical interest in chlorine's acidity, Berthollet passed the gas into a solution of caustic potash and observed the formation of a previously unknown salt, potassium chlorate ($KClO_3$), together with potassium chloride (KCl). Recognizing a high proportion of oxygen in the new salt, Berthollet interpreted the reaction, in essence correctly, as involving a redistribution of the oxygen of the oxymuriatic acid, so as to form on the one hand what he regarded as super-oxymuriate of potash, and on the other hand the ordinary muriate. The reaction in fact can be summarized in modern terms:[93]

chlorine + water = hypochlorous acid + hydrochloric acid
hypochlorous acid + potassium hydroxide
        = potassium hypochlorite + potassium chloride + water

hypochlorous acid + potassium hypochlorite + potassium hydroxide
→potassium chlorate + potassium chloride + water

This discovery of potassium chlorate aroused considerable interest because the new compound promised to find application in the production of a powerful new explosive. The significance of the discovery for bleaching was that neither potassium chlorate nor the simultaneously produced chloride had any bleaching action, so that chlorate formation in a bleaching liquor meant loss of bleaching power. It seems to have been this which first turned Berthollet against the use of alkali. In a letter of 22 May 1787 he informed Watt that he had now abandoned its use having found that it caused the liquor to bleach much less cloth, through formation of the new salt. In a further letter of 11 March 1788 he considered that at least half the chlorine was lost in liquors such as that now being marketed by the Javel firm, so that on this count alone the cost was doubled.[94] It must have been with some satisfaction that he felt able to denounce the product of the plagiarists at Javel, on what seemed good scientific grounds.

Berthollet was to persist in this theoretical objection throughout his writings on bleaching, and evidently considered it an edifying example of the way in which scientific understanding could guide technological practice. The historian can rather see in it an example of the way in which the limited chemical understanding of the late eighteenth century could be misleading as often as useful. The matter is worth considering a little further, both in order to see why Berthollet was misled and also because it will begin to give some insight into the complexity of the chemistry which the pioneers of chlorine bleaching were up against. At a time when there was still disagreement and confusion among chemists regarding the chemical nature of a solution of chlorine in alkali, Berthollet, in fact, with his characteristic feeling for the subtleties of chemical reaction, showed for the most part an outstandingly good understanding of the situation. Already in 1787 he had realized that there was only partial conversion to chlorate, and that the reactions occurring varied in a complex way with conditions. By 1803 his understanding had become more precise,[95] and he saw that combination of oxymuriatic acid and alkali produces largely oxymuriate (hypochlorite) in the first instance, together with a smaller quantity of super-oxymuriate (chlorate) and muriate (chloride). The oxymuriate, like free oxymuriatic acid itself, held its oxygen only loosely and so retained its bleaching action; but it slowly reacted, over the course of days, to form further useless super-oxymuriate and muriate. Berthollet also recognized another mode of decomposition of the oxymuriate, by the action of light with the evolution of oxygen, but that does not concern us here. He appreciated that chlorate formation was markedly influenced by the concentrations of alkali and chlorine, and suspected that it varied with temperature. Unfortunately, he rather overestimated the significance of chlorate formation, and in particular he failed to appreciate the important fact that it is greatly retarded by an excess of alkali. In reality, his objection to the use of alkali was largely groundless provided the alkali was employed in suitable quantity.

Modern understanding of the behaviour of chlorine solutions is based on the variation in their chemical constitution with the degree of acidity or alkalinity, as measured by the pH value. This is shown in Fig. 3.1. A plain solution of chlorine in water might typically have a pH value of about 2, the chlorine being present in the form of molecular chlorine and hypochlorous acid in roughly similar proportions. If alkali is progressively added to such a solution its constitution changes, as further molecular chlorine is converted

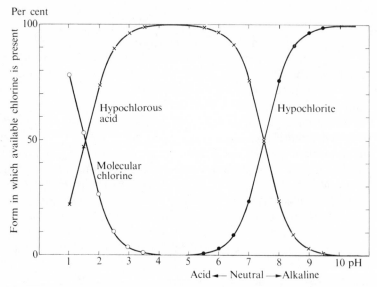

Fig. 3.1. Variation in composition of a chlorine solution with degree of acidity or alkalinity. For 0·1 per cent chlorine water at 25°C. Plotted from data in Kirk and Othmer (1963–72), vol. 4, 911.

to hypochlorous acid, and as the hypochlorous acid is then in turn converted to hypochlorite. The formation of chlorate, by reaction of hypochlorous acid and hypochlorite, occurs with maximum speed towards the neutral point (pH 7), where the proportions of the reacting species are optimal. In more alkaline conditions the rate greatly falls, as hypochlorous acid largely disappears from the solution. The effect of adding alkali to chlorine water is thus the reverse of what common sense would expect, for the natural supposition would be that if the addition of a little alkali resulted in harmful chlorate formation, the addition of more would be worse. Such a common sense view was voiced by O'Reilly, for example, echoing Rupp,[96] and in that misguided belief these two writers prescribed for the liquor a minimal dose of alkali such as would very probably have produced optimal conditions for chlorate to be formed!

The Javel manufacturers, on the other hand, do seem to have been vaguely aware that a high proportion of alkali was somehow beneficial, since Berthollet, in revealing the secret of *eau de Javel* in 1789, gave a recipe for its imitation which would have produced a secure excess of potash. No reason was given for employing potash in such quantity and Berthollet did not appreciate that it countered chlorate formation: his view indeed was that chlorate formation was promoted both by a high concentration of chlorine and by a high concentration of potash.

Clearly, what was needed was precise information on the extent to which chlorate formation did in fact occur, or the extent to which alkalis did affect bleaching power, and it is disappointing how little was published on this question. Berthollet himself made some rather vague indirect inferences, but the only direct study to be published was one by Rupp in Manchester, in 1798, a study which was widely quoted but fairly slight. The paper was translated by O'Reilly in his *Annales des arts et manufactures*, and was discussed by him in his *Essai*; it was cited with approval in 1804 by Berthollet, who described the work as 'decisive'. Rupp's experiment consisted in measuring the volume of a standard dye solution which was decolourized by a sample of chlorine water, and then the volumes decolourized by similar samples to which progressively larger numbers of drops of a potash solution had been added. He found that the addition of alkali increasingly reduced the decolourizing power, the reductions found varying between 9 per cent and 26 per cent. The result was highly misleading. Berthollet was impressed by the demonstration that such significant losses occurred even at the low concentrations involved in this experiment, and hinted that at the higher concentration of *eau de Javel* liquors the loss could be presumed to be greater still. It was not until 1817, when Welter made a careful study of bleaching powder the results of which were extended by inference to alkaline liquors too, that the former fears regarding chlorate formation were clearly seen to have been misguided.[97] And it appears to have been only with the work of Balard in 1834 that the role of excess alkali in retarding chlorate formation found explicit recognition.[98]

If the scientific objection to the use of alkalis was misguided, the economic objection was nevertheless substantial, for the expense of the alkali in itself greatly increased the cost of the liquor. Descharmes, on the basis of 1791 prices, reckoned the increase at about 25 per cent. O'Reilly in 1801, following Rupp but substituting French values for English, arrived at a similar figure. These were in reality underestimates since they assumed only a minimal quantity of potash. For *eau de Javel*, Berthollet prescribed four times Descharmes's quantity of potash and six times O'Reilly's, making the increase in cost some 100–150 per cent. He himself in 1804 wrote that in liquors of the *eau de Javel* type the cost was nearly tripled, his figure presumably bearing in mind the supposed chemical loss too. These estimates can be compared with the actual manufacturing costs incurred by the producers of *eau de Javel* in Paris in the mid-1820s.[99] The cost of the potash then constituted over 64 per cent of the total (including labour, fuel, and interest on capital); to put it the other way round, the use of potash increased the cost over that of a plain chlorine solution by 180 per cent. The force of the economic disincentive may be judged from the fact that even the long-suffering Descharmes, who in his book advocated the use of potash as his main answer to the odour problem, nevertheless when he came to calculate bleaching costs proposed its use only for half the chlorine baths.

The answer ultimately proved to lie in the use of lime in place of the alkalis. This again can be traced to Berthollet's early scientific studies, in which he found that chalk ($CaCO_3$) and slaked lime ($Ca(OH)_2$) reacted with an aqueous solution of chlorine in a similar way to the alkalis, though more weakly.[100] As with the alkalis, hypochlorite bleaching solutions are formed.[101] The economic objection lost most of its force in the case of chalk and lime, since these were very much cheaper than the alkalis: in the mid-1820s, for example, the average price of the potash employed in Paris for the manufacture of *eau de Javel* was 87 fr. per 100 kg, while the price of the lime by then in use for the manufacture of bleaching powder was 6 fr. per 100 kg, and the price of chalk was 1 fr. per 100 kg.[102] Berthollet himself early saw in chalk a viable means of overcoming any fumes problem in practical bleaching. In May 1788 he informed Watt that if powdered chalk was added to the chlorine water it reduced the odour without apparently harming the bleaching power;[103] he later concluded that its use did entail some loss of chlorine as chlorate, but that the effect was negligibly small.[104] One of the first to adopt this method was Descroizilles, at whose works it was certainly in use by 1795.[105] In his descriptions of 1795 and 1804 Berthollet suggested the use of chalk in the form of a suspension, to be added to the chlorine bath from time to time during bleaching, if alleviation of the odour was considered necessary. A similar method was mentioned by Descharmes. Berthollet preferred the use of chalk to that of lime, partly no doubt because it was even cheaper, but also through a mistaken belief that lime would cause greater chlorine losses through chlorate formation (see p. 159). It may have been in part his theoretical misconceptions which blinded Berthollet to what was to prove the nineteenth century's definitive answer to the problem of chlorine use, for it was to be lime, not chalk, which proved the superior agent. And whereas Berthollet seems in principle to have preferred to use any palliative agent as an additament to chlorine water, in the belief that at such dilution chlorate formation was minimized, it was in fact as an absorbent for the gas that lime came rather to be employed: either in solid form to yield bleaching powder, or in aqueous suspension to form a lime bleach liquor. Bleaching powder will be discussed later; here we shall consider the development of lime bleach liquors, which at first were the more important of the two.

The use of lime instead of alkali in the water of the receiver, so as to obtain an inexpensive imitation of *eau de Javel*, might seem a very obvious development. That in reality it was not so obvious was due to the fact that lime has only a very low solubility, which seemed to limit the extent to which its solution could be expected to absorb chlorine gas. Effective use of lime depended on the realization that despite its low solubility, if the lime was kept in suspension while the chlorine gas was delivered then extensive reaction did occur, producing a soluble bleaching compound (calcium hypochlorite), and so giving rise to a strong bleaching liquor. This discovery is particularly associated with the name of the Glasgow bleacher Charles Tennant, for although there were earlier experimenters with lime in Britain during the 1790s, some of whom may have anticipated him in some degree, it was he who was primarily responsible for bringing such liquors into common use.[106] In January 1798 Tennant obtained a patent for the preparation of a bleaching liquor by passing chlorine into an agitated suspension of lime in water. Many bleachers in Britain having earlier resorted to *eau de Javel*-type liquors, despite the expense, the much cheaper lime liquor now quickly found favour. Tennant at once began selling licences for the use of his method, and many bleachers adopted it without licence, challenging Tennant's patent on the grounds that the use of lime was not truly original. The patent in the event proved ineffective and the method rapidly came into general use in Britain. In Northern Ireland, for instance, there were said to be thirty plants preparing bleach liquor in this way by 1799.

On the development of practices among the generality of French bleachers information is exceedingly scanty. It does seem, though, that there was a longer dalliance with chlorine water in France than in Britain, and a slower recognition of the outstanding value of lime. Plain chlorine water was still presented as the ideal by Berthollet in 1804, and in skilled hands, such as those of Welter and Widmer, it succeeded well. Against objections that the fumes were insupportable, Berthollet maintained that they need not be troublesome if the solution was properly used. The employment of chlorine water persisted in some works for a surprisingly long time. Muspratt's *Chemistry*, published in 1860, observed that it had been practised until about twenty years earlier in many parts of France, especially in the department of the Oise, though the practice was now abandoned.[107] It is doubtful, however, whether the great majority can have succeeded with it on a large scale. If experimenters were prepared valiantly to suffer some discomfort, workmen were less sacrificially inclined. Thus, by 1790 even Bonjour, while still continuing to use chlorine water, had found it necessary to reduce its strength considerably, since that with which he had at first worked proved offensive to the workers; the change had meant that a larger number of treatments had become necessary, and the bleaching process was lengthened.[108] Descroizilles early adopted the use of chalk, because even saturated chlorine water was thus made less disagreeable to his workers than the weakest solution without it.[109] And a government team working on the application of chlorine bleaching to paper manufacture, in about 1794, reluctantly turned to an *eau de Javel* liquor when they found that chlorine water proved objectionable to the workers, and that chalk was an insufficient remedy.[110] O'Reilly in 1801 spoke of a large works near Paris, evidently using chlorine water, where he had seen the workers rolling on the floor 'dans l'excès de la douleur',[111] but one wonders how long any such conditions could have lasted. Most bleachers must soon have had resort to one remedy or another. In the early years of the nineteenth century, however, there seems still to have been no general consent as to what was best. The use of bleaching liquors prepared with lime was made known to French readers in 1801 by O'Reilly's *Essai*, which reproduced an account by William Higgins of the method of preparing such liquors in Ireland, but this lime liquor does not seem to have been seized upon in France with anything like the same alacrity as in Britain. Berthollet in his 1804 account did not even mention it. Chaptal in 1807 did, but without any special emphasis:

In almost all works it has been attempted to correct the insupportable odour of this acid [chlorine], sometimes by carbonates of lime or pure lime, sometimes by alkalis; but this is always to the detriment of the acid's virtue. Of all substances, that which alters its properties the least, and which nevertheless corrects its bad odour, is chalk.[112]

Chaptal added that chalk 'is therefore in all factories mixed with the acid at the moment of using it', but one may doubt whether there was any such uniformity of practice as this implies. The lime bleach liquor was evidently in common use by the early 1820s[113] but it is not known whether it had found extensive adoption in the Napoleonic period or only after the resumption of normal communications with Britain in 1815.

Problems of control

Experience early taught that too strong a chlorine solution attacked and weakened the fabric. Bleaching liquors therefore needed to be of controlled strength, and some means of knowing the strength was required. The bleacher's nose could provide a rough guide, of course, and in his 1795 description Berthollet relied on this when he gave the following rather vague directions for the preparation of the liquor: the appropriate strength was to be attained by diluting a stronger solution 'until the mixture has a fairly lively but supportable odour; a stage which practice easily teaches one to recognize'. With experience, such reliance on smell might indeed suffice for many purposes. As Berthollet pointed out, however, it was desirable to have some more precise measure, and he indicated a method of estimation invented by Descroizilles, which was already becoming known through his earlier advocacy and which in due course was to come into standard use.

In devising his method, Descroizilles would have been aware of the hydrometer, an instrument which was finding growing use in the later eighteenth century as a ready means of determining the concentration of solutions from their specific gravity; he would have recognized, however, that such instruments are of no help in the case of chlorine water, since this is far too dilute for its specific gravity to differ appreciably from that of water itself. Descroizilles therefore had recourse to a chemical method, and in so doing opened up the development of a major new branch of chemical analysis: volumetric analysis.[114] Taking the decolourizing action itself as his basis, Descroizilles simply adopted the technique of measuring how many volumes of a standard dye solution (0·1 per cent indigo) were decolourized by one volume of the chlorine water. To facilitate the test he devised a special instrument which he called a berthollimeter. In function this was an early type of burette, since it served to measure the volume of the indigo test solution employed, although in form it was really a graduated cylinder, a kind of instrument which similarly was hitherto practically unknown. The cylinder was marked off in equal volumes, the scale reading from bottom to top. To make a determination, one put the chlorine water sample into the cylinder, bringing the level up to the first point on the scale, marked zero; then the dye solution was progressively added until it ceased to be decolourized. The number of volumes of dye decolourized was then given directly by the reading on the scale. Descroizilles came to adopt the convenient practice of denoting the strength of the chlorine water in terms of degrees, rather after the manner employed in connection with the hydrometer. Thus, a sample which decolourized eight times its own volume of the dye he described as having a strength of 8°; this in fact was the strength he found for water saturated with chlorine (at a temperature of 10° [Réaumur]), and so solutions used in bleaching would show a lower reading in proportion as they were more dilute.

Descroizilles must have developed this technique very early in his work on bleaching, for its principle was employed by Berthollet in some experiments of September 1788. Berthollet described the technique briefly in his 1789 account of chlorine bleaching, and again in his book on dyeing of 1791, and in his 1795 bleaching instruction. Descroizilles himself described it in detail in a paper in the *Journal des arts et manufactures* in 1795, reissued as a brochure in 1802. The method was described, too, by Descharmes and by O'Reilly, and that it was coming to be considered standard by the beginning of the nineteenth century might be judged from the fact that Berthollet in 1804 gave his directions for the strength of the bleaching liquor in terms of Descroizilles's degrees: the bleaching baths for the first immersions were to be at a strength of 2°, and those for later immersions at 1°. The gain in precision over his 1795 instructions is obvious. Besides finding application in the control of the bleaching operations the technique was also, of course, important for work on the development of bleaching methods. Rupp's experiments on the effect of alkalis in bleaching liquors, for example, would hardly have been possible without it, misleading though the conclusions drawn in that instance were. Descroizilles's method did have its weaknesses, deriving in particular from his use of indigo for the test solution. Commercial indigo was rather variable, and also the reading obtained proved to depend somewhat on the exact manner of operating: on whether, for example, the dye was added quickly or slowly to the chlorine sample. The test was thus rather inaccurate and improvements were to be introduced from the 1820s, notably by Gay-Lussac. The inaccuracy was more objectionable in scientific than in industrial use, however, and so long as a bleacher employed the berthollimeter in a reasonably consistent fashion it must have proved a valuable guide.

Descroizilles is generally considered the chief founder of volumetric analysis. Though anticipators might be named, it was Descroizilles who first brought volumetric methods into popular use and who introduced the basic instruments. His berthollimeter was the first burette to become commonly known, and he also introduced into chemical analysis the pipette and the volumetric flask. Hitherto, analysis had been essentially gravimetric analysis, based on weight determinations. This could give results of great accuracy, but called for chemical skill, patience, and time. In contrast to this, Descroizilles's berthollimeter typifies the virtues of volumetric analysis in general: it made up for its comparative lack of accuracy by its speed and simplicity, and it yielded a reading which could be directly meaningful to the practical user, without requiring complicated calculations. These virtues were to make volumetric analysis of particular value in commerce and industry, and Descroizilles himself was soon to make a further important contribution in this direction when in 1806 he introduced what he called the alkalimeter, another burette device, graduated for estimating the purity of commercial alkalis.

It is perhaps of interest to give some indication of the concentrations at which chlorine seems in fact to have been employed. The strength of 8° which Descroizilles gave as the maximum found for chlorine water might be assumed equivalent to a concentration of about 0·45 per cent, the approximate practical solubility of chlorine at the temperature he indicates.[115] The solutions of 1–2° which Berthollet prescribed for bleaching in 1804 would thus have had a chlorine concentration of some 0·05–0·1 per cent. A figure approaching 0·1 per cent is also suggested by the preparation details Berthollet gave in 1789. *Eau de Javel*, as it was manufactured in the 1820s, had an available chlorine content of about 1·5 per cent[116] (in other words, its bleaching action was equivalent to that of a hypothetical plain chlorine solution of concentration 1·5 per cent); preparation and titration details from earlier writings would suggest values in the region of 0·5–1·5 per cent for *eau de Javel* liquors prepared from the 1790s.[117] Since these were generally to be used after dilution with 10–12 parts of water, the concentration of the resulting bleaching baths would have been of the order of 0·05–0·15 per cent available chlorine. By way of comparison, in modern practice concentrations of 0·05–0·3 per cent available chlorine are used for bleaching cottons. Thus, the concentrations employed in early bleaching appear to have been within the range still used, and towards the lower end of that range.

While early bleachers thus appreciated the significance of chlorine concentration, and by employing Descroizilles's technique could exercise a reasonable measure of control over this factor, there was also another important factor governing the activity of bleaching liquors which was much more difficult of comprehension. This was the degree of acidity or alkalinity of the liquor, as measured by its pH value. We have already seen that the chemical constitution of a chlorine solution varies with its pH value and so it is no surprise that its bleaching characteristics vary too. In particular, the intensity of the oxidizing action proves to be at a pronounced maximum in the region of the neutral point, with the result that bleaching proceeds with especial rapidity there (a reason suggested is that the mechanism involves the formation of a complex between hypochlorous acid and hypochlorite). Use of alkali in the liquor can thus either enhance or retard the speed of bleaching according to whether a lesser or a greater quantity is employed. It might be thought that it would be most advantageous to employ a neutral solution, so as to maximize the speed; it has come to be recognized, however, that neutral conditions promote not only the bleaching but also degradative attack on the fibre cellulose itself, and so this neutral region is now considered a danger zone to be avoided, and bleaching is conducted either in acid or more generally in alkaline solution. The alkaline side of neutral is normally preferred, partly because an acid solution is less stable, inconveniently producing chlorine gas; and partly because such degradative attack on the cellulose as occurs in acid conditions tends to produce aldehyde and ketone groups, which constitute one cause of yellowing on ageing, whereas in alkaline conditions the result is primarily carboxyl groups which do not present the same drawback. The provision of a suitable pH is complicated by the fact that the bleaching reactions themselves produce acidic products, so that the pH value tends to fall as bleaching proceeds. To prevent it falling into the neutral region, it is usual to add a so-called buffer agent, which serves to counteract the acids produced and stabilize the pH at an alkaline level. The buffer in fact takes the form of excess alkali, preferably in the form of the carbonate.

It would obviously be interesting to know how much understanding and control early users of chlorine acquired over these pH effects. Unfortunately, the contemporary literature contains only scanty indications and so our discussion must be somewhat conjectural and tentative; an examination of the British literature would probably throw further light on the question. Certainly there was early awareness in France of both the quickening and slowing effects which alkali could produce. Berthollet himself, in his earliest bleaching experiments, observed an increase in speed when he added potash to the chlorine water (evidently because he added only enough to bring the solution near to neutral).[118] The fact that addition of alkali not only reduced the odour but also seemed to speed the bleaching in this way, was a further reason why Berthollet adopted the practice in his early experiments. In 1804 he explained the effect in terms of the affinity of potash for the muriatic acid present in oxymuriatic acid (chlorine), this facilitating the release of oxygen. The opposite effect of alkali in diminishing the bleaching activity was early observed by Widmer and by Haussmann, in their application of chlorine liquors to the clearing of the grounds of printed goods.[119] Employing for this purpose solutions of the *eau de Javel* type, they both found that the proportions of chlorine and potash required careful adjustment to give a liquor of such activity as would bleach the grounds without harming the pattern: a higher proportion of chlorine they found to increase the bleaching action, whereas a higher proportion of potash weakened and slowed it. Although both the enhancing and retarding effects of alkali were thus described in the early literature, it is not clear to what extent there was a systematic appreciation of the action of alkali, in producing either effect according to its proportion. Berthollet, for example, in 1804 still described *eau de Javel* preparations as having a more prompt bleaching action than chlorine water, although Chaptal, for instance, clearly described their action as being slower,[120] which is what one would expect to have been in fact the case for liquors prepared with an excess of alkali. A few pages later, Berthollet did quote Widmer on the effect of alkali in attenuating bleaching activity, but without noting the apparent inconsistency with his earlier remarks, and without offering any explanation of the slowing effect. There was probably some tendency to confuse the slowing effect of alkalis with the supposed loss of chlorine through chlorate formation, such a confusion being apparent in O'Reilly's *Essai*, for example.[121] The retarding action of excess alkali might have further contributed to the scientific prejudice against the use of such additives; and seen in this light, an advantage of chalk over lime would be that excess chalk would produce a nearly neutral solution, with a faster bleaching action than the alkaline liquor yielded by lime. A further aspect of the pH effect of which there was an early awareness was the heightening of the bleaching activity when sulphuric acid was added to a liquor prepared from bleaching powder (an alkaline solution of calcium hypochlorite); this Berthollet in 1804 attributed incorrectly to a supposed re-formation of chlorine by decomposition of chlorate. A similar effect would also be expected with *eau de Javel* liquors, of course, but this does not seem to have been recorded. To summarize, it seems that there was an early awareness of various aspects of the pH effect on speeds of bleaching, but little systematic understanding and no very helpful theoretical insight.

It can be imagined that the variation in bleaching characteristics with pH must have contributed appreciably to the difficulties of early experimenters and early adopters of the new bleaching. The use of alkali would be likely to produce somewhat variable results according to the amount added, and this would be exacerbated by the considerable variability in composition of the commercial alkalis themselves: prior to the introduction of Descroizilles's alkalimeter, from about 1806, there was no ready means of ascertaining the real alkali content of the impure commercial materials. Variability in liquors prepared with alkali would occur more particularly if the alkali was used only in the rather small quantities indicated by Descharmes and O'Reilly; if employed in excess, as seems to be the case in Berthollet's directions for *eau de Javel*, one would expect the result to have been a buffered alkaline solution of relatively consistent and stable pH, similar to the bleaching liquors now employed except in containing many impurities from the commercial alkali. Chalk and lime would similarly have given variable results unless used in excess. British bleachers considered lime to give more reliable liquors than alkalis,[122] partly no doubt because, since it was cheaper, they were more likely to employ it in large amounts, and partly perhaps because of a somewhat greater purity and consistency.

For ordinary bleaching, some degree of latitude in the activity of the liquor could no doubt be tolerated, but one application demanding more precise control was the clearing of grounds in textile printing. We have already seen how Widmer and Haussmann were both impressed by the control needed if grounds were to be satisfactorily bleached without affecting the pattern, a problem made the more difficult by the varying fastness of the colours which patterns employed. In 1791 Widmer commented shrewdly:

What I should like above all for the perfection of this process, would be an agent which indicated at the same time the acid strength [i.e. the chlorine concentration] and the alkaline strength of the liquor; I could then easily confide the bleaching of printed goods to a workman, whereas at present I am obliged continuously to supervise these operations myself.[123]

In the event, the difficulty Widmer here mentions—that the control required was such as to demand his own personal attention—eventually resulted in the use of chlorine for clearing grounds being abandoned at Jouy, despite its early promise.[124] Chlorine, of course, continued to be used there for the less delicate task of bleaching raw cloth prior to printing. Haussmann in 1791 similarly commented on the considerable supervision the process required on his part, as a result of which he continued to prefer the traditional method outside the winter months.[125] Berthollet in 1804 wrote that he did not know what success others might have had with the method, or whether anyone had overcome the problem which had caused its abandonment by Widmer. The use of chlorine for clearing grounds had come into common use by the 1820s but information on its progress is not to hand; it seems to have developed more quickly in Britain than in France.[126]

It is clear that such understanding and mastery of the pH effects as bleachers may have acquired was essentially a matter of empirical trial and error. When to these complexities one adds the further fact that the bleaching reactions, like all reactions, are temperature sensitive, a variation in air temperature of 8 °C (15 °F) being sufficient to halve or double the speed of bleaching, it will be appreciated that the handling of the chlorine liquors was no less an art than the old operations inherited from the traditional bleaching process.

We may finally note that if the control of bleaching liquors had its difficulties control of the action of gaseous chlorine was even more problematical, and this largely obstructed the application of chlorine to bleaching in gaseous form. Early experiments on gas bleaching were conducted towards 1789 by Chaptal, who exposed cloth to the action of chlorine in a lead chamber temporarily borrowed from sulphuric acid manufacture, and who considered his results to be moderately encouraging. Berthollet, however, was in no doubt that the future lay with chlorine solutions.[127] As he pointed out, the gas itself acted more powerfully than its solution, and so presented even greater dangers of uneven bleaching and degradation of the fabric. Only one bleacher in France is known to have had any success with gas bleaching. This was Boulanger, of Saint-Julien near Troyes, who practised the method to a modest extent between 1802 and 1805, and patented plant for the process in 1806; his patent described the exposure of goods in a stone or brick chamber, 24 feet long, 14 feet wide, and 10 feet high, and it appears that he had two such chambers.[128] Another who attempted the use of gas was Montagnat, in a venture near Ambérieu, but this was abandoned in 1805 after Montagnat's efforts had repeatedly resulted in degradation of the cloth (see below).

Commercial bleaches

The first commercial bleach to be developed was *eau de Javel*. We have already described the origins of this product and have indicated that its high cost prevented its success as a commercial bleach for industrial use. *Eau de Javel* did, however, rapidly acquire a smaller-scale popularity for stain removal, particularly among washerwomen. It came into general use in Paris in the early years of the nineteenth century, and was sold by most of the capital's grocers by 1809.[129] By 1821, the washerwomen of Paris had become so enamoured of it that they were applying it not only to stain removal but to general cleaning, too, as a substitute for soap; a leading soap manufacturer complained that over three-quarters of them were employing it in this way, spoiling the linen by its excessive and uncontrolled use.[130] The production of *eau de Javel* was taken up by many others besides the original Javel firm, developing as a scattered, fairly small-scale activity. There were twenty producers in Paris in the mid-1820s, some of them chemical manufacturers, others drysalters, combining the preparation of *eau de Javel* with their trade in soap, alkalis, and similar goods.[131] The total production in Paris then amounted to 1 350 000 kg a year (greater, probably, than France's annual output of sulphuric acid before the Revolution). It sold at 24 fr. per 100 kg, as compared with the price of about 70 fr. per 100 kg at which it had originally been launched by the Javel company. *Eau de Javel* was made in the provinces too. The early-nineteenth-century product had an available chlorine content of about 1·5 per cent (see p. 152), which compares with modern figures of 5·5 per cent for household bleaches of similar character, and 12–15 per cent for those of commercial grade. It would gradually lose its bleaching power on keeping, partly by chlorate formation but probably very largely by slow evolution of oxygen:

$$2KOCl \rightarrow 2KCl + O_2$$

The catalysis of this decomposition by metallic impurities gives the poorest commercial grades today a half-life of a few weeks, and one would expect early *eau de Javel* to have had a stability of a similar order.

While *eau de Javel* thus found its outlet for laundry use, the product which was to find commercial success as an industrial bleach was, of course, bleaching powder. This had its origins in the discovery made by Charles Macintosh in Scotland that slaked lime ($Ca(OH)_2$) in the dry state would absorb and retain a large quantity of chlorine; and that the resulting powder could then be used to prepare a bleaching liquor when required, simply by stirring it in water. The chemical nature of the powder has been a subject of debate virtually ever since its discovery. Regarded by its original developers as oxymuriate of lime, it is now recognized to be of complex and variable constitution, represented approximately by the formula $Ca(OCl)_2.CaCl_2.Ca(OH)_2.2H_2O$; in water, an alkaline hypochlorite solution results. Like *eau de Javel*, the powder gradually lost its bleaching power on keeping but had sufficient stability to be commercially distributed. At the time of his discovery, Macintosh was an associate of Tennant in the exploitation of the lime bleach liquor which Tennant had patented in 1798. The new discovery of a solid bleach was patented in Tennant's name in 1799,[132] and the manufacture of the powder was begun that year by Tennant, Macintosh, and other partners, in a newly established works at St. Rollox, near Glasgow. Despite the convenience of bleaching powder, the majority of bleachers for many years continued with the direct preparation of their own bleach liquors, on grounds of economy. The production of the powder did steadily grow in importance, however. Output at St. Rollox rose from 52 tons in 1799–1800, to 293 tons in 1810, 333 tons in 1820 and 910 tons in 1825, the growth continuing similarly thereafter.[133] The St. Rollox works progressively took up the manufacture of sulphuric acid, soda, and soap as well, and in the second quarter of the century came to be generally considered the largest chemical enterprise in the world. As the manufacture of bleaching powder spread, it was destined towards the mid-century to take its place alongside soda and sulphuric acid as one of the great staple products of the nineteenth-century heavy chemical industry.

The advantages of bleaching powder were various and considerable. Its solid state made transport easier than that of *eau de Javel*, and its higher chlorine content made transport cheaper. Ure in 1822 indicated an available chlorine content of about 22–8 per cent for commercial bleaching powder made in Britain,[134] and the French product was evidently of similar strength: that made in Paris seems to have been of about 20 per cent in the early 1820s and 25–30 per cent from the mid–1820s.[135] The modern figure is about 35 per cent. For a given weight, bleaching powder thus contained some 13 to 20 times as much chlorine as *eau de Javel*. Probably its most important advantage, however, was in price: the powder made in Paris in the mid–1820s sold at 120 fr. per 100 kg, and so cost only one-quarter to one-third the price of *eau de Javel* for an equivalent chlorine content.

Despite these advantages, the French seem to have been slow to appreciate fully the virtues of bleaching powder and were certainly slow to take up its manufacture in any significant way. This was despite the fact that they did have some early acquaintance with the St. Rollox product, since Tennant exported it to France during the peace of Amiens (1802–3). It was presumably this imported article which Bosc in the spring of 1803 described as being on sale in Paris at the depository of Bardel *fils*.[136] Unfortunately, with the resumption of war in May 1803 supplies from St. Rollox ceased; the firm informed a continental inquirer that the King had now prohibited its export.[137] When Alyon published a paper on bleaching powder in January 1805, it was based on a rare sample of Tennant's product which had come his way by chance, a bleacher from Brussels having brought to Paris a few ounces remaining from a batch he had bought when supplies were still available.

One factor contributing to the neglect of bleaching powder in France was probably the lack of exact information on the subject. Tennant's patent was reported by O'Reilly in 1800, in his *Annales*, but the details given were erroneous. O'Reilly wrongly described the product as being made from carbonate of lime, and though he did mention its preparation by action of gas on solid, he dwelt at greater length on a supposed preparation by stirring carbonate of lime into chlorine water. In his *Essai*, the following year, he now rightly indicated the use of lime but again described preparation in the wet way: by passing chlorine into lime water the product was supposed to be precipitated as a paste.[138] In fact any precipitate would be dibasic calcium hypochlorite ($Ca(OCl)_2 . 2Ca(OH)_2$), which does have bleaching action but which has never been developed commercially because it is extremely difficult to dissolve. Berthollet in 1804 correctly described the preparation of bleaching powder as being by absorption of chlorine gas in dry slaked lime. Alyon in 1805 evidently felt that there was more to it than that. He published an analysis of Tennant's powder remarking that the British were keeping its composition a secret. From the analysis he concluded that Tennant made it by absorbing chlorine in a mixture of two parts slaked lime and one part salt; he recommended that better proportions were three parts slaked lime to eight parts salt. This admixture of salt, particularly in such quantity, is something of a mystery (though it may be remarked that Labarraque in the 1820s prescribed an addition of 5 per cent salt as facilitating the absorption of the gas).[139] Chaptal, in his standard textbook of applied chemistry in 1807, mentioned bleaching powder only obliquely and said nothing of its preparation.[140] Parmentier in the same year considered it difficult and expensive to prepare, and objectionably unstable.[141] None of these early writers showed any awareness of the precautions needed for the production of good bleaching powder, and it is likely that the products obtained by early experimenters were more or less defective. In particular, it proves to be important to conduct the chlorination of the lime in such a way as to prevent the heat evolved producing an excessive temperature rise. Higher temperatures promote loss by formation of chlorate and chloride, and there is also a transition temperature at 40 °C (104 °F) above which the product assumes a different chemical character, becoming deliquescent and losing some of the stability of the properly made article. This need to moderate the temperature was recognized by the 1820s.[142]

Bosc, and particularly Alyon, spoke with some approval of bleaching powder, but Berthollet's verdict was less favourable. He cited comparative trials by Welter which had shown that when chlorine was absorbed in lime it produced only one-tenth of the bleaching effect of chlorine simply dissolved in water, an extraordinarily misleading result. Berthollet, of course, attributed the loss of bleaching power to chlorate formation, which he believed from his own researches to be considerable;[143] and he concluded that the product could be of value only in special circumstances, where economy was a secondary consideration. This objection was to be clearly dispelled only in 1817, when a careful study by Welter showed that when chlorine was absorbed in lime it retained the same bleaching power as when dissolved in water, no chlorate being formed.[144] He maintained a discreet silence on the discrepancy with his earlier result.

The earliest known commercial production of bleaching powder in France was that mentioned by Alyon in 1805, when he advertised the product as being available from the chlorine bleaching works of Foucques, on the île Saint-Louis in Paris. This was probably only a small affair and the powder presumably had the curious composition which Alyon prescribed. By 1816 bleaching powder was being made by Widmer at Jouy, but again presumably only in a fairly small way, for the firm's own use; the lime was reacted with chlorine in a revolving barrel, into which the chlorine entered through a hollow axle, an apparatus which Ure considered ingenious but unsuited to very large-scale production.[145] The continuing neglect of bleaching powder in France was such as to draw comment from a writer in the *Annales de chimie* in 1819; remarking that Tennant was producing it at the rate of 3000 kg a day, the writer considered it astonishing that the product was not more widely used in France.[146] It seems indeed that significant commercial production was then just beginning. It was listed that year among the products of the leading chemical firm of Darcet and Chaptal *fils*, at la Folie near Paris. The firm were large producers of Leblanc soda and their bleaching powder no doubt utilized by-product hydrochloric acid from that manufacture. It was to be through the development of such an association with soda manufacture that bleaching powder was to acquire its major importance as the century wore on. In view of this, and in view of the dramatic growth of the soda industry in France from about 1810, the slowness of manufacturers to take up bleaching powder is the more surprising; it is no doubt to be explained partly by the technical difficulties involved in capturing the hydrochloric acid for use. Even Darcet and Chaptal *fils* evidently had only a limited output, for a later report was to describe the production of bleaching powder in France as having developed in a major way only from 1823. In that year the greater part of France's consumption was said to have been supplied by the 30 000 kg then made in the Seine department (i.e. the Paris region). By 1827, however, production in the department had reached 300 000 kg, a leading part in this development having been played by the manufacturers Anselme Payen, Ador, and Bonnaire, working in association.[147] A statistic published in 1826 gave production in the department as 100 000 kg, with manufacture conducted in four works.[148] These figures may be compared with Tennant's own output at St. Rollox of 910 tons (about 910 000 kg) in 1825. It might further be remarked that the price at St. Rollox of £27 a ton[149] (65 fr. per 100 kg) was then little more than half the price of the French product (120 fr.). The growth of production in France from 1823 perhaps owed something to an improved knowledge of British manufacture, derived from an article published by Ure in 1822; it seems to have been the method indicated by Ure which was generally followed by French manufacturers in the mid-1820s, the lime being exposed on trays to the action of chlorine gas in chambers of stone or plastered wood.[150]

Bleaching plant

In turning now to discuss the plant in which bleaching liquors were made, it will be convenient first to consider how the chlorine gas was generated, and then how it was dissolved.

The generation of the chlorine—by distilling a mixture of sulphuric acid, salt, and manganese dioxide—in itself posed no great problems, since the quantities involved were not very large. The earliest experimenters used glass matrasses or retorts, and glass continued to be employed by some workers on a commercial scale. The vessels most commonly adopted for commercial working, however, were stoneware carboys, which were less fragile than glass and available in larger sizes. Berthollet in 1795 and 1804 recommended egg-shaped carboys of about 14 to 16 inches diameter, these taking a charge of some 80 lbs. (40 kg). Larger ones were also available, Dumas in 1828 referring to carboys taking 62–75 kg.[151] The number employed varied with the bleacher's needs. Berthollet envisaged that one or two would generally suffice and this can be taken as an indication of the typical scale of working at the beginning of the nineteenth century. One of the largest chlorine bleachers in France at that time was Descroizilles, who in 1805 can be reckoned to have been distilling some 250–90 kg of materials a day,[152] calling for perhaps six or seven carboys. By the 1820s ranges of up to six or eight carboys appear to have become not uncommon.[153] An alternative material for the distillation vessels was lead, employed as early as 1787 by Watt, and subsequently quite widely adopted in Britain.[154] Leaden vessels were less commonly used in France, however. Though they had the advantage of being resistant to fracture, a countervailing drawback was their need to be heated by a steam jacket, or water bath, if the danger of melting was to be avoided; French bleachers generally preferred stoneware which they could happily heat directly or on sand baths. The use of water baths was considered by early French writers to reduce the yield, Loysel, for instance, claiming that the yield was halved;[155] but this belief may have been mistaken for later texts are silent on the point. When leaden vessels were employed in France they would seem to have been similar in size to those of stoneware, for leaden matrasses of 15–16 inches diameter were referred to in the 1820s.[156] The additional advantage of lead, that it made possible stills of larger size, became a major consideration only when chlorine was to be generated on the much larger scale required in bleaching powder manufacture. In Britain, Ure in 1822 described sizeable stills for this purpose built of lead and cast iron, with steam jackets and mechanical stirrers.[157] The stills Ure described probably resembled those of Tennant, who in 1825 was reported to be using 15 or 20 leaden retorts at St. Rollox, each about 5 feet in diameter and weighing nearly 3 tons.[158] With the development of bleaching powder manufacture in France, large stills of lead and cast iron began to find some use there, too,[159] although stoneware carboys were to remain very common down through the nineteenth century, even for bleaching powder manufacture.

More problematical than the generation of the gas was its subsequent solution. The difficulty here arose because the prime aim of early bleachers was to dissolve the gas in plain water, and its solubility in this is low. It could be dissolved quite readily in an alkali solution, of course, to yield a liquor of the *eau de Javel* type. In that case one could simply receive the gas into alkali under an inverted trough (as was practised by some in the 1790s), or in stoppered bottles (as seems to have been the method generally used for making commercial *eau de Javel* in the early nineteenth century—see Fig. 3.2).[160] The efficient solution of chlorine in water alone, however, called for more elaborate arrangements.

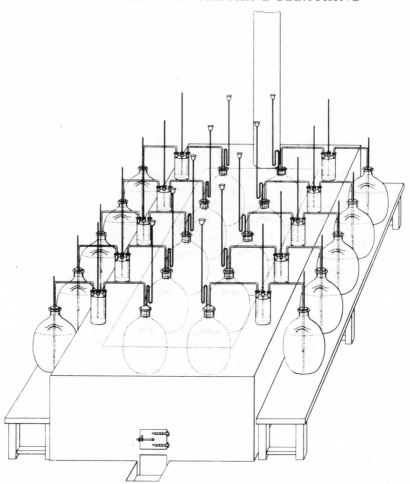

Fig. 3.2. Apparatus for commercial preparation of *eau de Javel* in early nineteenth century. (From *Dict. tech. Atlas*, vol. 1 (1835), 'Arts chimiques', Pl. 16. See also Julia de Fontenelle (1834), Pl. 1; Chevallier (1829), Pl. 3.)

As a general problem, the dissolving of gases in water had become a matter of interest to chemists from the mid-eighteenth century, when the investigation of gases developed as a main field of research. A convenient method published by Peter Woulfe in 1767 was soon popularized under his name, and was taken up from about 1773 by the French chemists, who considered it a major advance in laboratory technique.[161] It was this method which Berthollet used when he came to prepare his first bleaching solutions, and in this he was followed by the other early experimenters. The method consisted in bubbling the gas through a series of water-filled bottles arranged one after the other (Fig. 3.3); before the first collecting bottle it passed through a preliminary bottle to trap any acid rising from the still, which would otherwise contaminate the solution collected. As bleaching progressed from laboratory to workshop, there was some endeavour to scale up this apparatus for commercial working: in England, Rupp recommended a series of four to six casks arranged in Woulfe's manner, and no doubt similar methods were tried in France. From an early stage, however, Woulfe's apparatus was generally considered to be unsatisfactory for large-scale operation, and alternatives were sought.

The earliest alternative was devised by Berthollet's assistant Welter, and was first mentioned in March 1788.[162] Berthollet included a detailed description of it the following year in his first major publication on the new bleaching. By then the plant also incorporated unspecified incidental improvements by the young mechanician C. P. Molard (a friend of Bonjour

M.ʳ Bertholet's Apparatus.

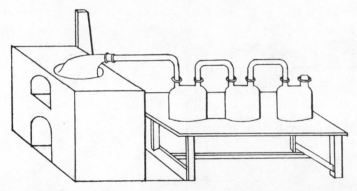

Fig. 3.3. Woulfe bottle apparatus for preparation of bleaching liquor. Berthollet's early apparatus as depicted by Bourboulon de Boneuil in his English patent specification of 1789 (patent no. 1678). (Patent Office)

and Welter, later to acquire prominence as the leading figure in the creation of the *Conservatoire des arts et métiers*). In this plant (Fig. 3.4), a first noteworthy feature is the safety tube now fitted to the preliminary acid trap—an invention of Welter. It consisted of a vertical open-ended tube dipping into a little water in the bottom of the trap, and it ensured that any fall of pressure in the plant during the course of the distillation would be relieved harmlessly by the entry of air, and not by the undesirable suck-back of water from the receiver. A simple but ingenious device, it was soon to become commonplace, not only in industrial plant but in laboratory apparatus too: Lavoisier in 1789 considered it to have greatly facilitated the conduct of distillations.[163] The chief interest, however, lies in the receiver itself with which Welter replaced the series of Woulfe bottles. This consisted

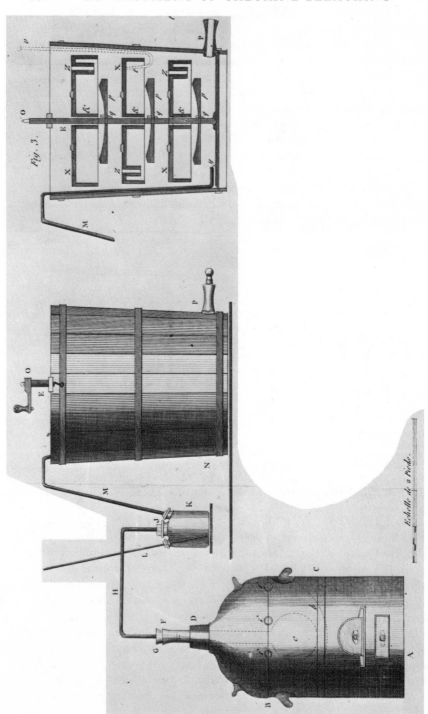

Fig. 3.4. Welter's plant for preparation of chlorine water, 1789. From Berthollet (1789b).

of a vat filled with water and fitted with two arrangements designed to promote solution by increasing the contact between water and gas. The first was a series of inverted troughs, arranged one above the other; the gas was conducted under the lowest trough and when that was filled bubbled up to fill the second, and then similarly the third. One can perhaps see here a first germ of the kind of absorption column of which a famous later example would be the Solvay tower in the modern soda industry. The second device to promote solution was a hand-turned agitator, which allowed the water to be stirred by rotating paddles between the troughs.

This receiver, advocated by Berthollet in 1789 and closely imitated in the book by Pajot Descharmes, had the drawback of being rather difficult to construct and was soon recognized to be unnecessarily complicated. In 1790 Berthollet remarked that the agitator was dispensable, and in his later publications he described a simplified vessel which just employed the inverted troughs. It was a receiver of this type, at Jouy, which was portrayed by Chaptal in his chemistry text of 1807, and this was perhaps the kind of plant most generally adopted in France, under the name *cuve à cuvettes*. There were also receivers in which paddle agitators continued to be employed, as an alternative to troughs; such receivers are found in texts of the 1820s under the name *cuves à moussoir*, but they were considered less convenient than *cuves à cuvettes*. Another type to be developed was the *cuve à serpentin*, in which the vat was fitted inside with a spiral inverted gutter.[164] The chlorine passed up the gutter, dissolving as it went. Alternatively, a spiral perforated tube might be used, or the gutter might itself be perforated, so as to produce streams of small chlorine bubbles. This device was perhaps inspired by the spiral coils which had long been used as condensers in the distillation of alcohol. An early example of its use is in a receiver with which Widmer experimented at Jouy in the early 1790s, in which spiral gutters were combined with inverted troughs, the gutters serving to conduct the gas from one trough to the next.[165] He found this combined arrangement to be highly effective but presumably it later came to be considered over-elaborate. The corrosive nature of chlorine meant, of course, that not only the form but also the materials of the receiver called for attention when the small-scale glassware of the laboratory was left behind. Welter's original plant was made of wood, but this was attacked to some extent and Berthollet in 1790 pointed to the desirability of finding an alternative. In the event wood was to continue as a main material, for it was soon found that it could be satisfactorily protected by a coating of mastic or varnish. The chief further materials to be adopted were masonry and stoneware, early employed by Widmer and probably by Descroizilles. Of the various metals with which there was some early experimentation, only lead came into common use, mainly for piping.

In designing plant, the main aim of early experimenters was to assist solution by maximizing and prolonging contact between the gas and water. It was known, of course, that pressure was a further factor which would promote solution, and this was the basis of a proposal by O'Reilly to use a series of Woulfe bottles in which the pressure was to be increased by giving them the form of tall cylinders, each seven or eight feet high.[166] This is an example of O'Reilly's armchair theorizing, however, for experienced workers knew well enough that one of their most troublesome practical concerns was the securing of gas-tight joints, and that any increase in pressure could only exacerbate this. A type of plant which approached theoretical perfection, in that it achieved a greater surface contact than any previously known, while at the same time subjecting the gas to negligible pressure, was the absorption column patented by Clément in 1821, in which the upward stream of gas was dissolved by a downward counter-current of water dispersed over a packing of small glass or porcelain balls. We have already mentioned this earlier, as the inspiration of the Gay-Lussac tower in the sulphuric acid industry. Here we may note that the preparation of bleaching solutions was one purpose for which Clément and others advocated its use.[167] The device does not, however, seem to have found any practical adoption among bleachers. Its sophistication was no doubt incommensurate with their fairly modest needs, and in any case the solution of chlorine in water was a matter of declining interest in the 1820s, as lime bleaches came into more general use.

While multiple troughs, spiral gutters, and paddle agitators all found some application, it appears that simpler receivers could also suffice, especially when solution was facilitated by adding a little lime or chalk to the water. Thus, the plant described by Dumas in 1828 amounted to just a single inverted trough, given an annular form so that further water and lime could be added through the central opening (Fig. 3.5).[168] Vitalis in 1823 even prescribed a straightforward covered vat, of no special construction, and evidently considered this quite adequate whether or not chalk was added to the water;[169] he was closely acquainted with the dyeing industry of Rouen and so such plant was presumably in practical use. It must have been highly inefficient but escape of chlorine gas into the workshop was prevented by dipping the exit pipe from the vat into an alkaline absorbent. Since it was customary to pass the gas from the several stills into a single receiver, the growing scale of bleaching operations is reflected in the size of the receivers employed, and it is interesting to compare the dimensions indicated by Vitalis and Dumas with those of bleaching texts a generation earlier. The vessel described by Pajot Descharmes had been a mere 3 ft. in diameter and $1\frac{1}{2}$ ft. deep, while that depicted by Berthollet in 1795 and 1804 was 6–7 ft. in diameter and 3 ft. deep. In the 1820s Vitalis and Dumas were speaking of diameters of 5–8 ft. and depths of 8–10 ft.

The bleaching liquor having been prepared, its application in turn proved less straightforward than might be imagined. The object here was to ensure that the goods were satisfactorily penetrated and evenly bleached, and various methods were described. According to Berthollet's original directions, the liquor was simply to be run into tubs in which the goods had previously been lightly arranged, but experience evidently showed this to be insufficient. In 1795 he prescribed that in the case of yarn the hanks were to be placed in baskets which were then to be repeatedly lowered into the liquor by means of a crane. For cloth, the pieces were again to be immersed in baskets, but repeated dipping was in this case insufficient and so the goods had instead to be trodden in the liquor by workmen. An alternative procedure, apparently employed by Widmer in the early 1790s, was to wind the cloth periodically over a dyer's reel so as to renew its contact with the solution, and it was a method using the reel which Berthollet indicated in 1804; the pieces of cloth,

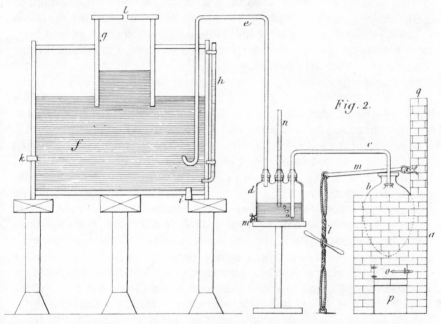

Fig. 3.5. Plant for preparation of bleaching liquor, 1828. From Dumas
(1828–46), Pl. 6, with description in vol. 1, 55–7.

sewn end to end, were now to be wound over reels through a series of four bleaching tubs. These various methods all employed the liquor in standing baths and this was the general practice. A different technique was adopted by Descroizilles, however. Instead of passing the goods through the liquor, he circulated the liquor through the goods,[170] presumably after the manner of the bucking operation, and one can see here the origin of the circulation techniques which have come to displace standing baths in twentieth-century practice. Manipulation of goods in open tubs obviously exposed workers to chlorine vapours and this led Rupp in Manchester to propose that bleaching be conducted in closed vessels, in which the cloth was to be mechanically wound back and forth between rollers.[171] This idea is not known to have been seriously pursued in France, however. Welter did devise a closed bleaching vessel with an agitator,[172] but bleachers generally remained content to work with open tubs. It was no doubt easier to alleviate the odour by throwing in a little chalk than by venturing into the mechanical complications of special plant.

Bucking and steam bleaching

While the replacement of grassing by chlorine brought important improvement to one part of the bleaching process, the other major operation in the bleacher's routine—treatment with hot alkaline lyes—remained inefficient as traditionally performed (see p. 114). This bucking operation began to receive attention at the turn of the century, and although subsidiary to our main concern, some brief account is here appropriate of the improvements that were then essayed.

Of central importance to these developments was the proposal by Chaptal in 1799 of a new mode of operation using steam.[173] Instead of repeatedly circulating lye through the goods, as in the old procedure, the goods were to be simply impregnated with alkali and then exposed to the action of steam in a stone chamber built over a boiler. This steam treatment aroused much interest and came to be almost as closely associated with Chaptal's name as the use of chlorine with that of Berthollet. It was not in fact altogether new, having been practised in a primitive way in India since time immemorial and having recently begun to find some use in the south of France. It was Chaptal, however, who publicized it and suggested its extension from cottons to linen and hempen goods. In Paris, the process was soon taken up with encouraging success by Bauwens, the noted cotton manufacturer, together with the chemical manufacturer Bourlier. And when Chaptal became Minister of the Interior, he responded to the evident interest in the matter by engaging O'Reilly to write an extended account of the process, the result being the latter's *Essai sur le blanchiment*. Copies of this were officially distributed throughout France in about June 1801, and inspired quite widespread experimentation.

That Chaptal's process aroused the degree of interest it did owed much to the exaggerated expectations to which it gave rise. In his 1799 paper, Chaptal had described the whitening of cotton by no further treatment than alkaline steaming followed by a couple of days on a bleach-field, and he contrasted the ease and economy of this with the Berthollian method. Many were thereby misled into seeing in the use of steam a method of bleaching which was virtually complete in itself, a major advance which promised to dispense with the difficult and expensive use of chlorine. The announcement of 'steam bleaching' thus introduced some confusion to the scene. One manufacturer, for example, in the midst of establishing a chlorine plant at Bordeaux, was stopped in his tracks by the news and made inquiries of the Ministry, fearful lest his venture be outmoded before it had even begun. The Ministry's reply, in April 1801, shows that the administration by then had a more realistic appreciation of the true value and role of steam treatment, and O'Reilly in his *Essai* properly described it as a complement to the use of chlorine rather than a replacement; his book may nevertheless have continued to foster exaggerated hopes by the exalted light in which the method was still presented, and by the eccentric theory he developed, which envisaged oxidation of colouring matter by alkaline vapours and so tended to obscure the distinction which Berthollet correctly saw between the oxidizing action of a true bleaching agent, like chlorine, and the solvent, cleansing action of alkali.

If steam treatment proved to be not quite the marvel which had been hoped for, it did still have considerable virtues even in its more modest role as an alternative to traditional bucking. It eliminated the tedious labour of ladling hot lye from boiler to bucking tub for hours on end; it cut down the considerable heat losses entailed in that procedure, so saving fuel; and by achieving a higher working temperature it significantly promoted the action of the alkali, for in the old bucking tub the goods could at best be brought only to within 10 °C of boiling point. Despite this, experimenters ended up for the most part disappointed and the steam treatment did not find common adoption. Simple and innocuous as it appeared, it turned out to have its own difficulties and dangers, and O'Reilly emerged as a less than adequate guide. One problem, for example, was that goods were liable to be 'burned'. Attempts to promote similar treatment for laundry use met with little more success. Though there were to be sporadic flirtations with steam later in the nineteenth century, only in our own times has Chaptal's method been taken up by the bleaching industry in a major way, and made the basis of modern continuous processes in which the scouring action of alkali is greatly accelerated by use of high-pressure, high-temperature steam.

This episode was not without some more immediate fruit, however, for it seems to have contributed to improvements that began to be introduced into ordinary bucking. When it became clear that his process was less straightforward than it had seemed, Chaptal charged three of his Ministry's technical advisers (Bardel, Molard, and Joseph Montgolfier) to set about researches in the hope of perfecting it. He seems to have continued to hope that it could be developed to a point where it would render the use of chlorine unnecessary. Investigations were duly pursued at the *Conservatoire des arts et métiers* from the end of 1801. These evidently never reached a wholly satisfactory conclusion, since in August 1803 they were said to be still in hand, and the promised publication never came. But a short interim report appeared in 1802, and this is of interest for its description of a new plant devised by Bardel.[174] In this there was a reversion to the practice of circulating lye through the goods, but with the noteworthy novel feature that the lye was raised to the top of the bucking tub automatically, by means of steam pressure. Another to contribute to the improvement of bucking was

Widmer, of Jouy, who had himself earlier investigated steam treatment. In 1804 Berthollet described bucking plant in operation at Jouy in which considerable gains in thermal efficiency had been achieved by mounting a bottomless bucking tub directly over the boiler, the two being separated only by a wooden grill, on which the goods rested; the boiling lye was raised through a central pipe by means of a pump.[175] The plants of Bardel and Widmer contain between them the basic principles of the chief types later to come into general use, and in particular of the familiar 'kier with puffer-pipe', a combined tub and boiler (like that of Widmer), in which the boiling lye was spontaneously raised up a central pipe by the pressure of steam

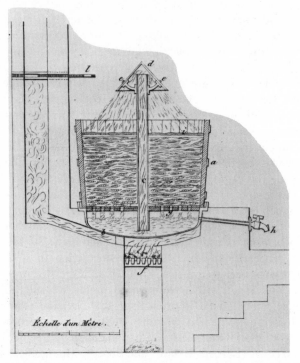

Fig. 3.6.   Early kier with puffer-pipe. (From *Bull. Soc. enc.* **14** (1815), Pl. 118.)

beneath the goods (see Fig. 3.6). According to Herpin, just such a kier was indeed invented by Widmer in 1805, and was then being used by Berthollet for his domestic washing.[176] If so, Widmer did not publish it, and it was as an English innovation that the kier with puffer-pipe was drawn to the attention of French readers in 1815.[177] Perhaps inspired partly by the French work, the improvement of bucking had begun to attract interest in Britain at about the same time, and general adoption of the new techniques was to be rather quicker in Britain than in France, where traditional bucking methods seem to have still prevailed in the 1820s.[178]

The advantages of chlorine bleaching

To assess the advantages of chlorine bleaching we must particularly consider its economic merits as compared with the traditional treatment,[179] and here we might begin by noting that at first the general economic viability of the new process was a matter of some doubt. Observers were struck by the high cost of chlorine itself, and one of the early objections was that this made the new bleaching expensive. This was a point conceded even by advocates of the new method, and Berthollet himself in 1790 advised its application only to fine textiles, not to common goods. This was because fine materials required less chlorine for their treatment, while at the same time their higher market value meant that any extra cost could more readily be absorbed.[180]

In seeking to lower costs, the attention of early workers focused on the raw materials and in this respect some significant improvements were soon seen. The chlorine having at first been generated by distilling manganese dioxide with hydrochloric acid, one advance came with the replacement of this acid by a mixture of sulphuric acid and salt. As a material for which there had hitherto been no very large-scale demand, hydrochloric acid was expensive, its price being given by Lavoisier in 1786 as 40–5 *sous* a pound[181] (408–459*l*. per 100 kg). That sulphuric acid could be advantageously used instead seems to have been a discovery made by Descroizilles, in his early trials at Rouen. This at least was the interpretation which Berthollet put on a rather vague press report of April 1788 to the effect that Descroizilles had made a notable substitution in the raw materials. The report prompted Berthollet to reconsider his own earlier rejection of sulphuric acid as unsatisfactory, and he asked Welter to examine the matter. When the latter quickly found the secret of success to lie simply in diluting the sulphuric acid with water, Berthollet then proceeded to publicize its use as an important economy.[182] It can be estimated indeed that it more than halved the original cost of the distillation materials. The significance of this must not be exaggerated, however, since in fact hydrochloric acid was very soon to come down in price, and it continued to be used by some bleachers.

More important was the question of the price of salt. Intrinsically salt was a cheap commodity, but it was made dear in many parts of France by the infamous salt tax known as the *gabelle*. Varying in a complicated way from one area to another, the effect of the *gabelle* was such that while salt might be had at $\frac{1}{4}$–$1\frac{1}{2}$ *sous* a pound ($2\frac{1}{2}$–15*l*. per 100 kg) in those exceptional areas fortunate enough to escape the tax, in areas where the tax was levied at its maximum rate the price was inflated to some 60–62*l*. per *quintal* (122–7*l*. per 100 kg).[183] In consequence, chlorine bleaching was less readily viable in some places than in others. The works of Bonjour and Welter were advantageously placed in a tax-free region, which is no doubt partly why they chose it; but other early centres of interest in the new bleaching, notably Paris and Rouen, had the misfortune to fall in regions of maximum taxation.

In Paris, the hindrance presented by the high price of salt received early attention from Lavoisier. Towards the end of 1786 he was investigating the possibility of applying to bleaching purposes the impure salt which issued as a waste by-product from the refining of saltpetre. At that time the salt was simply being dumped in the river and Lavoisier presumably hoped that its use might be permitted free of duty.[184] The outcome is not known. Eighteen months later the question of the salt tax was taken up by Descroizilles in Rouen, on whose behalf a petition for special exemption was presented to the central administration by the Normandy *Bureau d'encouragement* for the promotion of agriculture and commerce.[185] They suggested that any danger of fraudulent use might readily be avoided by adulterating the salt with copperas. That Descroizilles attached some importance to his request might be judged from his offer, in return, to make known all his processes and throw open his laboratory to all artisans wishing to learn about the new bleaching. His petition was not successful; it appears that he was offered the use of certain impure by-product salt[186] but that this was unusable. In the event the Revolution soon removed any need for special exemption, for one of the early Revolutionary reforms, in 1789–90, was the general abolition of the *gabelle* as one of the most roundly detested taxes of the *ancien régime*. The effect of this on bleaching, in former areas of maximum duty, can be reckoned to have been a further halving in the cost of the distillation materials.

In the further interests of economy Berthollet suggested that bleachers should make their own sulphuric acid, and should utilize their distillation residues for the manufacture of soda, but in practice these kinds of integration seem to have proved rare.

Of course, while chlorine was undoubtedly an additional expense for the bleacher, its use did also bring him countervailing economies and clearly what really mattered was how the total costs of the old and new bleaching compared. The use of chlorine considerably reduced the number of alkaline lyes required, perhaps to half or less, with consequent savings in alkali, fuel, labour, and plant; it shortened the grassing treatment (and for cottons eliminated it entirely) so bringing savings in labour and land; and it made possible the continuation of bleaching through the winter months, which was not only a matter of great convenience but also avoided the economic waste of idle plant. Unfortunately, precise comparison of old and new is frustrated by the complexity and variability of the processes, and contemporaries found the question as difficult as the historian. Berthollet's writings contain no attempt at any detailed comparison of costs, and such estimates as one finds elsewhere are sketchy. Pajot Descharmes in the 1790s gave calculations purporting to show chlorine bleaching to be fully competitive with traditional, but his estimates are very incomplete. Bosc, in 1803, on the basis of trials at Troyes, reckoned the new bleaching to be 20 per cent more expensive than the old, but again one may doubt the adequacy of his sums. Writers were still vague or silent on the subject in the 1820s. Blachette then guessed that if one took into account only materials and labour, chlorine bleaching perhaps emerged the more expensive, but that if one reckoned in the saving in interest on capital which resulted from the greater speed of the new bleaching, then it could be seen to have the advantage.[187] In general, it seems safe to conjecture that in its early years chlorine bleaching tended to be more costly than its rival, but that as it improved and became established the economic balance swung in its favour. Certainly, critics then took to complaining cynically that bleachers only adopted the new method because it cost them less. What does seem clear is that the cost advantage was not a striking one.

Turning now to consider the new bleaching from the viewpoint of the client, rather than that of the bleacher, a first point to be made is that it does not seem to have brought any reduction in the price he had to pay. In the 1820s the price of the new bleaching was said to be the same as the old, and in the 1790s and early 1800s the indications are that it was, if anything, slightly dearer.[188] This does not mean, though, that chlorine bleaching was without advantage for the client. On the contrary, he perhaps benefited more than the bleacher himself. A chief virtue was its elimination of all the problems attendant upon the cessation of traditional bleaching in the winter months. In Rouen, for example, it was said to have previously been common for textile workers to be thrown out of work once every four or five winters, in consequence of the seasonal interruption of bleaching.[189] This occurred when manufacturers exhausted their working capital in the production of raw goods, and then had to await the bleaching and sale of their goods in the spring before they could make more; it occurred too when weavers ran out of white thread and had to await the new bleaching season for fresh supplies. We are told that it was to provide alternative employment on such occasions that the town's fine promenades and carriageways had been built. Berthollet, too, commented on this seasonal problem, describing how peasant weavers were often deprived of their due rewards in the winter, when they found themselves obliged to sell their cloth prematurely—in a raw state and at a low price—being unable to wait for its bleaching in the spring.[190] By ending the seasonal irregularity of bleaching, chlorine overcame these consequent crises of capital. A related and perhaps even bigger benefit derived from the greater speed of the new bleaching. In general, chlorine seems to have cut the bleaching time for linens to a fifth or a quarter of the time of the traditional method, and for cottons to perhaps as little as a twelfth.[191] As a result of this, producers and merchants were able to profit from a correspondingly faster return of their working capital. The contributions which the new bleaching thus made to France's textile industry, by speeding and regularizing the turnover of capital, were especially important in view of the general shortage of funds which afflicted the country during and after the Revolution. In Rouen, chlorine bleaching was credited with having allowed textile workers to be maintained in employment during the difficult Revolutionary years, and with having thus contributed to the relative calm there. Whatever the truth of this, the process was sufficiently valued by municipal leaders for Berthollet to be greeted by laudatory deputations when he visited the town after the Revolution.[192] A larger claim for the new bleaching was made by Descroizilles in 1818.[193] He then considered that its economic benefits had been one of the main factors underlying the dramatic growth of the cotton manufactures in Europe since the Revolution. No doubt there is some truth in this, though the point is weakened by the slowness with which chlorine in fact came into general use.

Besides these economic benefits, chlorine also offered advantages for the quality of the bleaching. From the beginning it was considered to have a less wearing effect on the fabric than the more laborious traditional treatment, so that the goods lost less weight and retained greater strength (Blachette in 1827 compared a loss in weight of 26–7 per cent for linen bleached by chlorine with one of 33–5 per cent for traditional bleaching).[194] Chlorine was also said to yield a more striking white, and moreover to provide a better basis for any subsequent dyeing or printing: Descroizilles claimed that chemically bleached cottons took the Turkey-red dye better than ordinary goods, while Widmer found that in printing they took the colours more strongly and more firmly, so yielding a brighter pattern.[195] These virtues were by no means universally recognized, however. Popular prejudice in favour of the old 'natural' bleaching was long to continue, supported by the notion that since chlorine was a corrosive chemical it must necessarily attack and weaken the fabric; many textile printers, for their part, long disputed the merits of chlorine for the bleaching of cloth intended for printing. Such scepticism, of course, was not without some foundation, since the bleaching was not always well conducted: Widmer in 1806 laid particular blame on the English for having sent large quantities of badly bleached goods to the Continent. Prejudice and scepticism were evidently still quite widespread in the 1820s.

One final advantage claimed for the new bleaching was that it would benefit agriculture, by the return of bleach-fields to agricultural use.[196] In truth, though, this was more an appealing argument than a significant reality (at least when considered at a national as distinct from a local level). The area of land involved was extremely small. It can be reckoned that in the United Kingdom the vastly increased quantity of cotton and linen goods being manufactured by the early twentieth century would have required for their bleaching an area equivalent to no more than 2–4 per cent of the land then in agricultural use.[197] It seems, moreover, that land used for bleaching was not always without some agricultural return too: Bonjour in 1790 justified his continued use of a bleach-field partly on the grounds that if turned over to hay it would yield no more than it yielded in cut grass as a bleach-field (the product in the latter case being all the more abundant for the regular waterings).[198]

The growth of chlorine bleaching after the Revolution

Following the disruptions of the Revolution, the new bleaching began to spread and establish itself from the late 1790s. Information on its growth is scanty and so only a rather fragmentary sketch can be given.

It was perhaps in Normandy, and especially in the region of Rouen, that the new bleaching now made its greatest progress, no doubt thanks to the lead given there by Descroizilles. Descroizilles's own works won a silver medal at the industrial exhibition of 1801, as 'the finest Berthollian bleaching works in France', and in 1802 it went on to win a gold.[199] With a workforce of between 25 and 40[200] Descroizilles was not a particularly big employer, but the annual capacity of his works was probably at least equal to that of France's largest traditional bleachers: in 1806 he claimed to be bleaching some 250 000 kg of goods a year.[201] His nomination in 1810 to the *Conseil général des fabriques et manufactures* in Paris shows that he was rated among his department's leading industrialists. Though by 1806 chlorine was in use in several other local works too, Descroizilles claimed to be bleaching more goods than all the rest put together.[202] His chief rivals were Prevel *frères*, also of Lescure. They in fact were now the occupants of Descroizilles's original plant, for he himself had moved on to set up a second establishment in the same district.[203] From Descroizilles we learn that the new bleaching was also now being practised by Thuilot and several other bleachers in the valleys near Rouen (among them perhaps the cloth finishers Pottier *frères*, who were elsewhere mentioned as using chlorine in 1801),[204] and that there were also subsidiary chlorine plants in some of the local dyeworks, among others in those of de Fontenay *frères* & Cie, Maillère & Delamare, and Édouard Sévenne. At Évreux, some distance from Rouen but still in Normandy, chlorine was being successfully used in 1806 at the textile works of Messrs. Robillard, 'one of the first who, for the bleaching of yarns adopted the Berthollian lye'; it was probably also in use at Évreux by the manufacturer Désormeaux.[205]

Besides Descroizilles, the other early pioneer to continue in bleaching was Welter, who left Paris to return north in April 1797. He now established chlorine bleaching in the Belgian town of Menin, at that time part of France. This was at the linen manufactory of Vandermesch and Van Ruymbeke, and it appears to have been in effect a transference to Menin of the works Welter had earlier established at Armentières. The Menin works was said to have set an example which was followed by others in Flanders, though it should be added that according to the account given by Briavoinne (the early historian of Belgian industry) the use of chlorine was in fact very limited in that country before the late 1820s. The works continued, with Welter as a partner, until 1815, when the Belgian campaigns forced its closure. Following the separation of Belgium from France, the works was moved at the end of 1816 to Saint-Denis near Paris, where it operated under the name of Gombert *fils aîné* and Michelez.[206] It won a silver medal at the industrial exhibition of 1819 and at the same time Welter received the decoration of the Legion of Honour.[207]

Elsewhere in the north, chlorine bleaching continued in its early centres of Lille and Valenciennes. In 1804 an official in the Nord spoke of its 'infallible progress' in that department, though this is not to be taken too literally: the bleacher Pluchart *fils* at Cambrai, after working with chlorine for several years, had by January 1802 abandoned it because of its difficulties and had reverted to the traditional method.[208] There is nothing in truth to indicate any remarkable progress; the department was said in 1804 to have 'several' chlorine works for yarn in Lille and a single works in Valenciennes for cloth.[209] At Saint-Quentin, in the neighbouring department of the Aisne, the new bleaching began to be adopted from 1801.[210] This was at the personal instigation of Chaptal, who as Minister of the Interior visited the town in the company of Bonaparte at the turn of the year 1800–1, and among other manufactories inspected one of its four large bleaching works. Following his visit Chaptal enjoined the town's bleachers to set aside their traditional methods in favour of chlorine, and this led all four of them, independently and in some secrecy, to conduct trials. Though for the most part their results were disappointing, one of them did succeed. This was Dupuis, who was also a manufacturer of sulphuric acid and so may be presumed to have been the most experienced of the four in chemistry. By the beginning of 1802 he was reported to be overcoming the commercial prejudice against the new bleaching, and it was he who was to be the leading figure in its establishment at Saint-Quentin. In Paris a notable figure to adopt chlorine, in 1800, was Richard-Lenoir,[211] whose cotton spinning and weaving works were to grow into the largest industrial enterprise in France under the Empire. By 1802 chlorine was perhaps also being used in Paris at the spinning works of Fournier, noteworthy as one of the first French works for the mechanical spinning of flax.[212] In the textile town of Troyes, the new bleaching was making faltering progress in the opening years of the nineteenth century:[213] by the winter of 1802–3 one venture had already collapsed (that of Bosc presumably), but another was getting under way in the hands of a young newcomer to the town called Boulanger.[214] This came to be well spoken of by the local chamber of commerce, from whose remarks it would seem to have been the only chlorine works in the district in 1805. Boulanger worked at only a modest level, however, bleaching a total of 9000 pieces in the course of the three winters up to 1804–5. He laboured under various difficulties. Though manufacturers were well enough satisfied with his results, they were nevertheless reluctant to leave their usual bleachers and took their custom to Boulanger only during the winter; more seriously, the termination of an unsatisfactory partnership had left him without backers and he lacked capital. In 1806 he was bankrupt and his appeals to the Minister of the Interior for 30 000 fr. assistance were unavailing.

The opening of the nineteenth century also presents signs that interest was growing in the east of France. At Saint-Dié in Lorraine, Philpin *frères* & Cie, manufacturers of ribbons, were evidently using chlorine in 1801, though they found it difficult and were not entirely satisfied.[215] In the same year, Rigouel, a dyer at Dun, was experimenting with local linen and hempen cloths, and reported 'passable' success.[216] At Mulhouse, the new bleaching was receiving attention from Nicolas Dollfus and Alexandre Jaegerschmid, who in July 1801 patented a new means of making chlorine.[217] At Strasbourg, Vetter, the director of a spinning works, was conducting large-scale trials on the bleaching of yarn by chlorine in conjunction with Chaptal's new steam lyeing process. He was said to have succeeded perfectly and experiments by Carondelet—reported in 1801 to the department's *Société d'agriculture*—found his yarn to be superior in strength to that bleached traditionally.[218] In 1810, Vetter (described now as a manufacturer of sail cloth in Strasbourg) was credited with having made improvements to Berthollet's process.[219] By 1806 there was also a chlorine bleaching works at Tomblaine,[220] near Nancy, and by 1811 the printed-cotton manufacturers Ziegler, Greuter, & Cie had plants at Guebwiller and Rouffach.[221]

In the south, Aix had a chlorine works by 1800,[222] and the new bleaching was reported at Castres in 1801/2.[223] At Bordeaux, a works was being set up in March 1801 by the manufacturers Gamblond & Cie.[224] At Ambérieu, a first chlorine venture was attempted in about 1802 by Montagnat, to compete with eight small traditional bleachers; this never attained satisfactory results, however, and in 1805 the property was converted to a further traditional works.[225] At Toulouse, it was reported in June 1804 that a first chlorine plant had recently been established, by Trébos & Cie.[226] Finally, at Moulins, a small manufacturer of cotton and silken goods by the name of Tallard *aîné* was said in 1806 to have given a lead in the use of chlorine in the department of the Allier.[227]

It is not possible to chart the growth of the new bleaching in quantitative terms, but it was certainly slower than is often imagined. A wholly misleading impression is given by such contemporary remarks as the assertion in 1807, in the *Bulletin* of the *Société d'encouragement*, that 'the method of Berthollet . . . is adopted almost universally'.[228] In fact, even in Rouen the local trade directory in 1817 still listed 11 traditional bleachers as compared with 3 using chlorine,[229] though admittedly the traditional works were probably smaller. In a French text published in 1828 one reads that for the bleaching of calicos most manufacturers still held to the traditional process. Even in the important cotton-printing region of the Upper Rhine, the ten bleaching works in 1828 still treated 176 360 pieces of cotton on the field (along with some 50 000 of linen), as against 219 738 by chlorine.[230] The unreliability of much contemporary comment is illustrated by an official publication from Rouen, which in 1823, reproaching those textile printers who continued to prefer naturally bleached calicos, held up as an example the departments of the

Upper and Lower Rhine, where the printers (so the writer alleged) used nothing but chemically bleached cloth.[231] In sum, it seems very likely that the goods being bleached traditionally in France continued to outweigh those bleached by chlorine until at least the third or fourth decade of the nineteenth century. The time-scale leads one to wonder how far the new bleaching owed its growth to the conversion of established bleachers and their clients, and how far rather to the influx of new blood, as the older generation disappeared, and as the textiles industry itself grew. In this latter connection, it is interesting that chlorine found readier adoption for the treatment of cottons than for linens. It was remarked in the Aisne department in the 1820s, for example, that it had been with the tremendous growth of the cotton industry that Berthollian bleaching had come into general use, while an observer elsewhere noted that it was particularly against its application to linens that prejudice continued.[232] The new bleaching thus had an especially close connection with the youngest and most dynamic sector of the textile industry. This was partly for technical reasons, the application of chlorine to cottons being easier and yielding clearer benefits, but it is no doubt also relevant that the production of linen and hempen goods in France retained a predominantly artisanal and domestic character into the mid-nineteenth century, longer than any other branch of textiles.

The adoption of the new bleaching was clearly slower in France than in Britain, and for this the untimely intervention of the Revolution can only partly be blamed. We have earlier noted that already before the advent of chlorine, French bleaching manifested a relatively conservative character. The adoption of sulphuric acid sours in eighteenth-century Britain not only provides an early pointer to a more innovatory spirit in this country, but may also in some small way have itself helped prepare the way for chlorine, by beginning to erode prejudice against 'chemical' treatments and also by familiarizing bleachers with the handling of this dangerous chemical, one of the basic ingredients in the Berthollian process. Conversely, the delayed adoption of lime bleaches in France, particularly bleaching powder, not only reflects but must also have contributed to her backwardness, and here a small share of responsibility perhaps belongs to Berthollet himself, for having so long championed plain chlorine water as the ideal bleaching agent. Far more important, though, must have been the differing economic circumstances on the two sides of the Channel. The speeding of textiles production in Britain by mechanization may be presumed to have heightened the incentive to espouse a new process which could similarly speed their bleaching. Moreover, the especial early association of chlorine with cottons was probably significant, for of course cotton manufacture was a far larger element in the British textile industry than in the French. Finally, it would be interesting to compare costs in the two countries; data here is lacking, but the higher price of bleaching powder in France in the 1820s hardly suggests that the comparison would be in her favour.

Chlorine bleaching in the paper industry

It was a natural step to extend the application of chlorine from the textile industry to the manufacture of paper, since it was from textile sources that paper was at this time made.[233] The modern manufacture from esparto grass and wood pulp dates only from the second half of the nineteenth century. Until then, rags constituted the paper-maker's prime material: rags of linen and hemp (the staple clothing fabrics of French peasants and artisans), and also, increasingly in the nineteenth century, of cotton. Before the adoption of chlorine, the colour of the paper obtained closely depended on that of the rags employed. In the sorting which was the first stage of manufacture, the best white rags were set aside to make the finest white paper; rags of a somewhat inferior quality, retaining a degree of natural colouration, were destined for lower grades of white writing and printing papers; while the coarsest rags, together probably with most of those which had been dyed or printed, and along with such further materials as sack-cloth, old ropes, and fishing nets, were relegated to the manufacture of brown wrapping papers. A chief attraction of chlorine was that it permitted the production of a higher grade of paper from a given grade of rags, and so added to the value of the product. At the same time, by allowing good white papers to be made from rags hitherto disqualified by their colour, it helped to alleviate the recurrent shortfalls in the rag supply. A subsidiary, less striking benefit was that it contributed to the abandonment of the old paper-making practice known as *pourrissage* (rotting).[234] This consisted in piling the sorted rags in heaps, wetting them, and leaving them to ferment for up to five or six weeks. The primary aim was to soften the fabric and so assist in its subsequent reduction to paste, under the pounding of the heavy wooden mallets of the stamping mill. In fact, in the newer cylinder machines which were beginning slowly to displace the old stamping mills, this preliminary softening was unnecessary, since the new machines were more effective, shredding the rags by means of rotating blades. Nevertheless, the rotting process continued to find favour in France, even in conjunction with the cylinder machines, since it was also believed to have secondary value as a modest bleaching treatment. It was considered to solubilize colouring matter by oxidation, and so to promote its removal in the course of the conversion of the rags to paste. The introduction of chlorine obviously made any other bleaching treatment redundant, and the abandonment of rotting resulted not only in a considerable saving of time but also in a small but significant improvement in yield (perhaps by some 10 per cent), since rotting had been accompanied by notable losses.

The first to propose the use of chlorine in paper-making was Chaptal.[235] Already by 1789 he had experimentally applied a chlorine liquor to the bleaching of 100 lbs. of coarse rags (or perhaps paste), intended for the production of blotting paper; he estimated that for an increased cost of 7 per cent the value of the product had been raised by 25 per cent. These are presumably the experiments which Chaptal in 1819 described as having been conducted at the paper works of the Montgolfiers (at Annonay). They are not known to have resulted in any regular adoption of chlorine there, but Chaptal's short article did receive wide attention and he was soon followed by other experimenters both in France and in Britain. Further work in France is encountered in 1792. In May of that year a certain Christophe Potter of Paris was led by experiments of his own to take out a patent for the bleaching of rags and paste, essentially imitating Berthollet's procedure of alternating alkaline lyes with bleaching baths.[236] Nothing further is known of him. A little more can be said, however, about the experiments which Pajot Descharmes conducted at about the same time in the paper works at Courtalin (near Faremoutiers-en-Brie, to the east of Paris).[237] This was one of France's more important paper works, considerably developed before the Revolution by Réveillon, as part of his renowned wallpaper concern; it was now in Government service, engaged in the manufacture of *assignats* (the new paper currency). It appears that the success of the works which Descharmes had set up at Bercy in 1791, to whiten dyed and printed textiles for re-use, had led to the idea of applying similar treatment to rags, so as to allow dyed and printed rags to be used in the making of white paper; incentive was provided by the general shortage of rags at the time. With the authorization of the finance and interior ministers, Descharmes went to Courtalin in the summer of 1792 and carried out trials to the satisfaction of Government representatives. He later claimed to have established a workshop there able to treat 3000–4000 lbs. of rags a day. If so, it would seem to have been short-lived, for less than two years later we find a Government team again tackling the problem of applying chlorine to paper manufacture, with no mention of Descharmes.

This Government project of 1794[238] was the most important early venture in France. Again the concern was with the manufacture of *assignats*, but besides the attractive prospect of obtaining high grade banknote paper from lower grade rags, there was also now a more especial motive: at a time when the forging of *assignats* had reached disastrous proportions, it was hoped that chlorine would yield a distinctive paper which it would be difficult to counterfeit. The project was in the hands of the *Comité des assignats* of the Convention, and in particular of Pierre Loysel, one of its members and a man of some chemical knowledge (before becoming politically active during the Revolution he worked at the Saint-Gobain glass-works, and from 1786 was a non-resident member of the Academy of Sciences, as Darcet's correspondent).[239] Care was taken to obtain the best advice: the Committee consulted Berthollet, Fourcroy, and Guyton, and also drew on the practical experience of Welter, Athénas, Alban, Carny, Ribaucourt, and a certain Marchais. Even so, the use of chlorine proved to be not entirely a simple matter. Apart from the difficulties with the odour, already mentioned, some experimentation was also needed before a satisfactory product could be obtained. It was found, for example, that if the rags were bleached before being reduced to paste, the bleaching was superficial and the resulting paste was not a good white; if, on the other hand, the bleaching liquor was applied to the paste itself, it did not penetrate satisfactorily and the bleaching was uneven. The investigators therefore settled on the application of the bleach at an intermediate stage in the disintegration of the rags, a practice which came to be that most generally followed by the later paper industry. The project did successfully culminate in the adoption of chlorine for the regular manufacture of *assignats*. When the notorious financial collapse of the mid-1790s proceeded to rob the *assignats* of all monetary worth, there was some consolation in the thought that at least the paper on which they were printed was a tribute to French chemistry. The superiority of the paper, so Fourcroy observed, was plain for all to see.[240] The works where the paper for *assignats* was made, besides that at Courtalin, were the important manufactory of Didot at Essonnes (where in the later 1790s Robert was to invent his revolutionary paper-making machine), that of Delagarde *aîné* at le Marais (Jouy-sur-Morin), and a works at Buges, near Montargis.[241] It may be that in some of these works the use of chlorine survived the suppression of *assignat* manufacture in 1796, for the Courtalin works was honoured at the 1801 exhibition for its output of papers difficult of counterfeit, while at Jouy-sur-Morin chlorine was certainly in use five years later.[242]

The work just described remained unpublished until 1801, when Loysel gave details and suggested that chlorine would be of benefit to the paper industry at large. The industry was not quick to take it up, however. This was partly in the belief that it would be too expensive: if the bleaching of textiles for the market had at first seemed economically dubious, how much less viable appeared the application of chlorine to the whitening of rags and tatters.[243] Such a view was probably reinforced by an early notion that rather laborious processing was called for, resembling that for textiles; later experience was to show that in fact relatively simple treatment sufficed. The point might also be made that when chlorine bleaching did in due course establish itself it proved to be a not unmixed blessing, since if it was not carefully handled the gain in whiteness was bought at the expense of some degradation in strength and other qualities.

The paper-makers known to have taken up chlorine before the end of the Empire are soon enumerated. In addition to its use at Jouy-sur-Morin, chlorine was employed by 1806 at Troyes, by Fléchey, Moreau & Cie; in June 1812 it was reported to have been successfully adopted at Annonay, by the great paper-maker Johannot, who was hoping it would allow him to lower the price of his products; and in the spring of 1813 scarcity of rags led to the commencement of bleaching experiments by one of the leading paper-makers in the Vosges, Claude Krantz at Arches.[244] In about 1814 chlorine was adopted by the paper industry at Angoulême: according to a later writer it then spread with 'incredible rapidity', coming into use in all the paper mills there within less than a year.[245] If so, this was not typical of the country as a whole, for it seems to have been only in the 1820s that chlorine began to find fairly general use. By the end of the 1830s it was thoroughly established.[246] To an even greater degree than her textile industry, France's paper-makers lagged behind the British in adopting chemical bleaching, as can readily be illustrated by the fact that in Scotland alone chlorine is thought to have been in use at no fewer than nine paper mills by 1795.[247] France's delay can probably partly be explained by a lesser severity of rag shortages on her side of the Channel: before the disturbance of trade by the Revolution France had been an exporter of rags while Britain was an importer.

Chlorine as a disinfectant

The familiarity of chlorine and its compounds in bleaching is today matched by their role as disinfectants, and we can conclude this chapter by observing that historically the one application closely followed upon the other.

Disinfectant fumigation had been practised since ancient times, using the fumes of burning sulphur and the vapours of various other materials, most of which in truth did little more than mask evil smells. As the study of gases became a central concern of chemists from the mid-eighteenth century, new fumigants began to attract attention. Thus, in 1773 Guyton de Morveau successfully used hydrogen chloride, in 1780 Smyth in England employed nitrous gases, and in the 1790s chlorine was suggested—and sometimes used—by a number of individuals.[248] One of the first applications of chlorine may have been made by Bonjour in 1793,[249] though better known is its adoption by Cruickshank in England two years later. It was with the opening of the new century, however, that chemical fumigation, especially by means of chlorine, began to develop in a major way, thanks above all to the missionary zeal with which Guyton de Morveau then took up the cause. In his *Traité sur les moyens de désinfecter l'air*, of 1801, Guyton extensively reviewed an array of fumigants old and new, and was left in no doubt by his experiments that pride of place went to chlorine, 'the anti-contagient *par excellence*'. Through his advocacy the gas was soon to acquire popular renown as 'gaz Guytonien'. The theory was that the putrid airborne miasmas which were the presumed cause of infection were burned up and destroyed by acid gases, all acids, of course, still being supposed to contain oxygen; the especially high proportion of oxygen in chlorine (oxymuriatic acid) made it especially effective, and had the further beneficial effect of stimulating and fortifying one's constitution against attack. From the numerous articles which now began to appear in the *Annales de chimie* one can judge the enthusiasm with which the new fumigation was taken up, and the attention it received from public and medical authorities. It was applied in ships and lazarettos, in church crypts and cemeteries, in assembly rooms, dissection rooms and above all in hospitals and prisons, it was brought to the rescue of ailing silk worms and to the extermination of foot-rot in sheep, and small portable flasks were marketed which could be carried on one's person to generate disinfectant clouds at the first whiff of danger.

Fumigation by chlorine was effective but it did have its drawbacks, of course, being both troublesome for the user and uncomfortable for the beneficiary. More convenient for many purposes was to be the use of hypochlorite solutions. An early advocate of hypochlorite was Masuyer, a medical professor in Strasbourg, who in 1807 reported successful experiments with bleaching powder. In 1813 he made over half a ton of the powder at the chemical works of Meunier, and used it to disinfect troops arriving in the town on the occasion of the great typhus epidemic at Mayence.[250] General recognition of the disinfectant value of hypochlorites came only in the 1820s, however. In view of their convenience, and in view of the widespread domestic use before then of *eau de Javel* as a bleach, this is rather surprising. The explanation seems to lie largely in the misleading theoretical views which prevailed, for if the agent of infection was an aerial miasma, only a gaseous disinfectant could be expected to have much effect against it. The man who won recognition for hypochlorites was the Parisian pharmacist A. G. Labarraque, whose attention was drawn to the subject by a prize which the *Société d'encouragement* offered in 1819 for a means of rendering less unhealthy the trade of the catgut maker. This was one of the most nauseous occupations of the day, consisting as it did in the processing of animal intestines, its insalubrity clearly manifest in the abominable odours which emanated from the workshops. After experimenting with a variety of substances Labarraque found that the noxious odours could be suppressed by means of bleaching powder or *eau de Javel*, and he was awarded the prize for his methods in 1822. He went on to champion hypochlorites as general disinfectants, assuming a role similar to that of Guyton earlier and soon attracting official interest and support. A curious illustration of the value of such materials came in 1824 with the lying in state of Louis XVIII, for thanks to hypochlorite treatment the King's remains could be exhibited openly to the public without offence to their noses. This contrasted with the lying in state a few months earlier of the poet Byron, who, lacking these French advantages, had to be content to lie sealed in a leaden coffin.[251] The solution of sodium hypochlorite which Labarraque especially favoured came to be long known as *eau de Labarraque*, though in fact it did not differ essentially from *eau de Javel*.

NOTES

1. The standard biography is now Sadoun-Goupil (1977), which appeared after this chapter was written.
2. AN, $F^{12}2299$–2300, [2].

3. This account is based primarily on *Enc. méth. Man.*, vol. 1 (1785), 59–82 (article 'Blanchissage', by Roland); Hellancourt (1790).

4. Chassagne (1971), 212; there are some inconsistencies in the figures given.

5. Ballot (1923), 528.

6. AN, $F^{12}1327$, [1], [2]. For earlier regulations imposing an even longer ban see: Bonnassieux and Lelong (1900), 252; Chaptal (1819), vol. 2, 256–8.

7. On Scottish developments see: Clow (1952), Ch. 9; Edelstein (1955); [O'Neill] (1876–7).

8. Musson and Robinson (1969), 252.

9. Henderson ((1972), 21) attributes the translation to John Holker, but the Bibliothèque Nationale catalogue ascribes it to P. Henri Larcher.

10. [Brisson (1780)]; AN, $F^{12}1327$, [3]; $F^{12}2415$, [2].

11. AN, $F^{12}2299$–2300, [5].

12. Pajot Descharmes (1798), 117.

13. Berthollet (1785); id. (1788a); Partington, vol. 3, 503–6.

14. AN, $F^{12}1329$, [1], [2], [3]; Sadoun-Goupil (1974).

15. Smeaton (1962), 11.

16. From this there was deducted 600*l.* for the benefit of Macquer's widow.

17. On Berthollet's relations with Watt, and on British developments generally, see the excellent account by Musson and Robinson (1969), Ch. 8.

18. On this body see: Pigeonneau and Foville (1882); Baker (1973), 217–19.

19. The following account is based on: Pigeonneau and Foville, 94, 99–100, 158, 280–1, 308–11, 323–4, 407; *J. de Normandie*, 21 Nov. 1787, 2 and 12 April 1788.

20. Pigeonneau and Foville refer to correspondence about the new bleaching process between Berthollet and the agriculture department, in AN, H 1516, but I have found nothing on the subject there.

21. A group of scientists, including Berthollet, did gather at the Arsenal that day, but for trials on another matter.

22. AN, $F^{12}2415$, [3].

23. *Enc. méth. Man.*, vol. 2, supplement, article 'Blanchissage'.

24. AN, $F^{12}2299$–2300, [5].

25. J. Bonjour (1853); *Biog. univ.*, new edn., 'Bonjour'; Smeaton (1959).

26. F. Welter (1910).

27. Musson and Robinson, 273.

28. *J. de Normandie*, 2 and 12 April 1788.

29. This account is based primarily on AN, $F^{12}2299$–2300, [5]. See also: AN, $F^{12}2415$, [1]; Berthollet (1789b); Lemay (1932) (quoting an important letter of 30 July 1788 from Berthollet to *Bureau du commerce*).

30. Bonnassieux and Lelong (1900), 455, 460.

31. If this is taken to mean that bleaching was at the rate of 150 pieces a day, it would amount to a through-put of some 45 000–50 000 pieces a year; if, as seems more likely, it means that at any given time 150 pieces were in course of being bleached, the figure would be about 7000, since bleaching took about eight days.

32. F. Welter (1910).

33. Berthollet (1789b).

34. AN, $F^{12}1329$, [3].

35. *J. de Normandie*, 13 Oct. 1787, 20 Dec. 1788.

36. The following paragraph derives mostly from the account by Grandcour himself in *J. de Normandie*, 12 April 1788.
37. Lemay (1933–4).
38. ADSM, 3 J 168, [1]; Simon (1921), 31–2.
39. On Descroizilles see particularly Simon; also Lérue (1875).
40. AN, $F^{12}937$.
41. Beugnot (1835).
42. AN, $F^{12}1329$, [3].
43. Berthollet (1790*a*).
44. Lemay (1933–4).
45. AN, $F^{12}1329$, [3].
46. Chaptal (1789*a*); Lavoisier and Berthollet (1789).
47. O'Reilly (1801), 114.
48. AN, $F^{12}1508$, [17]. His enterprise also included workshops for the singeing of cloth and the printing of cottons.
49. Loysel (1801).
50. Berthollet (1789*b*).
51. Thiéry (1787), vol. 2, 644.
52. *J. de Paris*, 29 Dec. 1788 (quoted by Dorveaux (1929)); *J. de Normandie*, 7 Jan. 1789, Supplement.
53. Musson and Robinson, 287.
54. Ibid.; Berthollet (1789*b*).
55. Lemay (1933–4).
56. See Musson and Robinson, Ch. 8, *passim*.
57. Bosher (1970), 187.
58. Chapman (1969), 68.
59. Berthollet (1789*b*), 186.
60. AN, $F^{12}1329$, [3].
61. Basic information on Jouy is from: Clouzot (1928); Fages (1860); Labouchère (1866); *Biog. univ.*, new edn., 'Widmer'.
62. Berthollet (1789*b*), 187.
63. Berthollet (1791*b*).
64. Clouzot, vol. 1, 21.
65. Leuilliot (1951–2); Soc. ind. Mulhouse (1902), vol. 1, 308–9.
66. Haussmann (1791).
67. AN, $F^{12}1329$, [3].
68. Ballot (1923), 533.
69. Some details in AN, $F^{12}1508$, [1].
70. Darcet *et al.* (1797*a*), 113. Ribaucourt in 1793 directed an early alum venture at Lacroix-Saint-Ouen in the Oise, until in 1794 being drawn into the wartime saltpetre programme; after the Revolution he for some years directed the alum and copperas works at Saint-Georges in the Aveyron (AN, $F^{12}1508$, [34]; $F^{12}2234$, [8]). He published on technical chemistry and was formerly an apothecary and chemistry teacher in Abbeville (Pancier (1937–8), 40–54).
71. Pajot Descharmes (1824).
72. Berthollet (an III [1795]), 193.
73. Richard (1922), 647.
74. Richard, 491–2, 495.

75. AN, $F^{12}1508$, [17].
76. Huet de Coetlizan (1803–4), 144.
77. AN, $F^{12}2299$–2300, [3].
78. *Enc. méth. Chymie*, vol. 2, 590–615 ('Blanchiment').
79. The ministries inherited from the *ancien régime* had finally been abolished in April 1794, to be replaced by a dozen *commissions exécutives*, each headed by one or two commissioners, with a similar number of *adjoints*. Berthollet's appointment to the *Commission d'agriculture* was as a commissioner, and he was described as head of the commission by C. A. Costaz (a member of the administration). See: Gerbaux and Schmidt (1906–37), vol. 3, 277; C. A. Costaz (1818), 53–4; Du Verdier (1972); Parker (1965b).
80. Berthollet (1804), vol. 1, 212.
81. Berthollet (1785), (1788a).
82. Berthollet (1789b). For glimpses of his work in 1787–8, as revealed particularly in correspondence with Watt, see Musson and Robinson, 262–305 (*passim*).
83. Berthollet (1789b), 158–9.
84. Ibid., 172–3.
85. Berthollet (1790b); id. (1791a), vol. 1, 47–68; id. (1803), vol. 2, 190.
86. Berthollet (1789b), 159; Musson and Robinson, 273.
87. *J. de Normandie*, 7 Jan. 1789 (Supplement).
88. Quoted by Lemay (1932), 81.
89. Berthollet (an III [1795]), 193.
90. Pajot Descharmes (1799), 64–5.
91. Berthollet (1785), (1788a).
92. Berthollet (1788b).
93. More precisely, the essential reaction is:

$$2HOCl + OCl^- \rightarrow ClO_3^- + 2Cl^- + 2H^+$$

94. Musson and Robinson, 271, 287, 292
95. Berthollet (1803), vol. 2, 183–207; id. (1804), vol. 1, 267–73.
96. O'Reilly (1801), 116–17; Rupp (1798).
97. See p. 159; also *Dict. tech.*, vol. 3 (1823), 'Blanchiment', pp. 153–4.
98. Balard (1834), 294.
99. *Recherches*, vol. 3 (1826), Table 108.
100. Berthollet (1785), (1788a), (1788b).
101. The reactions can be represented:

$$CaCO_3 + 2Cl_2 + H_2O \rightarrow 2HOCl + CaCl_2 + CO_2$$
$$2\,Ca(OH)_2 + 2Cl_2 \rightarrow Ca(OCl)_2 + CaCl_2 + 2H_2O$$

102. *Recherches*, vol. 3 (1826), Tables 108, 109, 115.
103. Musson and Robinson, 297.
104. Berthollet (1804), vol. 1, 273.
105. Ibid., 267; Descroizilles (an III [1795]), 270.
106. Musson and Robinson, 321–7; Tennant (1947).
107. Muspratt [1860], vol. 1, 310.
108. AN, $F^{12}2299$–2300, [5].
109. Descroizilles (an III [1795]), 270.
110. Loysel (1801).
111. O'Reilly (1801), 99.

112. Chaptal (1807), vol. 3, 113–14.
113. *Dict. tech.*, vol. 3 (1823), 154.
114. See particularly Madsen (1958); also Szabadvary (1966), Christophe (1971).
115. Sconce (1962), 33. A calculation based on the equivalent weight of indigo (Madsen, 132, 145–6) would indicate a figure of only about 0·3–0·35 per cent, but the variability possible in the titration is sufficient for this not necessarily to be inconsistent with the figure we have assumed.
116. Calculated from data in A. Payen (1826).
117. Berthollet (1791*b*); id. (an III [1795]), 234–5; id. (1804), vol. 1, 267.
118. Berthollet (1785); id. (1804), vol. 1, 266.
119. Haussmann (1791); Berthollet (1791*b*).
120. Chaptal (1807), vol. 3, 114.
121. O'Reilly (1801), 116–17.
122. Musson and Robinson, 323.
123. Quoted by Berthollet (1791*b*), 256.
124. Berthollet (1804), vol. 1, 280.
125. Haussmann (1791).
126. Blachette (1827), 173.
127. Chaptal (1789*a*); Lavoisier and Berthollet (1789).
128. *Description*, vol. 8 (1824), 255–60; AN, $F^{12}1007$, [1]. See also p. 176.
129. Magnien and Deu (1809), vol. 1, 314.
130. Decroos (1821), 45.
131. *Recherches*, vol. 3 (1826), Table 108.
132. Hardie (1952); Tennant (1947).
133. Mactear (1877), 23.
134. Ure (1822), 18.
135. *Recherches*, vol. 3 (1826), Table 109; A. Payen (1829–32), vol. 1, 156.
136. Bosc (1803).
137. Alyon (1805).
138. O'Reilly (1801), 57–8.
139. Labarraque (1826), 171.
140. Chaptal (1807), vol. 3, 114.
141. Parmentier (1807).
142. *Dict. tech.*, vol. 3 (1823), 154–5.
143. Berthollet (1804), vol. 1, 272–3; id. (1803), vol. 2, 199–200.
144. Welter (1817).
145. Ure (1822), 14.
146. *Ann. chim. phys.* **10** (1819), 425.
147. A. Payen (1829–32), vol. 1, 156.
148. *Recherches*, vol. 3 (1826), Table 109.
149. Mactear (1877), 23.
150. Ure (1822); Chevallier (1826).
151. Dumas (1828–46), vol. 1, 59.
152. AN, $F^{12}2245$, [16].
153. Blachette (1827), 147; Dumas (1828–46), vol. 1, 59.
154. Musson and Robinson, Ch. 8 (*passim*).
155. Loysel (1801).
156. Vitalis (1823), 165.

157. Ure (1822), 15–17.

158. Alcock (1827), 133.

159. Chevallier (1826).

160. Haussmann (1791); Loysel (1801); Julia de Fontenelle (1834), vol. 1, 75.

161. Campbell (1957); Partington, vol. 3, 301.

162. Musson and Robinson, 295.

163. Lavoisier (1789), vol. 2, 453–4, Pl. iv.

164. On these various receivers see *Dict. tech.*, vol. 2 (1822), 14; Blachette (1827), 155.

165. Berthollet (an III [1795]), 243.

166. O'Reilly (1801), 90.

167. See e.g. 'Description' (1822).

168. Dumas (1828–46), vol. 1, 56.

169. Vitalis (1823), 165–70.

170. Chaptal (1807), vol. 3, 113.

171. Rupp (1798); O'Reilly (1801), 99.

172. Loysel (1801).

173. The following account is based primarily on: Chaptal (1800b); id. (1801); O'Reilly (1801); Bosc (1803); Herpin (1839); Edelstein (1958); AN, $F^{12}2299$–2300.

174. [O'Reilly] (1802).

175. Berthollet (1804), vol. 1, 242–6.

176. Cf., however, Rouget de Lisle (1851), 30.

177. *Bull. Soc. enc.* **14** (1815), 18–19.

178. Blachette (1827), 232–42.

179. Cf., for an economist's discussion of chlorine bleaching in Britain, Wolff (1974).

180. Berthollet (1790a); AN, $F^{12}658^A$, [1].

181. Pigeonneau and Foville (1882), 323. Such a price is confirmed by other sources (AN, $F^{12}2244$, [4]; Gerber (1925–7)).

182. *J. de Normandie*, 12 April 1788; Berthollet (1789b), 162–3; Musson and Robinson, 296, 305.

183. Marion (1923), 'Gabelle'; Cochois (1902), 60, 97–8; G. T. Matthews (1958), Ch. 4.

184. Pigeonneau and Foville, 323–4.

185. AN, $F^{12}658^A$, [1].

186. According to a source quoted by Ballot ((1923), 532) it was 'sel de morue, de nitrières et d'autres salaisons'.

187. Pajot Descharmes (1798), Ch. 17; Bosc (1803), 68; Blachette (1827), 129.

188. *Annuaire* (1823), vol. 1, 203–4; Duchemin (1893), 137; Ballot (1923), 533.

189. 'Note' (1803); AN, $F^{12}2245$, [16].

190. Berthollet (1789b).

191. Bosc, in 1803, writing on the bleaching of hosiery (linen presumably) at Troyes, compared a time of 15–16 days by chlorine with one of 70–80 days by the traditional process (Bosc (1803), 68). At Rouen, it was said in 1823 that linens which would formerly have taken 4–5 months by the traditional method were now bleached by chlorine in a month (*Annuaire* (1823), vol. 1, 203–4). In Britain, Tennant in 1799 spoke of chlorine liquors as having enabled cotton manufacturers to bring their goods from the bleachery to the market in a third of the time formerly required ('Specification' (1800)), and soon it seems the time for cottons

in Britain was reduced to about a tenth (Wolff (1974), 156–7). In France similarly, cottons (for printing) were said in 1831 in the Upper Rhine to be bleached in a week instead of three months (Penot (1831–2), 359).

192. 'Note' (1803).
193. Descroizilles (1818), pp. v–vi, 70–2.
194. Blachette, 128.
195. Berthollet (1790a), (1791b); *Moniteur*, 16 May 1806.
196. Berthollet (1789b).
197. A rough order-of-magnitude calculation based on data in Mitchell and Deane (1962), 78–81, 179, 182–3, 202.
198. AN, $F^{12}2299$–2300, [5].
199. *Seconde exposition* [1801], 19; *Exposition* [1802], 53.
200. Thirty workers were indicated in 1796 (Duchemin (1893), 137), up to 40 in 1803 ('Note'), and 25 in 1810 (AN, $F^{12}937$).
201. AN, $F^{12}2245$, [16]. For comparison, a bleacher at Saint-Quentin in 1802 claimed an annual capacity of 40–50 000 pieces for each of the four bleaching works there (AN, $F^{12}2299$–2300, [3]); according to the local authority, these works were (in about 1790) 'les plus belles et les plus vastes de France' (AN, $F^{12}1549$, statistical note of 1806). The weight of cloth pieces in France varied considerably but was commonly in the region of 1–10 kg, with the fine lawns and *batistes* which were the speciality of Saint-Quentin falling towards the bottom of such a range.
202. These and the following details on Rouen are largely from AN, $F^{12}2245$, [16].
203. The early history of Descroizilles's enterprise is obscure. According to Arvers (ADSM, 3 J 168, [1]), Descroizilles's initial undertaking with the de Fontenays had been only a pilot scheme, which the de Fontenays had abandoned once it had served its purpose in enabling Descroizilles to stand without them. Descroizilles's first major industrial enterprise had then been established in the hamlet of Lescure, in association with Prevel *frères*. The latter had continued in ownership of this works when their partnership with Descroizilles ended, and Descroizilles had then set up a new works, at the château de Lescure, in association with Chatel *fils* [son of the nearby sulphuric acid manufacturer], Delamarre *aîné* and others. From Descroizilles himself we have only a rather unclear reference (in 1806) to the works of the 'Sieurs Prevel frères, qui l'ont acquis des sieurs Descroizilles, dépossédés de leur premier établissement par la tourmente révolutionnaire' (AN, $F^{12}2245$, [16]). The second Lescure works must presumably have arisen after January 1796, for a report of that date does not mention it (Duchemin (1893), 136–7). Descroizilles moved to Paris in about 1810 and his works was then run for some time by his son Paul. Some brief details on this and on its subsequent history are given by Poussier. The property was perhaps still standing in 1924, when Poussier proposed that a commemorative plaque be affixed there.
204. AN, $F^{12}2299$–2300, [3].
205. *Notices* (1806), 79, 82.
206. F. Welter (1910); Héricart de Thury (1819), 18, 214–16; Briavoinne (1839), vol. 1, 343–4. The Gomberts were related to the Vandermesch family, and Gombert *père* had a successful cotton-spinning works in Paris with which Welter had been associated in the mid-1790s.

207. L. Costaz (1819), 119, 370.
208. AN, F^{12}2299–2300, [3].
209. Dieudonné (1804), vol. 2, 158.
210. AN, F^{12}2299–2300, [3]; Pingret and Brayer (1821), note to Plate 34; Picard (1865–7), vol. 2, 32–3.
211. Ballot (1923), 105–6, 532.
212. Troisième exposition [1802], 34; Ballot (1923), 230.
213. Bosc (1803); Dict. univ. commerce (1805), vol. 1, 499.
214. AN, F^{12}2299–2300, [4]; F^{12}1007, [1].
215. AN, F^{12}2299–2300, [3].
216. Ibid.
217. Description, vol. 4 (1820), 18.
218. AN, F^{12}2299–2300, [1].
219. Leuilliot (1959–60), vol. 2, 465–6.
220. AN, F^{12}1550.
221. Leuilliot (1959–60), vol. 2, 363.
222. AN, F^{12}1549.
223. Ballot (1923), 533.
224. AN, F^{12}2299–2300, [3].
225. Bossi (1808), 30–1; AN, F^{12}1549.
226. AN, F^{12}2299–2300, [4].
227. Notices (1806), 24.
228. Bull. Soc. enc. 6 (1807), 110.
229. Tableau (1817), 27. Arvers in 1818 named five chlorine bleachers: Descroizilles, Prevel, Selot (at Bapeaume), Monfray (at Petit-Maromme), and Houssaye (on the road to Quevilly). (ADSM, 3 J 168, [1].)
230. Doin (1828), 46–7; Penot (1831–2), 359–60, and Table 24.
231. Annuaire (1823), vol. 1, 204.
232. Brayer (1824–5), vol. 2, 291; Annuaire (1823), vol. 1, 204.
233. General sources on paper manufacture include: McCloy (1952), Ch. 5; Ballot (1923), Ch. 12; Lalande [1761]; Desmarest (1788); Le Normand (1833–4).
234. Comments on this in Bouillon–Lagrange [1798–9], vol. 1, 147–8; A. Payen (1825); Charpentier (1890), 52.
235. Chaptal (1789a); Lavoisier and Berthollet (1789); Chaptal (1819), vol. 2, 43.
236. Description, vol. 1 (1811), 214.
237. Tuetey (1917), 376, 533; Pajot Descharmes (1824); id. (1798), Ch. 25.
238. Loysel (1801).
239. Kuscinski (1916–19), 419; Index biographique (1954).
240. Enc. méth. Chymie, vol. 3 (1796), 586.
241. Gerbaux (1899), (1903).
242. Seconde exposition [1801], 25; Notices (1806), 317.
243. [O'Reilly] (1807).
244. Bull. Soc. enc. 4 (1806), 231, and 11 (1812), 137; Janot (1952), vol. 1, 123.
245. Lacroix (1863), 59.
246. Dict. tech., vol. 3 (1823), 184–5; Lacroix (1863), 360, 374.
247. Thomson (1974), 36–8.
248. Partington, vol. 3, 529–30; Smeaton (1957), 32–4; Guyton de Morveau (1801).
249. See above p. 135, which derives from Bonjour (1853), 5. It is surprising, however,

that Guyton (1801) does not mention it, and it may be that Bonjour actually used hydrogen chloride.

250. Parmentier (1807); Masuyer (1824).
251. Alcock (1827), 5. Alcock gives an extensive account of French developments based on the writings of Labarraque and others.

4

THE BEGINNINGS OF THE LEBLANC
SODA INDUSTRY

The natural alkalis

THE use of alkalis in the arts and manufactures long preceded the development of alkali production as a branch of chemical industry. This was made possible by nature's provision of a ready if not exactly bountiful supply of alkaline materials in the ashes of burned vegetation. 'Natural' alkalis obtained in this way were among the most heavily used chemicals of the eighteenth century, important as basic ingredients in the manufacture of soap, glass, and saltpetre, and as chemical agents in the cleansing and dyeing of textiles.[1] Two chemically distinct types of alkali were obtained, depending on whether the matter burned had grown inland or in salty coastal soils: inland vegetation gave rise to the product known in commerce as potash (in which the alkali was potassium carbonate), while coastal plants furnished commercial soda (whose alkali was sodium carbonate).[2] Along with the real alkali, these commercial products contained high proportions of impurities, which varied in quantity and character according to the species of vegetation burned and the time and manner of its burning. There were consequently marketed numerous different varieties of alkali, particularly of soda, generally named after the place from which they came or the plant which furnished them.

For the production of potash, materials such as dead wood and plant debris in general were burned in heaps or in pits to yield an ash containing perhaps 10–20 per cent potassium carbonate. This ash was to some extent used simply as such, but in general that intended for commercial distribution was next partially purified by leaching with water and evaporating the solution to dryness. Only then did the ashes strictly speaking acquire the name potash in the commercial terminology of the day, the name deriving, of course, from the method of production by boiling down in iron pots. The dark alkaline residue so obtained was also known in France more specifically as *salin*, to distinguish it from a purer variety of potash, known as 'pearl ash', produced by calcining *salin* in a reverberatory furnace so as to eliminate the coloured impurities which would be harmful in dyeing and in certain other applications. *Salin* and pearl ash generally contained between 40 per cent and 80 per cent potassium carbonate. The production of potash alkali in

France itself was only partly able to supply the country's needs, and so home resources were supplemented to an important extent by imports from more richly forested parts of the world, such as Russia, North America, and especially the countries of northern Europe.

For soda, too, France depended heavily on imports, in this case chiefly from Spain. It was from Spain, moreover, that the finest quality sodas of commerce came. Spanish soda was produced on the Mediterranean coast in the regions of Alicante, Cartagena, and Malaga, the best variety being that known as 'barilla' after the plant which furnished it. The barilla plant was specially grown for soda production and its cultivation was a jealously guarded Spanish monopoly: the removal of barilla seed from Spain was an offence said to be punishable by death. After being harvested the plant was burned in pits to yield a hard grey mass, which was broken into fragments and sold as 'stones of soda'. Unlike potash, soda was marketed without preliminary purification and so the alkali content was lower. Barilla generally contained about 20–33 per cent sodium carbonate. Besides her imports from Spain, France also bought large quantities of soda from Sicily, and to a lesser extent from the Levant, the soda from the Levant being of interest for its inclusion of an exceptional variety known as 'natron', which derived not from a vegetable but from a mineral source (it occurred as a natural deposit in Egypt). France produced some soda herself, but this supplied only a fraction of her consumption and was not of the highest quality. Although there had been trial sowings of barilla in France at the beginning of the eighteenth century, and again in 1782–4 by Chaptal, these were without result beyond demonstrating the possibility of its cultivation there. Of the sodas which France did produce, the best was that called *salicor*, or soda of Narbonne, containing about 14–15 per cent sodium carbonate. It was obtained from the *salicor* plant, specially cultivated for the purpose in various parts of the Languedoc, chiefly around Narbonne. In 1766 the diocese of Narbonne was said to produce some 12–15 000 *quintaux* a year (0·6–0·75 million kg); production was probably little higher than this in later years, and may indeed have been lower.[3] Another French soda was *blanquette*, or soda of Aigues-Mortes, produced from wild plants on the Mediterranean shores between Aigues-Mortes and Frontignan. This contained only about 4–10 per cent sodium carbonate but prior to the abolition of the salt-tax in 1790 it was valued by makers of green glass for the tax-free salt it contained, which they could separate and sell as a profitable by-product. The poorest of the French sodas was that called 'soda of Normandy', 'ashes of varech', or simply 'varech', which was similar to the kelp produced in Ireland and Scotland and was made by burning seaweeds gathered on the Normandy and Brittany coasts. This was very variable in composition and often contained scarcely any sodium carbonate (ranging between 0·5 and 8 per cent), but it was greatly used in Normandy in the manufacture of window-glass and in the glazing of pottery, applications in which its neutral salts as well as its small amount of alkali would be of service. In about 1730 some 800–1000 tons a year are said to have been produced in Upper and Lower Normandy, and in 1775 1200–1500 tons (1·2–1·5 million kg); it seems likely that production continued to grow in the latter part of the century, following an easing of government restrictions in 1772.[4]

We have no satisfactory year-by-year statistics for France's alkali imports in the eighteenth century, but some approximate indication can be given of the amounts she was receiving at the end of the *ancien régime*. Coquebert in 1794, for example, spoke of France's buying two-thirds of the soda produced by Sicily and by the chief soda regions of Spain, presumably referring to the state of affairs before the war. The production figures he gives—deriving from the French consuls—would point to imports from these countries of 293 000 *quintaux* (14·37 million kg). France's total soda imports in 1788, from detailed trade figures given by Magnien and Deu de Perthes, were 30 434 055 pounds (14·91 million kg). The total in 1792—according to an official report of 1794—was 277 768 *quintaux* (13·32 million kg), net of some very minor exports. We can conclude that on the eve of the Revolution France was probably importing something like 15 million kg of soda a year. Potash imports can be guessed to have been of the order of 2–4 million kg.[5]

France's heavy dependence on imports made supplies vulnerable in time of war, and it was chiefly this which stimulated a growth of interest during the final quarter of the eighteenth century in the possibility of procuring soda by chemical means from common salt.

The possibility of producing soda from salt

The roots of the soda industry are to be traced to an important scientific study of the chemical constitution of common salt which Duhamel du Monceau presented to the Academy of Sciences in 1737; for it was this which first established a chemical relationship between salt and soda.[6] Prior to Duhamel's work, although chemists were familiar with the acid contained in salt (hydrochloric acid, then called 'spirit of salt'), they remained ignorant of the base with which the acid was combined. Some believed the base to be an 'earth', but Duhamel formed the view that it was a 'fixed alkali salt', and undertook experiments to prove his opinion by isolating it. After failing in various attempts to separate the base from its acid by direct means—by heating salt with charcoal, for instance, or with iron—he at length accomplished his objective by an indirect technique, which consisted in first reacting the salt with vitriolic acid to form Glauber's salt (sodium sulphate), then calcining this with charcoal to obtain a hepar (sodium sulphide), next displacing the sulphur of the hepar with vinegar (forming sodium acetate), and finally driving off the vinegar by calcination (yielding sodium carbonate).[7] For good measure he also effected the decomposition a second way, by converting salt to cubic nitre (sodium nitrate) and then deflagrating this with charcoal, again obtaining sodium carbonate. Having thus secured what he regarded as the base of salt, Duhamel found that it was indeed an alkali as he had thought, and he concluded from his study that it was identical with the alkali contained in natron and in vegetable soda. He also ventured the essentially correct suggestion that it was by decomposition of salt from the soil that the alkali of soda plants was formed (the plants in fact produce organic sodium salts which then decompose to the carbonate on incineration).

Duhamel thus laid the scientific foundations for the manufacture of soda, demonstrating on a laboratory scale the practical possibility of producing soda from salt. The particular reactions he employed, however, would have been expensive for commercial exploitation, and Duhamel in fact said nothing about the potential industrial significance of his work. He was interested rather in the theoretical issues it raised, becoming engaged in a dispute with the German chemist Pott as to whether the base was truly to be considered an 'alkali salt', or an earth as Pott still maintained. It is only three decades later that we find the first stirrings of interest in the commercial possibilities which Duhamel's work had tacitly revealed, and this interest then first manifested itself in Britain.[8] The earliest endeavours seem to have been those of a distinguished group consisting of James Watt, Joseph Black, and John Roebuck, who over the course of several years from 1766 sought to develop a workable manufacturing method from a process of Black's devising. They were joined in 1770 by the chemist James Keir, following independent experiments of his own. In the event this scheme was without issue, but another company, of unknown identity, was making soda near London by 1771. In the later 1770s, with the disruption of alkali imports by the American War of Independence, interest grew. Thus, when on 22 May 1780 Alexander Fordyce petitioned the Commons to be exempted from salt duty for a soda manufactory at South Shields, he was followed within a month by no fewer than seven further petitioners, of whom two similarly claimed to have already established works. The activity at this time was further reflected in the award of five patents for alkali manufacture between 1779 and 1783. Henceforth soda production was to have a continuing place among the chemical manufactures of Britain, though for many years only in a very modest way, as a scattered activity of a few individuals using a variety of processes worked on a limited scale. Only in the 1820s, following the introduction of the Leblanc process from France, was it to become a major industry.

By 1779, news of developments in Britain had evidently reached the ears of John Holker in Rouen, for on 28 May that year, in his capacity as *inspecteur des manufactures étrangères*, Holker sought permission from the authorities in Paris to send an emissary to England in quest of information. The *intendant du commerce* Tolozan judged such a step to be unnecessary, however, at least for the time being, since by now there were experiments under way in France.[9]

The soda venture of Malherbe and Athénas, 1777[10]

The experiments in question were those of Malherbe and Athénas, whose venture, beginning in 1777, was the first attempt at soda manufacture in France.[11] Malherbe, the inventor of the process followed, was a Benedictine *abbé* at Saint-Germain-des-Prés in Paris, where he taught philosophy and was in charge of the apothecary's shop. Athénas is variously described as having been an apprentice apothecary in the house where Malherbe lived, and principal assistant in the apothecary's shop at the Abbey. He was taken into association by Malherbe at the beginning of the scheme, since Malherbe's religious calling prevented his own active participation in a manufacturing enterprise.

Malherbe had apparently begun his experiments in 1774, and he had discovered his process towards the end of 1776. He apprised the finance minister of his discovery the following July, in a memoir which further announced that a works was projected at le Croisic, on the Brittany coast. (Brittany was one of the exceptional regions in France where no salt duty was levied, and it was there that the cheapest salt in the country was to be had.) Interest had been shown by a company of English merchants, who had procured samples of Malherbe's trial product, and Malherbe was hoping to find sales not only in France but in England, too, if the state of war permitted. He was soon seeking Government protection for his intended venture, and his process was consequently examined early in 1778 by the chemists Macquer and Montigny, as commissioners for the *Bureau du commerce*. Their report of 13 March found that the process did result in a partial conversion of salt to soda. There were doubts, though, about its large-scale viability, and so before agreeing to an exclusive privilege the *Bureau* next demanded that large-scale trials be conducted. In 1779 Malherbe and Athénas accordingly travelled to le Croisic, where they set up furnaces and operated their process on several thousand pounds of materials under the eye of the inspector of manufactures Grignon. In his report of 16 August Grignon considered that, bearing in mind the imperfect facilities and the pioneering nature of the trials, the results had been favourable, and had demonstrated the possibility of obtaining soda from salt in very large quantity.

The process Malherbe had devised was partly based on that by which Duhamel du Monceau had achieved the first laboratory production of soda from salt. It resembled in this a number of manufacturing methods that were to be developed in the following years, including, as we shall see, that of Leblanc. These methods all began by converting the salt to sodium sulphate, and reducing this to sulphide by fusion with charcoal. Where they differed was in their manner of then eliminating the sulphur so as to obtain the soda itself. Duhamel's own technique had involved the expensive use of vinegar. Malherbe was the first to develop a cheaper alternative, employing iron instead. By fusing the sulphate with scrap-iron as well as charcoal he obtained a crude mass containing free soda in a caustic state (sodium hydroxide). On standing for some days this mass then crumbled to a powder as it took up carbon dioxide from the air, forming sodium carbonate. The soda was extracted from the crude product by leaching with water. Malherbe's inspiration for the process had been a technique then in standard use for extracting metallic antimony from its sulphide ore—by fusing the ore with iron nails—a technique depending on the known pre-eminent affinity of iron for sulphur.[12] As Malherbe's project progressed, two variants on his process were soon developed by Athénas. Athénas found that instead of iron one could also use iron ore for the reaction with the sulphide. And for the preliminary conversion of salt to sodium sulphate he devised a means of employing copperas (iron sulphate) instead of sulphuric acid: by heating salt with copperas, sodium sulphate was formed with the evolution of hydrochloric acid, and fusion of the resulting mixture with charcoal then again gave a crude mass containing caustic soda. The advantage of this variant, clearly, was the lower cost of copperas, and Athénas apparently devised it when his early hopes of making sulphuric acid cheaply, without using saltpetre, ran into difficulty. He was not to attempt its exploitation, however, being obstructed in this by the salt duty: the duty dictated that soda manufacture be conducted in Brittany, whereas the country's chief pyrites deposits (for the production of copperas) were in Picardy.

Following the favourable report from Grignon the partners were apparently promised an exclusive privilege in February 1780, though it was not actually dispatched. The next three years were then spent in endeavours to establish a manufactory. The large-scale trials had shown that le Croisic was not after all the best place for a works, and so after considering Gournay in Normandy (near the pyrites deposits of Picardy), the partners finally settled on port Lavigne (or port de Vigne) near Nantes. While the works there was in course of construction, however, the company was broken by the death in 1781 of its chief capitalist, a *négociant* of Angers who had promised 150 000 *livres* of its 180 000 *livres* capital. Despite this blow the project struggled on a little longer. On 13 March 1782 Athénas formed a partnership with individuals named Jourdan and de la Bernardière, and on 16 April he at last secured his exclusive privilege, for the usual fifteen years. Besides soda the works was also to make its own sulphuric acid. It appears from Malherbe that the project was finally abandoned as an expensive failure in 1783, having cost a total of 117 538 *livres*. The reason for the failure is not known. There was evidently some difficulty in attracting capital, but it may also be that the technical operations left something to be desired. Athénas seems subsequently to have made some further endeavours but without success, and in December 1786 he was said to be reduced to producing *eaux-de-vie* at Nantes. A local official reported in January 1787 that he had 'made use of his privilege only to betray the confidence of those who have lent him money'.[13] His later career as a manufacturer of sulphuric acid has already been indicated in Chapter 2.

Although the project of Malherbe and Athénas thus never reached fruition, the process they attempted was to have a longer history. Later concern with it in France will be considered below. Here we may note that essentially similar methods were patented in England in 1781 by Alexander Fordyce and by Bryan Higgins. It is just possible that these owed something to Malherbe, since one of Malherbe's associates complained of the le Croisic trials having resulted in a leakage of their secret. Many years later, in the 1850s, there was to be a notable revival of interest in the process when it was taken up by Kopp and for a time seemed likely to prove a credible rival to the then established Leblanc process. Experience showed it to have serious technical drawbacks, however, in causing rapid degradation of the plant and in giving only a very disappointing yield.[14]

The Academy's soda prize, 1781

During the period when Malherbe and Athénas were engaged in their project, the nation's alkali supplies came to arouse some concern owing to the disturbance of overseas trade by the War of American Independence (1776–83, France intervened in 1778). The war brought an interruption in shipments of potash from North America, and also made difficult the importation of potash by sea from northern Europe, while at the same time there developed shortages of Spanish soda. The resulting rise in alkali prices brought representations to the Government from users, and a number of steps were consequently taken to assist supplies.

There can be little doubt that the prime agent in these moves was the *Régie des poudres,* and in particular Lavoisier, one of its leading members. Since acquiring control of the country's saltpetre and gunpowder industry in 1775, the *Régie* had been making strenuous efforts to expand the production of saltpetre in France, and in this it was meeting with conspicuous success: saltpetre production rose from 1 906 000 pounds (crude) in 1776 to 2 727 000 pounds in 1781, and was continuing to grow steadily.[15] A main factor in this improvement, according to Lavoisier, was the use of potash in the extraction and refining processes. Although the use of potash was not altogether new, having found some place in production practices for centuries, it had previously often been neglected or imperfect since the chemical role of the potash was not understood. Lavoisier was perhaps the first to arrive at an understanding of its true function and importance, seeing that it served to improve the yield by producing additional saltpetre (potassium nitrate) from the various other nitrates which accompanied the natural saltpetre in the deposits exploited.[16] With this insight potash had acquired a more emphatic and more rational place in saltpetre production, and the *Régie* had become a large consumer at a time when the country's supplies were restricted.

As one step to improve supplies, the *Régie* sought to encourage the production of potash in France by issuing a detailed instructional pamphlet. This work, written by Lavoisier, was published in 1779 and was distributed in the provinces. The Government followed this up by imposing a ban on the export of potash (a ban also in practice applied to soda), by royal orders of 10 February 1780 and 26 April 1781. And the Government further decided to try to promote alkali production by offering a prize on the subject. The finance minister assigned this task to the Academy of Sciences, apparently leaving the exact subject to the Academy's discretion. It seems likely that Lavoisier played a large part in instigating this prize offer, and in then determining its shape, and it was he who drew up the prize programme which the Academy in due course published.[17] Not surprisingly the needs of the *Régie* for potash loomed large in the explanations which were there given. The Academy had not, however, decided to make potash production itself the subject of the prize. Recognizing that prospects in that direction were limited by France's comparative poverty in forest-lands, it adopted the indirect strategy of seeking instead to promote the production of soda. Some of the main consumers of alkali could quite well use either material, the kind customarily employed in any particular locality depending on their relative availability: glass manufacturers were in such a case, for example, as were those simply requiring alkaline lyes for the cleansing of textiles. If supplies of soda could be improved, a switch in consumption might thus ease pressure on the supply of potash for saltpetre production (where only potash would do). Particularly enticing was the long known theoretical possibility of obtaining soda from salt, for this would offer a virtually unlimited resource. The topic fixed was accordingly the 'alkalization' of salt. The prize programme, published in 1781, offered a prize of 2 400 *livres* to be awarded in November 1783 for a solution to the following problem:

To find the simplest and most economical process for decomposing marine salt on a large scale, [and] extracting the alkali which serves as its base, in its pure state, free of all acid or other combination, without the value of this mineral alkali exceeding the price of that obtained from the best foreign sodas.

The contest did not prove to be a great success. The prize went unawarded in 1783, and after being re-offered for 1786 and then again for 1788 it seems to have been finally withdrawn. So far as can be judged it attracted few entrants. There is evidence of only five submissions for 1783, one for 1786, and one in 1787. Moreover, two entries in 1783 came from the same author, who was also responsible for the communication of 1787, this merely reiterating a process he had indicated earlier. The various processes suggested in the surviving memoirs need not detain us here, since none of the authors gave any indication that he was planning commercial exploitation.[18] The prize was not a particularly large one, and we shall see that chemists with serious manufacturing intentions, rather than entering for the prize preferred the alternative course of approaching the Government for an exclusive privilege.

The prize contest, in fact, came into some conflict with the Government's parallel efforts to promote soda manufacture by granting exclusive privileges to inventors. Its announcement brought protests from Athénas and his associates, who demanded that the competition should not take place, and then when it did complained that it had prejudiced their venture by undermining public confidence in them. In a letter to the Academy of 11 April 1783, Jourdan complained that the proclamation of the prize had created the impression that their own methods must be inadequate, and had 'put off capitalists without whom one cannot mount an enterprise of consequence'.[19] From the Academy's point of view a greater embarrassment was the discovery that the process indicated in the best memoir received (which Macquer thought in August would probably gain the prize) was similar to one for whose exploitation Guyton de Morveau secured a privilege in June.[20] The granting of Guyton's privilege was attacked by some members of the Academy as an injustice to their prize entrant. It also transpired that several of the processes described in other prize memoirs were essentially similar to processes for which a further privilege was then being sought, by Hollenweger. This conflict between the prize contest and the granting of privileges was perhaps one reason for the non-award of the prize. The declared reason, apparently, was the imperfection of the methods submitted in achieving only an incomplete decomposition of the salt.

Guyton de Morveau, 1782

Guyton de Morveau, by the time of his soda privilege, was a chemist of some repute.[21] Although his professional career until 1782—when he retired from his post with a pension—had been that of an advocate in the law court at Dijon, he had also since the 1760s been interesting himself in chemistry, in both its pure and applied aspects. He had begun to attract scientific notice in 1772, with some researches on the calcination of metals, which were an important influence on Lavoisier, and in 1776 he had begun to deliver an annual chemistry course at the Dijon Academy. In 1778–9 he had ventured into the field of chemical manufacture, mounting a saltpetre works at Dijon in response to the Government's promotional efforts for that industry. By about that same time he had also become interested in the possibility of soda manufacture.[22] On hearing in 1779 of Malherbe's activities in Brittany, he discontinued his own researches for a while, but his interest in the subject was rearoused in 1781 by the Academy's prize offer, which he took as an indication that the Brittany trials had proved disappointing. By now, moreover, his engagement in saltpetre manufacture had made him all too aware of the inflated level of alkali prices.

Rather than enter the Academy's prize contest, Guyton preferred to make a direct approach to the Controller-General of Finance, in the belief, as he later explained, that this course would produce quicker results. In a letter to the Controller-General of 16 February 1782, he offered to establish a soda works for the Government, anywhere it might care to choose. Alternatively, he said, he would like a privilege to manufacture on his own account. Soon afterwards he submitted for the judgement of Macquer 'a memoir containing over 400 experiments, the examination of all known and possible methods, and the exposition of the two processes which I proposed as advantageous'. (Guyton was characteristically a thorough worker.) Macquer privately informed him in May that his processes were greatly preferable to that of Malherbe, and then on 31 July presented a highly favourable official report, concluding

[that] M. de Morveau has completely resolved the important and very difficult problem of the entire alkalization of common salt. That his processes are totally different from that for which dom Malherbe has obtained a privilege, and that they are above all much simpler, much more sure, much easier and consequently far more practicable and far more economical.

The Government did not wish to engage in soda manufacture itself, and so after some delay Guyton was granted an exclusive privilege for his processes on 3 June 1783. The administrators of the *gabelle* would not countenance any concession in salt duty for a works in the interior, and he was constrained to make his works in Brittany, where it would automatically enjoy duty-free salt, but where it would also, of course, be far from his home province of Burgundy. Despite this inconvenience Guyton on 22 August rented a property at le Croisic and immediately set about installing his works. By 24 September he could inform the *intendant du commerce* Tolozan that he had instructed a director, that the works was already equipped with vats, boilers, and five furnaces, and that the commencement of large-scale working awaited only the arrival of supplies, which were then being ordered.

We are not told what the two processes covered by Guyton's privilege were, but we can be virtually certain that one of them was a method based on the formation of soda as an efflorescence from a paste of brine and lime. Guyton indicated such a method in a sealed note which he deposited in May 1782 with the Dijon Academy, and this method was also the first among a number of processes of which he filed a description with the administration in 1788. According to his 1788 description it consisted in the exposure in shallow boxes, in a low, damp, and somewhat confined place, of a paste made from lime and concentrated brine. In these conditions sodium carbonate slowly formed as an efflorescence on the surface, to be periodically scraped off until the materials were exhausted.[23] Guyton considered this to be an imitation of the way in which nature sometimes produced soda as a mineral deposit (most famously in the natron lakes of Egypt). The method was essentially the same as that which Watt and Black had sought to exploit in

Britain some fifteen years earlier, though Guyton cannot have known this since the Scotsmen had kept their method secret. Guyton's knowledge of the reaction derived from Scheele, who had published it in 1779 after discovering it for himself.[24] The second method covered by the privilege was probably based on the formation of soda in a paste of brine and lead oxide (a manufacturing method discussed below, p. 207). Such a method was the second of those Guyton described in 1788, and an earlier interest in it is suggested by his remark in October 1783—when seeking assurance that Hollenweger's process was different from his own methods—that he had heard Hollenweger's process to consist in the use of calx of lead.

A number of factors conspired to prevent Guyton's venture developing as he had hoped. He felt disadvantaged from the outset by the location of his works at le Croisic. If it had to be in Brittany, it would have been much better placed near Nantes, but Guyton was expressly forbidden from forming a works within 10 leagues (about 28 miles) of Nantes, as a protection for Athénas, who had a prior privilege to work there. More important, however, was the blow he received in October with the news that a soda privilege had been granted to Hollenweger, for the prospect of a rival made Guyton and his backers apprehensive. On 18 October, less than a month after his cheerful report to Tolozan, Guyton was informing the Controller-General that he now found himself obliged to stop supplies to the works and operations there, and to suspend publication of his intended advertising prospectus. From a later letter we learn that the news of Hollenweger's privilege caused his prospective backers to withdraw their funds. Guyton continued to maintain at le Croisic a 'provisional establishment', where in December 1784 he had 'an agent (*commis*) and some workers, who work according to my very small means', but by then a new obstacle had arisen. A first batch of soda (three barrels), dispatched for the glass-works in Burgundy which Guyton had established that summer, had been held up by the customs authorities in Nantes. Since Brittany lay outside the central customs region of the *cinq grosses fermes*, the authorities were insisting (quite rightly) that it should pay import duty as foreign soda before being allowed to pass into the interior. Guyton refused to pay and the soda was still being held in August 1787.[25] Whether the works continued to operate after this new setback seems doubtful; if it did it can only have been in a small way, presumably for the local market. Insofar as soda was made there at all it seems to have been by the lime process.[26]

Hollenweger, 1783

Hollenweger had approached the Controller-General in April 1783, seeking an exclusive privilege for a works he wished to form on the Brittany coast.[27] He was a former pupil of the Saint-Gobain glass company, France's premier glass concern, and his scheme followed trials he had made at Saint-Gobain and at the same firm's Tourlaville works near Cherbourg. This firm must have been one of the largest alkali consumers in the country, employing over 1000 tons of soda a year in 1770.[28] It had refused to support its pupil in large-scale manufacture, however, lacking faith in his methods, and his complaints having led to his dismissal Hollenweger was now embarking on a soda venture of his own. Among those to take an early interest in the scheme was the physicist Meusnier (at that time employed as a military engineer in Cherbourg).

Hollenweger demonstrated his processes to the Government's adviser, Macquer, who then presented highly favourable reports in July and August 1783, speaking of Hollenweger's 'knowledge in the theory and practice of chemistry', and concluding that 'of all those who have undertaken up to the present, or who propose to undertake, to establish artificial soda works by the decomposition of marine salt, it seems to me that it is Sr Hollenweger who should obtain the most success, through his familiarity with chemical operations, his knowledge, and the resources of his mind'. Hollenweger's processes resembled the process of Malherbe in that they began by converting salt to sulphate and by then reducing this to sulphide. Exactly how he then proceeded to convert the sulphide to soda we do not know, but Macquer tells us that he had several methods, one of them very similar to that of Malherbe but more complete, the others quite different.[29] A very pure soda was obtained. Macquer was also impressed by a method Hollenweger had devised for making sulphuric acid without saltpetre (see p. 58), and considered that this would be a great advantage. A substantial enterprise was evidently intended, for besides the immediate production of soda, and of sulphuric and hydrochloric acids (the latter acid to be collected as a by-product), Hollenweger also had in mind for the longer term the manufacture of sal ammoniac, alum, and vitriols. He was impatient to start building, having formed a company with funds at the ready, and being already in the summer of 1783 in receipt of a salary as director of the proposed works. He was granted his privilege on 23 September, with the condition that he must form his works at least 16 leagues (about 40–5 miles) from Nantes.

There was some initial difficulty in securing a site, but a works was duly established between 1784 and 1786 in the commune of Batz (two or three miles south of le Croisic).[30] An inventory of the works drawn up in 1794, by which time it was idle, speaks of two buildings housing half a dozen furnaces, one of which, from the description given, was probably in reality some form of lead chamber. There were also raw materials: Glauber's salt, two barrels of saltpetre, and two or three tons of sulphur.[31] How long the factory worked, or indeed whether it ever did begin regular manufacture, we do not know. It had apparently still not begun in June 1788, for a local statistical report then mentioned that it should be in activity very shortly.[32] Its existence was evidently unknown to the Government when in August 1788 Hollenweger's privilege was revoked on the grounds that he had not established a works. The Guérande district council in 1794 described 'Aulveger' (Hollenweger) as an *émigré*. The mayor of Batz said that the director and entrepreneur (Hollenweger presumably) was dead, and that his associates had not continued with the works. A certain Cormeray, who in 1794 unsuccessfully sought to acquire the works to make sal ammoniac there, wrote that its founder, having failed, had moved to England and had there died.[33]

Guyton, de Bullion, Carny, and Géraud de Fontmartin, 1788–1789

At the beginning of 1788 there came a new petition for a privilege, from the *marquis* de Bullion,[34] who wished to form a works on the Atlantic coast to make soda, along with sulphuric and hydrochloric acids. His process was duly examined in May by Berthollet (Macquer's successor). Meanwhile, Guyton also still had ambitions to manufacture, and indeed had petitioned the Controller-General the previous August (without reply) for an improvement in the terms of his own privilege to allow him to set up works at Nantes. Guyton and de Bullion now agreed to a merger of their interests, to secure jointly a new and wider privilege which would assure them of adequate protection against competition and sufficient freedom in the location of their works. This they were granted by an *arrêt* of 23 August 1788. The privileges previously accorded to Athénas, Guyton, and Hollenweger were now revoked 'for want of their having formed their establishments in useful time', and a new fifteen-year privilege authorized Guyton and de Bullion to establish works, with exclusive rights to their processes, in six specified regions where salt was duty-free, these covering the greater part of the Atlantic coast (from Brittany to Guyenne), and also the coast of Flanders. To prevent hindrance from internal customs tolls, of the kind Guyton had earlier encountered, their soda was permitted to circulate throughout the kingdom free of internal duties. At least one works was required to be established within a year.

The new privilege covered seven processes.[35] The first two were the lime and lead oxide methods to which we have earlier referred. The third used double decomposition of salt and potash in solution. The fourth involved vitrification of salt with soda and felspar ($KAlSi_3O_8$), whereupon additional soda was formed. The fifth and sixth depended on double decomposition of salt with pyrolignate of lead or baryta to form pyrolignate of soda, which was then decomposed by heat to sodium carbonate. (The pyrolignates were salts of pyroligneous acid, a crude acetic acid obtained by the distillation of wood.) And the seventh process—presumably de Bullion's, since it is the only process to use sulphuric acid—again employed the pyrolignate method but applied it to sodium sulphate instead of directly to salt.

Despite their new privilege, Guyton and de Bullion had still not set up works when in the spring of 1789 yet another petition was presented, this time by J. A. Carny and Géraud de Fontmartin, who wished to undertake soda manufacture by processes of their own, and proposed to couple with it the production of mineral acids and sal ammoniac. Ambitiously, they hoped to finance their enterprise by floating a joint-stock company, for which they also sought permission. Carny was the principal partner here and we shall subsequently find him a prominent figure in the industry's later development. Son of a director of the mint at Grenoble, he had studied in Lyons and Paris and had then in 1778 joined the service of the *Régie des poudres*, in which he rose successively through the grades of *élève, inspecteur*, and *commissaire*. During his time with the *Régie* he had also occupied himself with various chemical researches and with chemical manufactures. While in Lyons as *commissaire* he had set up a works to produce potash from the waste materials of hat factories, and in 1784 or 1785 he had also established at Lyons a small mineral acid works (producing chiefly hydrochloric acid, it seems), which was to continue until the siege of Lyons in 1793, when it was destroyed and its manager shot dead. Carny left the *Régie* in 1788 in order to devote himself entirely to private industry.[36]

Carny and Géraud described eleven processes, all wet reactions performed variously on common salt itself or on the sulphate or sulphide.[37] The processes can be seen to reduce to three basic methods. One consisted in the formation of an organic salt which could then be decomposed to sodium carbonate by calcination; the organic salt was produced by double decomposition of common salt with lead or magnesium acetate, or of sodium sulphate with calcium acetate, or by boiling sodium sulphide with wood shavings. The second method consisted in the decomposition of salt, sulphate, or sulphide by boiling with a solution of lead oxide in soda. The third depended on the use of tartrates, the simplest variant involving double decomposition of sodium sulphate with calcium tartrate, followed by decomposition of the sodium tartrate with lime. There were also various elaborations for the recovery of reagents used.

Berthollet presented a favourable report on 13 May 1789, finding two of the processes to be entirely new: 'Messrs Carny and Géraud appear to me to have applied themselves to the problem . . . with much sagacity and a great knowledge of the chemical affinities whose play one must unravel and whose applications one must vary in the operations'. Since Carny and Géraud had new processes, the *Bureau du commerce* felt that a privilege could not be denied, yet it was loath to precipitate another nervous stalemate by multiplying privileges. The four interested parties were therefore urged to reach a private agreement, and this led to an association of the four to share the single privilege already accorded. The notarized act of their agreement, of 19 August 1789, was endorsed by an *arrêt* of the *Conseil d'état* of 25 October, which at the same time extended the privilege to allow the establishment of a soda factory in the salt-works at Peccais, in the Languedoc, again with duty-free salt. This was in response to a request for manufacturing rights in the south, where the partners were attracted by the immense soda market afforded by the soap industry of Marseilles and the scattered glass works of the southern provinces. The partners were required to form at least one works in each of the seven localities now covered within five years, and at least three within a year.

This grandiose scheme was never realized, probably because of the changed circumstances brought by the Revolution, already in its early weeks when the partners came together. Guyton is said to have now quickly become too absorbed in politics to concern himself with industry.[38] His three associates were evidently still interested in joint soda manufacture towards the end of 1790, when they petitioned the National Assembly for confirmation of their privilege, and for authorization to establish a works, but nothing seems to have come of their request.[39] The request was no doubt motivated by the plans then being prepared by the Assembly for the replacement of the old system of privileges by a more regular patent system. France's modern patent system was indeed soon afterwards instituted by a law of 7 January 1791, and it is possible that the privilege of Carny and his associates was among those which now lapsed under the terms of this law.

The early years of the Revolution, 1789–1793

Early in the Revolution, a number of institutional reforms swept away some of the hindrances which had played a part in frustrating efforts to mount soda works in the 1780s. The replacement of privileges by the modern patent system in 1791 eliminated the kind of *ad hoc* territorial restriction which in 1783 had prevented Guyton establishing his works at Nantes. The suppression of internal customs tolls in November 1790 removed a lesser obstacle of which Guyton had also complained. Above all the abolition of the infamous salt-tax, with effect from 1 April 1790, lifted the constraint which had driven early soda ventures inconveniently to Brittany, and opened up the possibility henceforth of conducting soda manufacture anywhere in France with salt at free market prices. At the same time added interest was by now being lent to soda manufacture by the development of chlorine bleaching, for the decomposition of salt could yield as twin products both sodium sulphate, applicable to soda production, and hydrochloric acid saleable to bleachers. The potential market for hydrochloric acid in the new bleaching was one circumstance which Carny and Géraud had urged in support of their soda project in 1789.

By the first years of the Revolution we find soda manufacture being conducted as a subsidiary activity in two of France's established chemical works: at Javel and Montpellier. The Javel manufacturers had been reported to be making soda already in 1787, when their product was said to be cheaper than the imported article. (In the same year soda manufacture was one of the concerns of Vallet and Bourboulon, after leaving Javel for England.) Before the Revolution, however, the manufacture might not have been regularly conducted, for soda does not figure among the firm's products as listed in an advertising prospectus of January 1789. The manufacture was at all events pursued in only a small way, using by-product sodium sulphate from the production of hydrochloric acid and *eau de Javel*. The process employed was essentially the same as that of Malherbe, involving fusion of the sulphate with charcoal and iron. In 1790–1 the firm complained that the elevated price of saltpetre—and consequently of sulphuric acid—prevented their giving the manufacture a greater extension.[40]

In Montpellier, Chaptal made soda from about 1789, using the lead oxide process to which we have previously referred. Later, in his applied chemistry text of 1807, Chaptal claimed to have conducted manufacture by this method 'très en grand'. The method was based on the reaction between litharge and brine, first observed in about 1770 by Scheele and published in 1775 by Bergman. As Chaptal was aware, the reaction was commercially exploited in England by James Turner, who besides soda obtained by it a yellow pigment; Turner had secured an English patent for the colour in 1781 and it had soon become well known as Turner's Yellow. The method of working followed by Chaptal was to make a paste of litharge and brine in large stoneware vessels, and then leave these to stand, with an occasional stir and further addition of brine; the paste stiffened and swelled, and after 24–48 hours caustic soda could be extracted by lixiviation with boiling water. A limitation on the method was the expense of the litharge employed, which made its economic viability dependent on profitable utilization of the oxychloride of lead formed as the second product. Chaptal indicated three uses for this, any or all of which he might have exploited: the production of a yellow pigment (Turner's Yellow) by calcination; the production of a white pigment by reaction with sulphuric acid; and reduction to metallic lead.[41]

The early Revolutionary period also saw several new projects with soda manufacture as their essential concern. Thus, in March 1790 partners called Bourgogne and Baudoin, of Hyères in Provence, requested an exclusive privilege for soda manufacture by a process they had developed, but their request was rejected without further examination and no more is heard of them.[42] More interestingly, in January 1791 the chemist Hassenfratz declared that he had found a large number of means of obtaining soda from salt and asked for a patent under the new legislation; he announced that he was engaged in establishing a works and that he had just conducted large-scale experiments in the department of the Allier. In 1792 he published papers on the decomposition of salt by lime, iron, and metallic oxides, but we have no further information on his venture.[43] There were two schemes, however, which are known to have resulted in the creation of works: a new venture by Carny, and, of course, the historic enterprise of Nicolas Leblanc.

Carny and Payen, 1791

Although Carny's soda project with his 1789 partners fell through, he himself in 1790 did set up a works in Paris, in the rue de Harlay, for the manufacture of hydrochloric acid. He claimed this to be the first works formed in France to manufacture hydrochloric acid on a truly large scale, though it seems in fact to have been only a fairly modest enterprise, with a production given in 1796 as 25–30 000 pounds a year. The acid was sold to bleachers, textile printers, and dyers. Carny obtained sodium sulphate by-product in some quantity, of course, but he did not use this for soda manufacture, instead simply selling it to hospitals as 'Epsom salt'.[44] Carny's desire to establish a soda works was realized in 1791, when he found a backer in J. B. P. Payen (father of Anselme Payen, the prominent nineteenth-century chemist and industrialist). Payen had studied at the *Collège de Navarre* in Paris, and had there shown a taste for the sciences, but his family had intended him for a legal career and on the completion of his education had endowed him with an office as deputy attorney in Paris. This having soon proved little to his liking, he in 1791 sold his post and on 6 July bought a property at Grenelle—on the banks of the Seine about half a mile north of the Javel works—where Carny then proceeded to install a works for the manufacture of sodium sulphate and soda.[45]

Carny patented his processes in January 1792, announcing that he intended to begin production very soon.[46] The works was then visited on 9 February by Berthollet, Parmentier, and de Servières (*Directeur des brevets d'invention*), to fulfil the formality of verifying the accuracy of the patent description. Carny demonstrated two processes to the visiting commissioners. The first began by converting salt to sulphate, with the novelty that instead of sulphuric acid, iron sulphate was employed: the salt was boiled with a solution of gently-roasted copperas (ferric sulphate), after which sodium sulphate was obtained by crystallization. The sulphate was then reduced to sulphide, and the final conversion to soda was accomplished by one of the lead oxide variants he had described in 1789, the sulphide being boiled with a solution of red oxide of lead in caustic soda, whereupon lead sulphide precipitated leaving an enriched caustic soda solution. This was then to be evaporated to dryness in iron boilers. Carny's patent description mentioned no use for the by-product lead sulphide but in 1789 he had claimed a method of treating it so as to obtain flowers of sulphur and white oxide of lead, which could then be reused in the process. Carny's second method was a new variant on the tartrate processes he had described in 1789. Salt was added to a solution of tartrate of potash (made by reacting the acid tartrate—cream of tartar—with potash), after which potassium chloride was separated by crystallization leaving a solution of the mixed tartrate of potash and soda. Addition of pyroligneous acid then precipitated acid tartrate of potash (to be reused), and produced pyrolignate of soda, from which soda was obtained by calcination in a reverberatory furnace. The potassium chloride by-product was converted to potassium sulphate, for which there was a market, by treatment with sulphate of iron. This second method, of course, was in effect just a roundabout way of accomplishing the double decomposition of salt and potash, and generated no new alkali but merely exchanged one species for the other. Carny's patent description also included a third process, which he did not demonstrate to the visiting commissioners and so perhaps was not intending to use. In this, sodium sulphide, made as in the first process, was converted to pyrolignate by treatment with pyrolignate of iron (made from the iron chloride by-product); the iron sulphide formed as second product was to be reconverted to iron sulphate by exposure to the air.

The venture appears to have enjoyed little success. Carny later, in 1797, described it as having been obstructed by the vicissitudes of price caused by the paper money (i.e. the depreciation of the *assignats*).[47] In 1794 Payen seems to have been making sodium sulphate at the works, but not soda.[48] After the Revolution he became important as a manufacturer of sal ammoniac, producing soda as a by-product, no longer by Carny's methods but by the Leblanc process which by then had become public knowledge.

It is to the historic enterprise of Leblanc himself that we must now turn.

The discovery of the Leblanc process

Leblanc,[49] like so many others, came to chemistry from a medical background. He was born in 1742, the son of a minor ironworks official in central France, but his father died when he was nine and his education was then cared for by a family friend who was a surgeon in Bourges. From about 1760 Leblanc himself studied surgery in Paris, and he then practised in the capital with sufficient success to have become by about 1780 a surgeon in the household of the Duke of Orleans.[50] It seems to have been only in the ensuing decade, when he was already entering his forties, that he developed a serious interest in chemistry. He then attended the lectures given at the *Collège de France* by Jean Darcet, who advised and encouraged him in his chemical researches, and to whom he wrote in 1788: 'I shall always be proud to own that it is to you that I owe the greater part of the knowledge I have acquired in chemistry'.[51] In the years 1786–8 he began to make a minor mark in the scientific world with some original researches on the crystallization of salts. These were presented to the Academy in a series of four papers which drew favourable reports from Darcet, Berthollet, and Haüy, and three of which were then published in the *Journal de physique*. The results of these and later investigations were in 1802 gathered together in his small book *De la crystallotechnie*. His crystallization studies constitute his only appreciable scientific work, but they are sufficient to secure for him a place as one of the lesser founders of the science of crystallography. At the same time as Haüy was developing his geometrical theory of crystal structure, Leblanc is remembered for having been the first to undertake the serious experimental study of the factors affecting the crystallization process, with the aim of explaining the variations in form which the crystals of a given chemical species could show, and of acquiring a practical mastery over crystal growth.

In general it seems fair to characterize him as an ordinarily competent chemist of the second or third rank, more dogged than distinguished.

Leblanc tells us that his interest in the soda question was first aroused in 1784, at the time of the Academy's soda prize, and that after experimenting with several known reactions and finding them to be unsatisfactory, he in due course discovered his own process in 1789.[52] The method he developed was a two-stage process. It began by reacting the salt with sulphuric acid to form sodium sulphate, and it then proceeded—in the second and essential stage—to convert the sulphate to crude soda by fusing it with charcoal and chalk. The crude product so obtained contained sodium carbonate which could then be extracted by leaching with water. The basic reactions are:

$$2NaCl + H_2SO_4 = Na_2SO_4 + 2HCl$$
$$Na_2SO_4 + 2C + CaCO_3 = Na_2CO_3 + CaS + 2CO_2$$

This is closely similar to the method of Malherbe, of course, save that where Malherbe used iron, Leblanc used chalk. It does not appear, however, that Leblanc devised his own process as a variant on that of Malherbe, with which he was probably unacquainted (the method was not yet commonly known, finding no mention, for example, among the half dozen alkalization reactions enumerated by Chaptal in his *Élémens de chimie* of 1790). The inspiration for Leblanc's work, as he himself explains, came rather from a suggestion published in the *Journal de physique*, by the editor Delamétherie. The suggestion came in the January 1789 issue, in the course of a discussion of British industry based on a visit Delamétherie had made to Britain the previous year. After mentioning that 'Milord Dundonas' (Dundonald) had succeeded in the large-scale production of soda from salt, Delamétherie went on to speak of the importance of finding means for such a manufacture. He referred briefly to the known decomposition of salt by lime and by lead oxide, and then he proposed the following method:

There is a manner of accomplishing this decomposition which would be very sure, but it would perhaps be too expensive. This would be, in suitable apparatus, to pour vitriolic acid on to marine salt; the marine acid would be disengaged and would pass into flasks, and the residue would be vitriol of natron or Glauber's salt. One would then decompose the vitriol of natron by calcining it with charcoal. The vitriolic acid would be disengaged in the form of sulphurous acid [sulphur dioxide], and the natron would be left pure.[53]

This was apparently just a speculative suggestion, perhaps deriving from something he had heard in England, or perhaps based simply on the known evolution of sulphur dioxide when free sulphuric acid reacted with charcoal. As an afterthought Delamétherie added that there might possibly be some hepar (sulphide) formed as well, but he considered that this would be incidental and proposed its removal by conversion to acetate followed by calcination.

We may note that Delamétherie was not alone in expecting that soda might be made by simply reacting sodium sulphate with charcoal. Such a method was indicated in 1783, for example, by one of the Academy's prize contestants, who had obtained some soda in this way in small-scale experiments (see n. 18). And the Italian chemist Lorgna, in a paper in the *Journal de physique* in 1786, described an experiment in which by fusing sodium sulphate on a charcoal block he claimed to have demonstrated its conversion to soda. On his understanding (and probably on the understanding of the Academy's prize entrant), the sulphate was reduced to sulphide and the sulphur of the sulphide then simply burned off. In the early 1790s two instances are known in which this kind of process was operated on a commercial scale. A certain Bernard, the founder of a chlorine-bleaching works near Lille, claimed in 1795 that from his sodium sulphate residues he had there 'constantly . . . extracted the soda by carbon', until the activity of his works was interrupted by the war.[54] And Ribaucourt in 1794 said that at his own bleaching works near Paris he had for nearly two years made soda by a method of this kind, and gave details of how the operation should be conducted. He had found the control of the process to be difficult, however, and this had eventually led him to make an addition of iron, as an agent suitable for separating the sulphur.[55] The fact of the matter is that the fusion of sodium sulphate with charcoal alone is not a viable manufacturing method. Although appreciable conversion of sulphate to carbonate can indeed occur—perhaps up to some 30 per cent, depending on conditions[56]—the major product is sodium sulphide, possibly accompanied by sulphite. The apparent success of Ribaucourt probably owed less to the real production of sodium carbonate than to the extensive hydrolysis which the more largely produced sulphide would undergo in water

$$Na_2S + H_2O \rightleftharpoons NaOH + NaSH$$

yielding an alkaline solution which he would no doubt have found serviceable as a lye.

It was Leblanc's addition of chalk which turned the unviable process proposed by Delamétherie into an invaluable manufacturing method, and which essentially constitutes Leblanc's discovery. Regrettably, Leblanc fails to tell us how he came to make this all-important addition, recording only that

Citizen Lamétherie inserted, in about the year 1785 [*sic*—1789], I believe, in the Journal de physique, some observations on the decomposition of sulphate of soda by incineration with charcoal: he did not doubt that one day new experiments would find the means of completely decomposing this sulphate, called Glauber's salt. I attached myself to this idea, and the addition of carbonate of lime perfectly accomplished my object. I informed Lamétherie; it was to his observations that I owed this first success, since they furnished the occasion for my latest work.[57]

To this account Delamétherie adds the subsidiary detail that besides reading the *Journal de physique* article Leblanc had also conferred with him personally on the matter.[58]

If Leblanc himself is singularly uninformative, we do have a detailed and graphic account of the circumstances of the crucial discovery from the pen of M. J. J. Dizé. Dizé told his remarkable story only after Leblanc's death, however, and its truth has always been a matter of dispute, which it remains impossible to resolve. At the time of the discovery Dizé was Darcet's twenty-five-year-old assistant, in charge of the laboratory at the *Collège de France*. He had known Leblanc for several years through Leblanc's attendance at Darcet's lectures, and indeed he tells us that he had been approached by Leblanc in 1787 with a view to their undertaking joint researches on the decomposition of salt, although in fact nothing had come of that suggestion. In 1790 he became Leblanc's industrial partner. Dizé published his story in 1810, in a letter to Delamétherie, prompted by the latter's recent reprinting in the *Journal de physique* of Leblanc's own brief remarks on the discovery. This seems to have been Dizé's only publication on the subject. We can supplement it, however, by an account penned by Boudet, who retold the story forty-two years later in the year of Dizé's death, drawing on oral testimony from Dizé and on his manuscripts, and introducing one or two detailed amplifications.[59]

According to Dizé's story, when Leblanc (in 1789) had found himself in possession of a process for the manufacture of soda, he had approached his patron, the Duke of Orleans, for financial backing to exploit it. As a first response to Leblanc's approach the Duke had then sought to have the process verified by Darcet, with whom he was well acquainted and from whom he had himself been receiving private chemical instruction. This much of the story, at least, seems fairly certain, since Leblanc himself speaks of his process as having been 'verified under the eyes of Mr d'Arcet . . . in the year seventeen hundred and eighty-nine by several experiments repeated in the chemistry laboratory at the Collège royal [the Collège de France]'.[60] Dizé continues by relating that Darcet had been too busy immediately to undertake the work himself, and so the task had been delegated to him. He and Leblanc accordingly set about trials in Darcet's private laboratory on the quai Voltaire. The process Leblanc brought forward simply involved fusion of the sulphate with charcoal, as indicated by Delamétherie, and when the two carried out their first tests they found, of course, that the product was not soda but largely sulphide mixed with some sulphite. In his own experiments Leblanc had apparently mistaken the sulphide for soda. Leblanc nevertheless insisted that he had obtained soda in this way, and maintained that success depended on achieving the right conditions. The two therefore continued the experiments for a fortnight, hoping to find conditions which would give them soda, but without success. In response to Leblanc's pleas, Darcet then persuaded Orleans to finance a series of further experiments—now at the

Collège de France—according to a plan they had drawn up, and in the course of these new experiments the idea came to them that it might be possible to decompose the sulphide by means of carbonic acid (carbon dioxide). The origin of this idea is not very clearly explained, but it seems to have arisen from the known decomposition of the sulphide by other acids, a property of which they made use in their analytical tests. By treating the sulphide in the wet way with a stream of carbonic acid gas they found that they could indeed obtain soda, a result which they found very striking and which afforded them their first ray of light. (Surprisingly, this simple means of decomposing the sulphide does not seem to have occurred to earlier workers.) The use of carbonic acid gas appeared impracticable on a manufacturing scale[61] and so they next tried to achieve the same result by reacting the sulphide directly with carbonate of lime (this being the source from which they had previously obtained their carbonic acid). All attempts at reaction in the wet way failed, but eventually they achieved success in the dry way. By heating to dryness and then fusing a strong solution of the sulphide mixed with charcoal and carbonate of lime, they found the sulphide content to be much reduced. And by then repeating the experiment with more complete fusion they obtained a product from which they were at last able to procure crystals of carbonate of soda. Having thus succeeded in producing soda from the sulphide they went on to attempt the same process starting with the sulphate: the sulphate was fused with charcoal and carbonate of lime, and of course the outcome was successful. After experiments lasting about three months the Leblanc process had been discovered.

Dizé's story has found some partisan acceptance among pharmacists,[62] who rise to him as one of their own, but more generally it has been rejected. Dizé was peremptorily dismissed, for example, in 1819, by the jury of the industrial exhibition in Paris, when he presented his case in the hope of an award.[63] More notably, his story was rejected in 1856, after a close investigation of the documentary evidence, by a commission of the Academy of Sciences headed by Dumas. The verdict was not unanimous, however, for Chevreul dissented from the majority view; he promised to discuss Dizé's case in his history of chemistry, but this work never appeared.[64] More recently, Dizé's story has been rejected by Gillispie, who has offered a conjecture of his own as to the manner in which Leblanc might have made his discovery. This pictures Leblanc as finding the inspiration for his use of chalk (or limestone) in the same article by Delamétherie as had provided the starting point for his researches:

La Métherie never mentions limestone in connection with soda. But he does do so in the preceding section of his essay, and necessarily so, for this section is devoted to the iron industry. . . . From iron La Métherie turns to the manufacture of soda. And it seems very probable that this juxtaposition is precisely what led to the Leblanc invention. If charcoal and limestone in a furnace separate iron from its ore, might they not somehow have a similar effect on Glauber's salt and liberate the soda?[65]

Gillispie's hypothesis has gained a certain currency but we do not find it persuasive. The juxtaposition in question is neither striking nor suggestive: Delamétherie says very little of iron manufacture, with no discussion of its processes and specifically no mention of limestone (*pace* Gillispie). There is in any case no true analogy between the extraction of iron from its ore and the 'extraction' of soda from its sulphate, and we can see no good reason why Leblanc should have imagined one. And if Leblanc *had* found this further inspiration in Delamétherie's piece, would it not have been natural for him to say so? We find Dizé's story more plausible.

Dizé's story does have its unsatisfactory aspects. His circumstantial and at times highly coloured personal narrative—from which we have here extracted only the bare outline—can be checked in a number of its details against other evidence and shown to be wrong or highly questionable. There are chronological discrepancies, for example, of which the Dumas commission made much. In particular the commission seized upon Dizé's assertion in 1810 that his researches with Leblanc had begun in August 1790, which is patently false. Objections on such grounds are essentially trivial, however, since inaccuracies of this kind are understandable in the recall of events that occurred twenty years before.[66] The one serious objection to arise stems from the legal documents which formed the basis of Dizé's partnership with Leblanc, and which were first unearthed by the Dumas commission. For it emerges that in these Dizé himself recognized the process as belonging to Leblanc. This powerful point is reinforced by the fact that Darcet, too, whom Dizé described as having been closely in touch with their researches, on several occasions recognized the process as Leblanc's. It seems clear that whatever Dizé may have said in 1810, twenty years earlier he had not considered himself to have had any major share in the discovery. It nevertheless remains possible that Dizé did assist Leblanc in his crucial researches, albeit in a more subordinate role than he later remembered or chose to describe. In the absence of any decisive evidence we are inclined to believe that there is some kernel of truth in his story, and to accept the technical details he gives. Dizé's account seems to savour more of imperfect and distorted remembrance than of total fabrication. One would have expected fiction to be more transparent than the curiously cryptic technical explanations he presented in 1810, and his erratic chronology would similarly seem more indicative of casual reminiscence than of artful skullduggery. Finally, there is some very slight technical evidence which might be cited in his support. Thus, his understanding in 1794 of the mechanism of the Leblanc process—as involving displacement of the sulphur from the sulphide by carbonic acid from the chalk (see p. 282)—can be seen to be nicely consistent with his later account of the process's origins. And his remarks in a paper of 1803 on the difficulties the experimenter encounters in attempting to produce soda by the agency of charcoal alone would seem to suggest personal experience of just such endeavours. The trouble is, Dizé explains, that sulphide is formed, which is less easy to destroy than one might think, particularly when dealing with large quantities:

The oft repeated exposure of alkaline sulphides to a moderate heat does not suffice to volatilise the sulphur; the alkali retains enough of it to render it improper for certain operations in the arts: moreover, the sulphur, in volatilising, burns in part and forms sulphurous acid which combines with the alkali. The operation, instead of becoming simpler, becomes more complicated since one obtains an alkali mixed with alkaline sulphide and sulphite.

The addition of carbonate of lime . . . as was practised in the first soda factory, established at Saint-Denis near Paris, was an inexpensive means of facilitating fairly promptly the decomposition of the sulphide of soda which was the result of the sulphate of soda decomposed by the charcoal.[67]

Leblanc's venture at Saint-Denis, 1790–1793

At the beginning of 1790 Leblanc and Dizé were called to a meeting with the Duke of Orleans in London, where the Duke was then living after having left France the previous October as a result of his involvement in the early events of the Revolution. In London, on 12 February, a preliminary agreement was drawn up before the notary James Sutherland, by which the Duke undertook to supply capital of 200 000 *livres* for the establishment of a works.[68] Besides making soda, by the process contributed by Leblanc, the works was also to produce sal ammoniac (as a by-product), and in addition 'white lead' by a process contributed by Dizé. This latter product was not in fact true white lead (basic lead carbonate) but lead sulphate, to be made by dissolving lead in nitric acid and then adding sulphuric acid to give the sulphate by precipitation. The intention to produce this 'white lead' is a little surprising since it has no natural connection with the manufacture of soda, and it seems likely that Dizé was included in the scheme not only for his 'white lead' process (which appears never to have actually been exploited) but also for his general technical skill. He himself later wrote that 'Le Blanc having no knowledge of speculative calculation, in no way accustomed to manufacturing operations, and naturally not very active, had me proposed by M. d'Arcet to enter into the projected partnership.'[69] Whatever the truth of this, we have other testimony that Dizé was indeed primarily responsible for the construction of the works (see p. 233). Also included in the partnership was Henri Shée, who was to be the Duke's representative at the works, in charge of its finances. Shée is described as having been colonel of the Duke's cavalry regiment and for some time prior to the Revolution his military secretary; he retired from the army at about this time (in 1791 it is said) for reasons of health.[70] In accordance with the terms of the agreement, when Leblanc and Dizé returned to France they deposited with the Paris notary Brichard, on 27 March, a sealed packet containing descriptions of their processes, certified correct by Darcet.[71]

On 9 August 1790 Shée bought a site at a spot called la Maison-de-Seine, just outside Saint-Denis, about six miles north of Paris,[72] and the erection of a works then seems fairly soon to have begun. Six months later, on 27 January 1791, the definitive act of partnership[73] was drawn up in Paris (the Duke had returned the previous July). The terms were generally similar to those of the preliminary agreement. Orleans provided 200 000 *livres*, which he was to be repaid, with interest at 10 per cent, from the first profits. Subsequent profits were then to be shared in the proportion nine-twentieths to Orleans, two-twentieths to Shée, and nine-twentieths to Leblanc and Dizé (to be divided between them in a manner specified in a separate agreement of 15 January).[74] Leblanc and Dizé were assured salaries of 4 000 *livres* and 2 000 *livres* respectively, from 12 February 1790, until their share in the profits reached these sums. The optimism of the associates is shown by the inclusion of an article to cover the eventuality of the annual profits' exceeding a million *livres*: in that case profits in excess of a million were to be shared slightly differently. Other articles mentioned the possibility of extending the Saint-Denis works in due course, and of establishing further works elsewhere. Leblanc, indeed, later described the works as having been established 'comme essai en grand', and Dizé wrote that it had been intended only as a pilot plant, the Duke's principal intention having been to set up a much larger factory at salt-works he owned near Marseilles, once the process had been successfully developed on an industrial scale.[75]

At the beginning of May 1791 Leblanc submitted a request to the Constituent Assembly's *Comité d'agriculture et de commerce* for a patent under the country's new patent system.[76] The system required a patentee to deposit a description of his invention at the patent office in Paris, where it would normally be made available for public consultation. There was provision, however, for the description to be kept secret in cases where political or commercial considerations warranted this, and Leblanc requested a secret patent in accordance with this provision, apparently being the first to do so. Such a declaration of secrecy required a special decree of the Assembly, and it seemed most unlikely that the Assembly would have time to consider his case in the very near future. Since Leblanc was anxious to secure his patent, the *Comité d'agriculture* therefore resolved on 2 September that his process should be declared secret provisionally, until the matter could be brought to the Assembly for a definitive ruling. Leblanc then submitted his description. In order for the patent to be declared secret, the law also required the accuracy of the description to be verified by commissioners, and so on 23 September a visit was made to the works by the Director of Patents, de Servières, in company with Darcet and Nicolas Desmarest (*Directeur des manufactures*). The commissioners were shown round by Leblanc and then filed a report which not only verified the description but also commented very favourably on the venture itself:

After having scrupulously examined the method employed by the said Sr Le Blanc . . . we have recognised that the invention was different and very superior to all that had hitherto come to our knowledge, as much for the economy, celerity, and sureness of the processes, as for the richness and purity of the results;...considering the inestimable advantages which should result from it for the supply of our soap-works, glass-works, and many other manufactures and arts, which hitherto have been forced to draw their soda at great cost from abroad . . . we consider that the discovery of Sr Le Blanc, on all political and commercial grounds, deserves the encouragement of the French nation, and that the secret of his discovery should be carefully guarded.

Leblanc received his patent, for fifteen years, on 25 September 1791.[77]

From Leblanc's patent description we can judge the progress that had now been made in adapting the process to an industrial scale. The method of working there described began by reacting salt with sulphuric acid in a large leaden pan, ten feet long and four feet wide, fitted with a cover from which a pipe carried off the hydrochloric acid gas to be dissolved in water in stoneware jars. It was not possible to bring the reaction to completion in this pan, since the lead would not withstand the necessary temperature, and so the material from the pan (containing the half-way product sodium hydrogen sulphate, $NaHSO_4$) was then transferred to a reverberatory furnace and heated more strongly to complete the conversion to sulphate.[78] The sulphate was then mixed with chalk and charcoal and fused—again on the hearth of a reverberatory furnace—to yield crude soda. This, the essential stage of the process, had now very nearly reached its final form. The substitution of a reverberatory furnace for the crucibles described in March 1790 meant that the reaction could now be carried out on a large scale, working several hundred pounds of materials at a time on a 6 ft. by 4 ft. hearth. The reverberatory furnace also had the important further advantage of allowing the progress of the reaction to be closely followed in a manner not possible when working with covered crucibles. The change from crucible to furnace operation had necessitated an amendment in the proportions of materials used,[79] and these had now been more or less optimized for furnace conditions at 100 sulphate: 100 chalk: 50 charcoal (as compared with the proportions of 100:50:25 earlier established for crucibles). Figures roughly similar to those determined by Leblanc were to remain in general use throughout the history of the industry. The crude soda obtained was said by Leblanc to furnish over 75 per cent of its weight of a 'soda of excellent quality' on leaching with water (equivalent to an Na_2CO_3 content of a respectable 28 per cent if we assume the 'soda of excellent quality' to have been $Na_2CO_3 . 10H_2O$). Ironically, it was not the soda-making stage itself but the preliminary treatment of salt with sulphuric acid which was proving the more problematical. The leaden pan being employed—a scaled-up version of standard laboratory apparatus—was still very unsatisfactory. Although it allowed a large quantity of salt to be treated at a time (200 pounds), its use was slow and troublesome, for it had to be assembled and all the joints luted at the beginning of each operation. Only one operation a day can have been possible. Moreover, the leaden cover and pipes were corroded by the hydrochloric acid gas, and the leaden pan

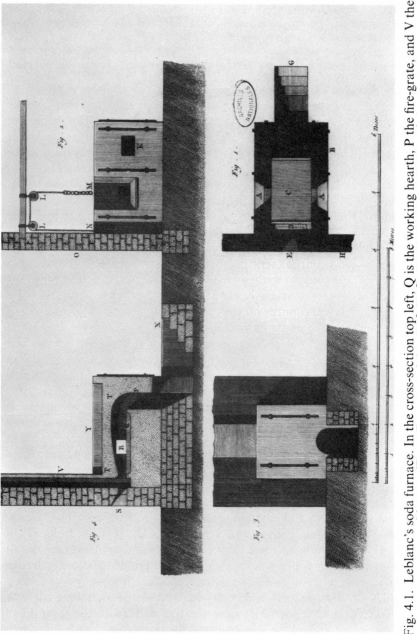

Fig. 4.1. Leblanc's soda furnace. In the cross-section top left, Q is the working hearth, P the fire-grate, and V the chimney. (From Darcet et al. (an III [1794]), Pl. 8.) (Archives Nationales)

itself, heated from below, was later said to have never survived four operations in succession without melting, despite being protected by iron plates from direct contact with the flame in an effort to prevent this. The difficulties were such that Leblanc was thinking of substituting stoneware or porcelain for lead, and of using smaller vessels. In short, by September 1791 the partners were able to carry out large-scale production after a fashion, but chiefly because of problems in producing the sulphate they were not yet in a position to start regular manufacture. Methods were still in the course of development and Leblanc remarked: 'There exists a multitude of means of improvement, on which I am daily making researches'.

Much further work, indeed, still lay ahead, and it seems to have been only in 1793 that the plant and methods acquired a satisfactory appearance. Shée wrote early in 1794:

It is only after more than three years of experiments of all kinds, of trial and error, of various constructions of furnaces, that experience has resulted in adherence to that which exists at present in the works at Franciade [the Revolutionary name for Saint-Denis]; not, assuredly, that I wish to claim that one could not do better; I believe, on the contrary, that the works is still susceptible of great improvements, and they are much to be desired, to avoid especially the frequent degradation of the furnaces; a better choice of bricks, of clay to join them, a surer touch on the part of the lime-burner, might render very useful economies in fuel, time and repairs; but it will be prudent to make no innovation without much circumspection.[80]

The works which had now arisen at Saint-Denis was situated beside the Seine, on a $2\frac{1}{2}$-acre site surrounded by walls, with its entrance opening on to the towpath.[81] About an eighth of the site was covered by buildings. There was a main building measuring 20 metres by 15, housing the furnaces (see Figs 4.2 and 4.3); and in addition there were living quarters, a number of large sheds, and a horse-driven mill for grinding the sulphate. A windmill nearby was rented for grinding the salt. Construction costs were said by Leblanc and Dizé to have amounted to 100 000 *livres*. The works was still not complete. There remained plans to mount a workshop for the purification of the soda, so that the firm could sell its product not only in the crude state but also in a refined form, either as soda crystals or (after calcination) as dry soda ash. There were also plans to erect a lead chamber to make sulphuric acid, for by making the acid themselves instead of buying from Javel the partners reckoned that they could save 25 per cent on this important item, which they required in large quantity and which formed by far the largest element in their manufacturing costs.[82] The essential part of the works, however, for the manufacture of soda itself, was now capable of regular industrial production.

The preliminary treatment of the salt with sulphuric acid, formerly so troublesome, was now also carried out in a reverberatory furnace, suitably adapted by lining the hearth with lead to form a shallow tray.[83] There were two such lead-lined furnaces, operated alternately so as to achieve a measure

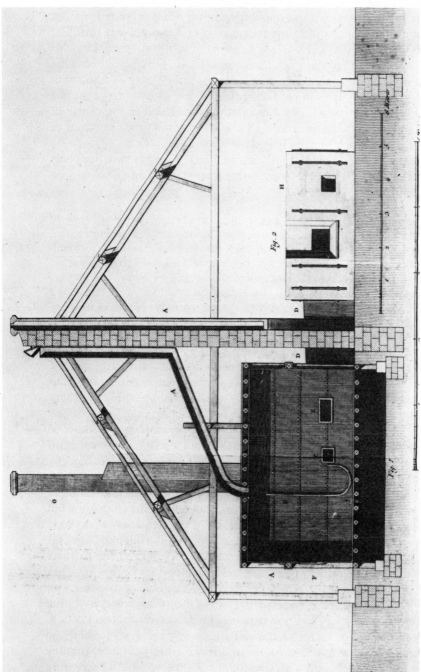

Fig. 4.2. Leblanc's plant for production of the sulphate. On the right, H is one of the two reverberatory furnaces in which the salt was reacted with sulphuric acid. The furnace gases, containing hydrochloric acid, could either be released through the chimney A, or passed via the conduit D into the lead chamber B, entering at E. (The lead chamber is shown in partial cross-section, to reveal the interior.) In the chamber the hydrochloric acid could either be simply condensed or combined with ammoniacal vapours (entering at C) to yield sal ammoniac. To promote condensation, steam was

of continuity. As in 1791, the half-way product obtained in the first instance was then heated more strongly in an ordinary brick-paved furnace to complete its conversion to sulphate. The substitution of lead-lined reverberatory furnaces for the covered leaden pan of 1791 was a great improvement, and such furnaces were to be adopted as standard when the industry became generally established. They avoided the need to assemble and then dismantle the apparatus at each operation, and allowed the same charge (200 pounds of salt per furnace) to be worked in a much shorter time: each of the two furnaces could work twice a day, or even four times if work were to continue through the night. The danger of melting the lead was also now averted, since heating was from above rather than from below. At the same time such furnaces did also present new disadvantages, though, and if they appear an obvious solution to the problem Leblanc and Dizé had faced, it was a solution which they probably adopted only with some reluctance. The reaction was no longer isolated from the fire of the furnace, and so the hydrochloric acid by-product now emerged mixed with fire gases, which would add considerably to the difficulty of its collection. Manufacturers were later to find this a great embarrassment. In Leblanc and Dizé's own plant the acid could either be wasted, by simply releasing the fumes through a chimney, or it could be collected by passing the fumes into a leaden chamber, measuring 6 metres long, 3 metres high, and 3 metres wide. In the chamber the acid was condensed, for sale to bleachers and others, or alternatively it was combined with ammonia gas to form sal ammoniac (a more profitable outlet). Condensation was promoted by blowing in steam from a boiler, and the ammonia was obtained by distilling animal materials in iron cylinders. One rather doubts the efficiency of these methods on any very large scale, and they were not to be imitated by the later industry.

The final stage of the process, the conversion of the sulphate to soda, seems to have undergone little change since 1791. The sulphate was mixed with chalk and charcoal, now in the proportions 100:100:55, and the mixture was fused in a fourth reverberatory furnace, whose hearth, as in 1791, measured about 6 ft. by 4 ft. It took a charge of 400 pounds. This stage called for some skill and care, the phenomena occurring in the furnace requiring close observation, since on this depended proper regulation of the fire and recognition of the point at which the material must be withdrawn. The mixture formed a pasty melt, which was to be vigorously stirred. It bubbled and its surface became covered with small candle-like flames. Then when the bubbling and jets of flame died down and the paste became more fluid, and when a sample withdrawn for inspection presented a suitably uniform texture, the operation was judged complete and the material was raked out, falling to the ground as a soft paste which hardened on cooling. The crude soda so obtained amounted to about 55–58 per cent of the weight of the mixture charged. It was broken into blocks and transferred to a store, where on standing it gradually crumbled to a powder as it took up carbon dioxide

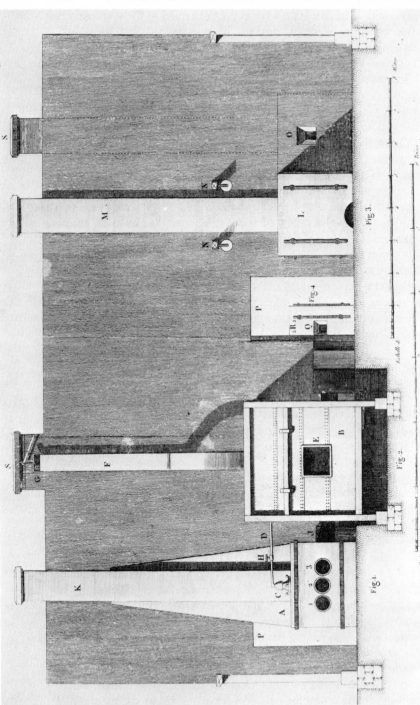

Fig. 4.3.   A general view of Leblanc's workshop. Through the doorway P in the centre can be seen one of the two furnaces (R) in which the salt was reacted with sulphuric acid. B is the lead chamber employed for collection of the hydrochloric acid by-product. On the left, A is the furnace in which ammoniacal vapours were generated by distilling animal materials in iron cylinders (1, 2, 3); the vapours passed into the chamber through the conduit J. C is an aeolipyle from which steam passed into the chamber through the pipe D. On the right, O is the opening of a furnace on the far side of the wall in which the product of the furnaces R was calcined to complete its conversion to sodium sulphate. The calcined sulphate was then

and water from the air, increasing in weight by some 15 per cent. Leblanc tells us that six charges could be worked in 24 hours,[84] and so from the above data the production capacity of the works can be estimated to have been in the region of 2–400 000 pounds of crude soda a year (weighed straight from the furnace), assuming a working day of 12–24 hours and 300 working days a year. The capacity of 6–700 000 pounds which Leblanc and his associates themselves claimed on several occasions[85] seems rather exaggerated, though they were no doubt referring to the soda in its market state, when it would have gained somewhat in storage.

By the summer of 1793 the works had produced a total of 30–35 000 pounds of soda in three or four trial runs, and 15–20 000 pounds of this had been sent out to Marseilles, Saint-Gobain, and London, and also to local wash-houses, to be tried in use.[86] As was first made clear by Gillispie, however, the traditional portrayal of the works as having become a flourishing industrial concern is entirely mythical. The fact is that it ceased operations without ever having begun regular production. The reasons for its stoppage are not very definitely known, but several likely factors can be suggested. Certainly false is the old story that the venture was struck down in full flight by the execution of Orleans and the factory's subsequent sequestration, for operations had in reality already been abandoned before these events.

Leblanc himself blamed the interruption of his venture on the Government's wartime emergency measures for the supply of raw materials to gunpowder manufacture. We have seen in Chapter 2 how a decree of 21 September 1793 declared all the sulphur in France to be henceforth at the disposal of the executive, and ended the sale of saltpetre to private industry by the *Régie des poudres*. These measures, we saw, led in 1794 to a virtual paralysis of sulphuric acid manufacture. It was presumably to the decree of 21 September that Leblanc referred when on 5 November he informed the Minister of the Interior that since sulphur and saltpetre were now requisitioned for the needs of war, it had become impossible for his works to continue; the sulphur and saltpetre which his works had possessed had been delivered to the *Régie* as soon as the law had become known, and 100 000 pounds of charcoal had also been offered.[87] We should note that Leblanc would have had good reason for wishing to blame his closure on imperious circumstances of this kind: the purpose of his letter was to seek assurance from the Minister that the non-exploitation of his patent would not affect its validity, and he was anxious to establish that it was through no fault of his own. There is nevertheless probably some truth in what he said. The real or prospective lack of materials might well have contributed to a final decision to close down the enterprise pending the return of better times. It is doubtful, however, whether this provides a complete explanation for the stoppage, for in practice the works had already been more or less idle since at least the previous July. On 26 July there had been introduced a law against hoarding which required those holding materials of prime necessity (among them salt, charcoal, and soda) to declare their stocks.[88] The partners at Saint-Denis made their declaration on 31 July, and it is noteworthy that whereas the law required manufacturers only to justify the use of their materials, the partners volunteered to offer their stocks for sale, at whatever price the local authority might fix. The materials were not in fact sold, but their offer suggests that the enterprise was no longer in very active development. It is in any case quite clear from the figures Leblanc gives that there was no significant further production at the works after this. The materials in stock when operations ceased included enough sulphate and acid for the production of 22 000 pounds of soda, and enough salt, chalk, and charcoal for the production of over 100 000 pounds more (given the necessary acid). There also remained some 10 000–15 000 pounds of manufactured soda and a quantity of sal ammoniac.[89]

A different explanation for the interruption of the venture was given in 1794 in a report on soda manufacture in France by a chemical commission headed by Darcet. The Darcet commission ascribed it to lack of funds[90] and it seems likely that this was indeed a main factor frustrating continued development in the summer of 1793. When the works closed, the partners still had 28 000–30 000 *livres* of their 200 000 *livres* capital,[91] but of course the works was not yet complete. The plan to build a sulphuric acid chamber had got no further than marking out the intended site, and the Darcet commission explained that it had been suspended not only because the war made sulphur and saltpetre unavailable but also through lack of funds. Also still unrealized was the plan for a workshop to refine the crude soda. We may presume that with Orleans in prison since 7 April the further capital which such additions would probably have called for was not to be had. In connection with the firm's finances it is also worth noting that payment of the partners' salaries ceased in April,[92] presumably as a result of Orleans' imprisonment.

Finally, we might make the subsidiary observation that the response of users to Leblanc's trial product had not been altogether encouraging. Although Loysel found it to serve very well for glass manufacture in trials at Saint-Gobain, those trying it for other purposes complained of the sulphide it contained, and of the hepatic odour (hydrogen sulphide) which it consequently gave with water, as Shée remarked in 1794:

my veracity obliges me to say that the advantages presented by the manner of operating in the factory at Franciade are not without drawbacks for soda destined for the use of washerwomen and soap-makers; they both of them complain that the soda contains too large a quantity of hepar [sulphide], which is harmful to their operations . . . Skilful chemists would perhaps find an easy means of eliminating this fault; but it is nevertheless true that as long as it remains these two kinds of consumers will prefer Spanish sodas, although in general less rich in mineral alkali. . . . I know that by reducing the soda to dried crystals one can avoid the drawback of the hepar: it remains to be seen whether what one would gain by reducing it to a much smaller volume, the greater facility of transport, the certainty and invariability of the proportion of alkali, would overcome the prejudices of routine, and cover handling costs.[93]

Leblanc and Dizé themselves both later commented on the drawback presented by the presence of sulphide in their product.[94] The problem was due in fact to the presence of sodium sulphide which on leaching with water dissolved with the soda (the calcium sulphide which formed a large part of the crude product was not itself directly troublesome since it was insoluble). When well made, Leblanc soda should contain only traces (less than $\frac{1}{2}$ per cent) of sodium sulphide, and so it seems likely that Leblanc's own product was still somewhat imperfect. Even in well-made Leblanc soda, however, there is a development of sodium sulphide after manufacture if the product is stored too long before use, or if its lixiviation is inappropriately conducted (by digesting with hot water, for example), and so it is possible that some part of the early trouble lay in use rather than manufacture.

The war crisis of 1793–1794

The emergency measures introduced in 1793–4 to assure gunpowder to the armies included not only the requisitioning of sulphur and saltpetre but also a great expansion of saltpetre production, and this aspect now calls for our attention since indirectly it resulted in soda manufacture again becoming a matter of interest to the Government.

The saltpetre industry depended on the natural formation of nitrates, including potassium nitrate (saltpetre), by the decomposition of animal and vegetable wastes. Nitrates formed in this way spontaneously accumulated in the soil of stables and similar places, and grew as an efflorescence on the walls of houses from the decay of refuse in the streets. Production could also be deliberately arranged by building beds of organic waste, but 'artificial' saltpetre plantations of this kind never acquired any great significance in France and it was on the collection of the spontaneous accumulations that the industry essentially depended. Licensed saltpetre collectors, under the authority of the Régie des poudres, gathered stable soils, wall-scrapings, and other such materials, from which they then extracted the nitrates by leaching with water, adding potash to convert other nitrates present to saltpetre. The crude saltpetre so obtained was then sold to the Régie, in whose refineries and powder works it was purified and compounded into gunpowder. The activities of the salpêtriers, with their right to collect materials on private property, were understandably unpopular, and the Revolution brought a decline in collection which added to the danger felt when the war increased the demand.

From August 1793 the organization of saltpetre and other war supplies was taken increasingly in hand by the Committee of Public Safety,[95] and in particular by Prieur de la Côte-d'Or, a young military engineer who was recruited to the Committee on 14 August and was then on 23 September given special responsibility for arms and munitions. As head of the Committee's *Section des armes et poudres*, Prieur was for the following year to have the role, in effect, of a minister of munitions, directing the national effort for the arming of the forces through the height of the war crisis, and presiding over that mobilization of scientists, as technical advisers and organizers of production, which was to be one of the most celebrated features of the supply programme and a chief factor in its success. The Committee's first major action on saltpetre came on 28 August, five days after the introduction of conscription by the famous *levée en masse*. At the Committee's instigation a decree of the Convention then removed restrictions on the collection of saltpetre-bearing materials, and declared all such materials to be henceforth at the disposal of the executive; the *Régie*'s collectors were requisitioned to stay at their work, provision was made for their number to be increased, and the price paid for their product was almost doubled. Some improvement in production ensued, but it soon became clear that the *Régie* and its collectors alone were incapable of fully exploiting the nation's resources and so bolder measures followed on 4 December. Prieur then introduced a decree in the Convention launching a so-called 'Revolutionary' programme of popular saltpetre extraction. The monopoly of the professional collectors was suspended and to supplement their work the population at large was exhorted to apply itself to the task: citizens were enjoined to extract saltpetre themselves from any suitable materials on their properties, by processes explained in a simple set of instructions quickly distributed to local authorities. About a month later the country was divided into eight regions to each of which there was appointed an inspector—the chemists Chaptal, Vauquelin, and Descroizilles among them—to stimulate and guide the new movement. In the early months of 1794 'Revolutionary' workshops sprang up all over France, mostly communal undertakings organized by local authorities. By July they were reported to number over 6000. To organize the refining of the saltpetre from these workshops, and its subsequent employment for gunpowder manufacture, a new central administration was formed under the name *Administration révolutionnaire des poudres et salpêtres*. This originated on 26 January as a Parisian body, whose responsibilities were then extended on 27 February to the country as a whole. It was to work in parallel with the *Régie des poudres*, which continued independently to operate its own works, supplied with saltpetre by the professional collectors. Both the *Administration révolutionnaire* and the *Régie des poudres* themselves now came under the authority of another new body, the *Commission des armes et poudres*, established in February as a general munitions executive under the direct control of the Committee of Public Safety. The Commission, with its several subordinate organs, was soon employing in its offices hundreds of secretaries, clerks, and messengers, while Prieur's own offices at the *Section des armes et poudres* acquired a staff of over eighty. At the head of the *Administration révolutionnaire* there was placed the chemist Carny, who in the weeks prior to its creation had attracted Prieur's attention by his proposal of expeditious new processes for saltpetre refining and gunpowder production, which were now to be exploited. Carny was joined at the beginning of March by Descroizilles, and early in April (after some persuasion) by Chaptal. The *Administration révolutionnaire*'s most spectacular achievement was to be the creation of a vast refinery in the former abbey of Saint-Germain-des-Prés in Paris, together with an equally impressive powder works on the outskirts of the capital at Grenelle, both based on Carny's processes. The refinery began work in June, treating saltpetre drawn from a large part of France and attaining ten times the output of any previous refinery, equal to all the *Régie*'s refineries put together. The powder works was in operation by May, and according to Chaptal employed the enormous labour force of some 2500 workers. The results achieved by the war effort were truly remarkable. The 'Revolutionary' workshops inaugurated by the decree of 4 December 1793 were by the end of July 1794 reported to have already produced 8 million pounds of saltpetre (crude), while the *Régie*'s workers in the same eight-month period produced over 4 million pounds. These figures are to be compared with the *Régie*'s production in the later 1780s of rather under 4 million pounds (crude) a year.[96]

The production of saltpetre required potash, of course, and so the Committee of Public Safety had to accompany its saltpetre programme with measures to secure supplies of this necessary alkali, particularly since imports were now interrupted by the war.[97] It began by providing for the requisitioning of available supplies: the decree of 21 September 1793, whose effect on sulphur and saltpetre we have considered above, also declared that all ashes, *salin*, and potash were to be at the disposal of the executive council, for the use of the *Régie des poudres*, which was to distribute the alkali to saltpetre-collectors in controlled amount (a pound of potash for every three pounds of saltpetre made). As the requirement for potash alkali grew in the course of 1794, progressive efforts were also made to increase its production. When in January the eight inspectors were appointed to organize popular saltpetre extraction throughout the land, they were directed to concern themselves at the same time with the production of *salin* and potash. Soon afterwards, on 1 February, a 'coupe extraordinaire' of wood was ordered in all the forests of the Republic, for the production of ashes and charcoal. And on 18 April the Committee introduced its major measure in the form of a general appeal for the burning of all useless vegetation. The Committee had hitherto refrained from such a general invitation lest it should serve counter-revolutionaries as an excuse to vandalize forests and pastures, but with requirements growing and the danger of counter-revolutionary subversion being considered to have retreated, citizens were now enjoined to undertake the collection and burning of all useless vegetation in the same way as they had already taken up the extraction of saltpetre. A simple set of instructions was to be issued explaining how to extract *salin* from the ashes. The principal effort was concentrated in and around the department of the Indre-et-Loire, where saltpetre production was especially active. There Vauquelin, Nicolas and Trusson were appointed commissioners to organize the burning of vegetation and the distribution and utilization of the ashes, and the inhabitants seem to have applied themselves to the task with some enthusiasm. In August the *représentant* Nioche reported to the Convention how in one commune

As soon as the needs of the Republic for *salins* and potash were known, to procure the saltpetre which is to exterminate all the enemies of our liberty, men, women, children, old people, took themselves *en masse* to the forest, more than a thousand in number, and on the same day we established eighteen combustion furnaces. I cannot paint the joy, the gladness and the truly Republican ardour with which we worked at cutting and burning all the useless branches which in this region cover thousands of acres.[98]

A detailed instructional brochure by Vauquelin and Trusson was published in September at the order of the Committee, which had already earlier in the year published a reissue, with additions by Pertuis and Sage, of the pamphlet written by Lavoisier in 1779. According to Richard the decree of 18 April was widely executed and procured a great deal of alkali.

Another possible means of assisting potash supplies for saltpetre production, of course, was by replacing potash in its other uses with soda. Unfortunately, soda too was in short supply, for France was at war with Spain, and indeed the soda shortage was itself causing some concern by the end of 1793 (we find the *Commission des subsistances*, for example, in December deciding to inquire severally of Darcet, Berthollet, and Chaptal whether both potash and soda could be dispensed with in the making of soap).[99] In this situation Carny drew the attention of the Committee of Public Safety to the potential resources which lay in the manufacture of soda from salt. In a memoir of 27 November[100] he presented observations on means of procuring to the Republic a soda production equal to its needs, and offered to communicate his own processes. His remarks inspired the Committee to take action on the matter in the new year.

By an *arrêté* of 27 January[101] the Committee appointed a chemical commission including Darcet to test Carny's methods along with such others as might be received, and then to draw up a report for publication. The *arrêté* also ordered that all who had begun works, or who had obtained patents for soda manufacture, were to inform the Committee of the state of their ventures. This latter order obviously had the Saint-Denis works particularly in mind, and in fact the chemist Guyton, on behalf of the Committee, wrote that same day to Dizé asking him to call to discuss soda manufacture at the Committee's *Bureau des poudres et salpêtres*.[102] Guyton was related to Prieur and was one of the several prominent scientists Prieur gathered in his offices as advisers. By a second *arrêté* of 27 January[103] the Committee also took preparatory action with a view to securing supplies of iron sulphate, which had been indicated, no doubt by Carny, as a valuable agent for the manufacture of soda. A student of the *École des mines* called Blavier was directed to go at once to the department of the Gard in southern France, where he was to assess the resources in copperas (and also in sulphur) which might be afforded by the pyrites exploitations around Alais; he was to find out how the materials there could most profitably and promptly be utilized to supply soda manufacture, and advise on buildings and other facilities which it might be useful to requisition. Four days later the Committee published a general announcement of its intentions in the *Moniteur*. It was explained that in order to help potash supplies for saltpetre production it was necessary to develop the manufacture of soda from salt, and that this called for extraordinary measures:

It is well recognised that repeated speculations have hitherto not even succeeded in reducing the importation of foreign sodas; it is evident that the best devised enterprises, left to their own resources, would provide only too-distant hopes, and would be useless for the present need. A single course remains; it will overturn all obstacles, for it will carry the revolutionary imprint: let us bring together all the enlightenment furnished by theory and acquired by experiment (*l'expérience*), and soon, by putting together and combining all this knowledge, with the help of the circumstances and of the high value they put on this industrial product, we shall see created a new art, destined first to contribute to the defence of liberty and afterwards to free us from a commercial dependence.

Anyone who had carried out experiments or who had collected any observations was invited to communicate them to the Committee, for examination by its soda commission:

A true republican does not hesitate to give up the very property of his mind at the call of the motherland which appeals for his aid.

The report of the soda commission, June 1794

The commission appointed to collect and examine processes for soda production consisted of the chemist Darcet, his former student B. Pelletier, and the mining engineer C. H. Lelièvre, an able trio, all members (until its suppression) of the Academy of Sciences. Another mining engineer, Alexandre Giroud from Grenoble, was called in from 15 March to lend assistance with analyses. Darcet was the senior figure and it seems to have been he who was chiefly responsible for the writing up of the findings.[104] The commission's report[105] was presented to the Committee of Public Safety on 20 June, having taken nearly five months to prepare, an unusually long time by 'Revolutionary' standards, reflecting the uncommon magnitude of the task. It was a substantial and detailed document, somewhat ill-organized, with signs of haste, but an extremely useful compendium of the processes then known. It described four chief methods: the Leblanc process, the Malherbe iron process (and a number of variants), Chaptal's litharge process (with variants), and Guyton's lime process. There were also briefer accounts of various other processes and suggestions, mostly contributed by Guyton and Carny and of more academic than commercial interest.

It was the Leblanc process and the various iron processes which had mainly occupied the commission's attention. Leblanc and his associates, we are told, 'with a noble devotion to the public good', had been the first to present their process to the commission. Darcet, of course, had already formed a high opinion of it in its early days, and its value in large-scale operation was now verified by trials at Saint-Denis on 28 March. We may observe in passing that the commission's relations with Leblanc and his partners may be presumed to have been very good: both Leblanc and Dizé always showed the very warmest regard for Darcet, and Leblanc was also personally acquainted with Pelletier and Lelièvre (he was a member of various administrative bodies himself and worked as a colleague with both in the *Commission temporaire des arts*). Malherbe's iron process, as developed in the late 1770s, was communicated to the commission by A. G. A. Jourdan, director of the glass-works at Munzthal and presumably Malherbe's former associate. There is no indication that Jourdan was currently exploiting the process[106] but essentially the same method had by now been in use for some time at Javel, and the commissioners were able to verify its efficacy, as worked by Alban, by large-scale trials at the Javel works on 17 April.

While the processes of Leblanc and Malherbe were thus proved to be effective, both methods, as they stood, presented the crucial drawback in the prevailing circumstances of employing sulphuric acid for the preliminary conversion of salt to sodium sulphate. The commissioners therefore made it an important part of their inquiry to examine the possibility of using iron sulphate (copperas) instead of sulphuric acid for this preliminary conversion. They attempted the reaction of iron sulphate with salt in the wet way—a method published in 1786 by Lorgna, and also indicated by Carny—but they found this to give unsatisfactory results. They did succeed with reaction in the dry way, however, using the variant on Malherbe's process which we have seen to have been early devised (though not exploited) by his partner Athénas:

$$\text{salt} \xrightarrow[\text{copperas}]{\text{calcination with}} \left[\begin{array}{c} \text{sulphate of soda} \\ +\text{oxide of iron} \end{array}\right] \xrightarrow[\text{charcoal}]{\text{fusion with}} \text{crude soda}$$

After satisfactorily testing Athénas's process at Javel, on 26 April, the commissioners then went on to develop a process of their own as a further variant. Reflecting on the time and expense involved in manufacturing iron sulphate from pyrites, they conceived the aim of employing pyrites directly, and found that calcination of salt with pyrites could indeed furnish sodium sulphate. Their experiments showed, furthermore, that the reaction could be usefully facilitated by adding coal, peat, or charcoal, whereupon it proceeded more quickly, and as a self-propagating combustion, so that instead of requiring calcination in a furnace the mixture could simply be formed into balls and burned on a grate. Lixiviation of the resulting combustion product gave a liquor from which sodium sulphate could be crystallized. The residue after lixiviation was brown iron oxide, which could then if desired be used with charcoal for the conversion of the sodium sulphate to soda.

The soda report did not recommend any single process as a definitive answer to the country's immediate problem. It reflected the stop-gap, improvisatory policies of the war effort in general in envisaging the adoption of a variety of different methods according to local conditions. The litharge process and Guyton's lime process were both thought likely to be of potential value in some circumstances. But the processes which the commissioners believed would be most generally useful were the iron processes (using iron sulphate or pyrites), and the chalk process of Leblanc (suitably supplied with sodium sulphate). The Leblanc process, the report said, was that which might be most generally adopted, because of the widespread availability of chalk and because it yielded a product which could be used in the crude state, when it resembled the customary natural sodas. Crude soda made with iron had necessarily to be refined since the iron impurity would be detrimental in use. On the other hand the commissioners considered that the iron process might have some superiority technically, in that theory suggested—and their experiments seemed to confirm—that the greater affinity of sulphur for iron than for chalk would result in a somewhat higher yield.

The commissioners concluded that by exploiting the various processes they described it would be possible for soda to be made in practically all parts of the Republic, at least on a sufficient scale to supply local needs. They ended their report with a geographical survey, enumerating the regions which would be particularly suitable for the formation of major works. It would be useful, they said, to establish manufacture in the departments of the Gard and the Ardèche, where pyritous deposits were abundant and where iron sulphate was already produced; soda made there would find a market in the neighbouring departments, and in particular could be sent down the Rhône to supply the great soap industry of Marseilles. Soda works would be well placed, too, in Picardy, and in the department of the Isère, in Dauphiné, where again there were abundant deposits of pyritous materials and where alkali was needed in bleaching and in the manufacture of linen. In the vicinity of Lyons it would be possible to conduct soda manufacture in conjunction with the copper works at Chessy and Sain-Bel, where the sodium sulphate could be made by adding salt during the roasting of the pyritic ore. The Vendée and the Loire-inférieure would again be good regions for manufacture, whether using salt or the sodium sulphate which a certain Daguin, of Angers, had reported to be obtainable there in large quantity from peat ashes; Daguin had patented methods for the extraction of this sulphate in 1791, and claimed to have produced 200 000 pounds in 1792. In the east the department of the Meurthe offered good resources, with salt, iron ores, pyrites, and coal all plentifully available, and with natural sodium sulphate easily procurable as a by-product of the salt-works there, as indicated by Nicolas (former chemistry professor at the University of Nancy): a factory could be established at Dieuze, for example, and would be particularly useful in supplying the glass-works of the region, which were said to be idle for want of alkali. Above all the commissioners thought it a matter of some importance that a large soda works be promptly formed to supply the needs of the capital, and to this end it recommended the revival of the factory at Saint-Denis.

The Saint-Denis works was now in the possession of the Nation. With the execution of Orleans on 6 November 1793, his soda works, like all his property, automatically passed to the State, and although in practice the works had remained for a little time in the hands of Leblanc and his partners, it was in due course placed under sequestration at the end of January 1794. Local officials visited the works on 27 January to draw up a general inventory, and four days later Shée wrote to Leblanc that he could no longer dispose of the smallest thing without a written permit.[107] By that time Leblanc and Dizé had both moved to posts in Paris, leaving the idle plant in Shée's care. Dizé was now embarked on a successful career as a military pharmacist: following a brief apprenticeship, he had served from early 1793 as a sub-assistant apothecary in the military hospital at Saint-Denis, and was then on 24 September called to work in the central medical stores at the *École militaire* in Paris.[108] Leblanc, for his part, had acquired an important post in the *Régie des poudres*: the growing work-load of the *Régie* had led the three existing *régisseurs* to request the appointment of a fourth, and Leblanc was named to fill this new position on 29 December, thereupon moving into residence at the Arsenal.[109] The factory at Saint-Denis, following its sequestration by the district authorities, was in March announced for sale. Dizé informed the Committee of Public Safety of this move, however, asking that the sale be postponed until he and Leblanc had been able to demonstrate their process to the commission, and suggesting furthermore that it would be preferable for the works to be retained until the commission had presented its findings. The Committee then responded swiftly, on 21 March, by ordering that the sale be suspended and charging the district authorities to see that the works was not disturbed, 'Considering that it is advantageous to profit from this establishment for the manufacture of soda, on which it will make further arrangements forthwith'.[110]

When the soda commission presented its report in June, the position of the works must thus have seemed ideal for its revival as a leading element in any Government programme for a national development of soda manufacture. The opportunity it offered was all the more notable in that to all intents and purposes it was the only soda factory France possessed (Hollenweger's old plant at Batz was remote, abandoned, and ignored).[111] Moreover, the commission had nothing but praise for its design and organization:

It would be difficult to assemble, in such a small space, more facilities and more conveniences than are to be found in this workshop; furnaces, mills, equipment and stores, everything is here arranged in the best order, all for the greatest convenience of service.[112]

The commissioners were so impressed, indeed, that they had detailed drawings of the entire plant prepared to accompany their report, and they commented that 'the factory at Franciade will be able, in a very short time, to be put into activity and serve as a model'.[113] Their observations merit extended quotation:

It would appear urgent to us to form without delay a large establishment for Paris: its daily needs, the continual consumption of soda by its manufactories, its wash-houses, etc., everything here commands a prompt manufacture. The establishment is ready-formed at Franciade; the citizen Dizé, one of the co-associates, has particularly directed its construction: it is mounted in such a way that it can serve equally for every kind of use and process of this sort; this is a justice which his co-associates render to him. This establishment is in the hands of the nation; and although it was formed to exploit marine salt by the agency of sulphuric acid, it is no less well disposed, in the absence of this acid, which the circumstances of the war do not allow it to procure, to operate by the agency of pyrites, of vitriols, of peats, in a word by such other material as its situation, and industry, might present as most suitable for reviving this manufacture. It is there, as also at Javelle, that the Committee can promptly put the extraction of soda into activity; it is there that the Committee can form a school, call pupils to receive the first lessons, to work there themselves and acquire, by their own experience, a knowledge of all the details of this new art.

The co-associates of this works, as also those who in different parts of the republic have made a sacrifice to the country of their knowledge and of their work, all equally full of confidence in the justice of the Committee, await, without disquiet, the just indemnities which are due to them, both for the advances of all kinds and for the time which they have devoted to creating and forming their establishment; they are ready, if the need of the country requires it, to devote themselves again to the same work, to which they are attached (for they do not conceal it) as one is attached to the fruit of one's industry and to the product of one's creation.[114]

Government response to the soda report

Following the report by the Darcet commission, the Committee of Public Safety took steps aimed at promoting a general development of soda manufacture in France, and its spokesman Barère, on 14 July, addressing the Convention on the triumphs of the munitions programme, spoke of the additional assistance the programme would soon be receiving from this new manufacture:

A considerable work on this new art, so important for the Republic, was ordered by the Committee; it is finished, and manufactories will be immediately devoted to its practice, through the attention and the encouragement which the Committee is hastening to give to it.[115]

In these hopes the Committee was to be deceived, however. Its plans for soda manufacture were in the event to prove totally fruitless. It succeeded neither in reopening the factory at Saint-Denis nor in securing the formation of any other works. We shall shortly consider the reasons for this in detail, but we may here observe in anticipation that a chief factor seems to have been a lack of 'Revolutionary' vigour in the Committee's own action on the matter. The Committee did not treat soda manufacture with the same urgency and resolve as munitions themselves, and this is evident from the outset in its response to the Darcet report.

The Darcet report did not present any precise plan for Government action but it did make several suggestions in passing and it seems clear that positive measures were expected. The commission suggested, for example, that Giroud might be sent to the copper mines near Lyons to examine the possibility of making soda there; it recommended Athénas and Daguin as men suitable to direct any work which might be undertaken in the Vendée and Loire-inférieure; it similarly recommended Nicolas for his particular knowledge of the Meurthe; and, of course, it strongly urged the revival of the Saint-Denis works and suggested that a school be formed there. There survives a draft for an *arrêté* of the Committee which sets out wide-ranging measures broadly in line with the commission's conclusions, and although the authorship of the piece is unknown, it is worth summarizing its proposals here as an indication of the kind of action which was evidently considered. Soda, it declared, was to be made in all places where the necessary materials were available, and the Darcet report was to be published so as to spread knowledge of the processes. Citizens wishing to undertake manufacture were within a month to inform the *Commission des armes et poudres* which would then procure to them the necessary facilities. The same Commission was to see that manufacture was immediately resumed at Saint-Denis, 'by procuring to the citizens Leblanc, Dizei [*sic*] and Shée all the facilities which are necessary to them'. It was also to promote soda manufacture at Javel. The Commission was furthermore to see that soda works were established at Alais, Laon, the Lyons copper works, and Croissy, and in the departments of the Isère, Meurthe, and Loire-inférieure. Each works established was to take four pupils below the age of conscription 'to train them in the knowledge and practice of the chemical arts necessary to the extraction of soda and the different products which might depend on this first salt'.[116]

The *arrêté* which the Committee actually adopted, on 2 July, was a good deal less specific and less forceful. It simply ordered the printing and distribution of the Darcet report and invited anyone wishing to set up works to make known his intention:

Citizens who would like to establish workshops for soda manufacture, by the processes described in the report or by processes of their own, are invited to make themselves known to the Committee, which will procure to their operations all the facilities of which they are susceptible.[117]

This brief order was published in the *Moniteur* on 11 July, without further comment. Before we describe the response it elicited, we should first consider the fate of the one works which did already exist, that at Saint-Denis.

Attempts to revive the Saint-Denis works

Although the Committee of Public Safety made no specific order with regard to the works at Saint-Denis, the question of its revival did receive some attention in the weeks following the Darcet report. The records are regrettably fragmentary but we do find, for example, a set of detailed proposals drawn up on 2 July by Dizé.[118] In these Dizé recognized that it would not now be possible to work with sulphuric acid, and proposed therefore that as a first step all the sodium sulphate in France should be requisitioned for the factory's use. The three furnaces intended for the production of the sulphate would not then be needed for that purpose and so could be applied to the manufacture of soda, increasing the plant's output fourfold (it would be necessary first, of course, to strip the lead linings from two of them). Meanwhile, in order to ensure continuing supplies of the sulphate, a new workshop should be built to make it by the Darcet commission's pyrites process, and to this end supplies of pyritous peat should be procured. Dizé estimated that by these arrangements the works would be able to produce 2 800 000 pounds of crude soda a year (probably a considerable over-estimate, a more likely figure would be about half this). He pointed out that the site was sufficiently large to accommodate extensions and that there would be easy access for building materials by way of the Seine. The proximity of the town of Saint-Denis, moreover, would be an advantage if a national school were established there. He suggested that the administration of the works should consist of two men to direct manufacture and one to keep the accounts (no doubt thinking of himself, Leblanc, and Shée), and also a representative of the people, who would facilitate supplies and would generally 'imprint a Revolutionary movement on the machine' (here perhaps thinking of the earlier allocation of a representative of the people to the great saltpetre refinery at Saint-Germain-des-Prés). He considered the matter to be of sufficient importance for the reorganization of the works to be effected by a decree of the Convention, which would have the additional advantage of publicizing the works throughout the Republic.

A glimpse of the discussions that were taking place is afforded by a letter from Leblanc to Dizé, of 11 July:

Each day I have been awaiting the results of your approaches; you had promised to inform me the following day of the course agreed with Darcet; I have learned, it is true, that he had fallen ill, but we could have consulted with Pelletier and Lelièvre for the proposed experiment. You were also to send me a copy of the note written by Fourcroy. It is essential for us to meet as soon as possible and agree upon our facts for the reply which we must make to Berthollet and to Fourcroy.[119]

Fourcroy—and probably Berthollet, too—was among the scientists assisting Prieur as advisers in the offices of the Committee's *Section des armes et poudres*. It was perhaps to their questions that Leblanc and Dizé were replying when they together addressed a note to the Committee on 14 July. They had here been asked how much soda the works could make if it were expanded within the limits of the site, and how much it would cost to increase its capacity to 3 million pounds of crude soda a year. Their response, in Leblanc's hand, was brief and hasty with some inconsistencies, but it presented the general conclusion that by making soda from sulphate in the two furnaces without lead linings, and by carrying out six operations a day in each (i.e. by working night and day), the works could produce a quarter of the 3 million pounds in question (a reasonable estimate). The partners declared themselves unable to estimate the cost of quadrupling the facilities, but pointed out as an indication that the existing works had cost about 100 000 *livres*. They added that it would be possible to begin manufacture there as soon as a sufficient quantity of sulphate and coal was procured.[120]

Nothing came of these early discussions and four months later the works was still lying idle. It is impossible to say exactly why this was, since during these months the archives are almost totally silent: we find recorded only a modest personal award to Leblanc, on 5 September, in recognition of his disclosure of his process to the nation and in consideration of his poor financial circumstances.[121] It appears that the Committee had taken no major action on the works, and that Leblanc himself had sunk into an equal inactivity. The summer, of course, had seen a profound change in the political situation. With the fall of Robespierre at the end of July, and the ensuing reaction against the Terror, the Committee of Public Safety was reorganized by the Convention so as to limit the scope of its responsibilities and ensure that it could never again assume dictatorial powers. In itself this does not seem to have had a great impact on the Committee's handling of the war manufactures, which with Prieur's departure from the Committee passed into the charge of his former advisers, Fourcroy and Guyton.[122] What does seem likely, however, is that as the coercive climate of the Terror retreated so did the zeal of Leblanc for the *bien public*, and we can imagine that amid the clamour against the recent tyrannies he was encouraged now to ruminate on the losses he himself had suffered from the execution of Orleans, and on the compensation that he might be able to claim.

The Committee evidently felt that Leblanc was being indolent when on 9 November, after a visit to the works by Fourcroy and Prieur, it wrote to him in the following impatient but rather vague terms:

The Committee, greatly occupied with the establishment of artificial soda works, wishing definitely to know how far it can count on the zeal of citizens who have begun establishments of this kind, invites you without delay to make proposals which you believe proper for assuring the large-scale manufacture of soda: you are aware that objects of this importance cannot be abandoned to a languishing indecision.[123]

The Committee at about this time seems to have been itself bestirring from a certain inactivity on the soda question (probably in association with the publication and distribution about now of the Darcet report). On 27 October we find it inviting the *Comité d'agriculture et des arts* to send two commissioners 'to the meetings and other experiments on the best processes for manufacturing soda'.[124] And it was no doubt with soda manufacture in mind that the Committee on 8 November ordered Besson—the representative on mission to the salt-works of the eastern departments—to instigate the extraction there of as much sulphate as possible, 'considering that the sulphate of soda which can be obtained from materials neglected in the salt-works can serve for the manufacture of other materials of which there is a shortage'.[125] The possibility of extracting natural sulphate in the salt-works of the Meurthe had been pointed out to the Darcet commission by Nicolas, and Nicolas had subsequently in the month of thermidor (July/August) again advocated the creation of workshops to exploit this source, suggesting that the sulphate then be sent to 'the national establishment which it is proposed to form at Franciade'. In a further communication of 25 August Nicolas had estimated the amount of sulphate which might be extracted in the Meurthe at over 2 million pounds a year.[126]

It was over two months before Leblanc replied to the Committee's inquiry of 9 November. The reason seems to be that he and Dizé were now deciding to press claims to an interest in the factory as former partners of Orleans, and were considering their best course of action.[127] In the rough justice of the Terror, the factory seems to have been regarded when first sequestrated simply as the property of the nation. The more precise legal position, as it now began to appear, was that the State and the three surviving partners had a joint claim on the enterprise in accordance with the terms of the original partnership, with the State now standing in the place of Orleans' heir. Cases of this kind were covered by a law introduced on 7 December 1794 as part of the Government's efforts to revive the nation's industry and commerce.[128] The new law allowed surviving partners in businesses whose property had been confiscated by the State to reacquire their property by buying out the State's share: those wishing to take advantage of the provision had to declare their intention within twenty days, whereupon the financial details would be settled by arbitration. Leblanc and his partners, however, did not make any application under this law, presumably because they lacked the necessary capital. When on 22 January 1795 Leblanc eventually replied to the Committee's note, he presented different suggestions.[129] These were cast in a characteristically oblique style and are very indistinct, but what he seems to have had in mind was the resumption of manufacture under the terms of the original partnership, with the State now in the place of Orleans. It would be necessary, he said, for the Government to ensure supplies of materials, and he pointed out that the only ways of obtaining sodium sulphate would be to extract it from salt-works residues, or to make it by the commission's pyrites process, which should be tested by a large-scale trial.

The Committee evidently found Leblanc's cryptic suggestions unacceptable and decided that the business should be liquidated. Since the partners did not intend to acquire it for themselves, while the Committee apparently did not wish to acquire full ownership for the State, it now began to think of selling the works to a third party. A request to buy the works had in fact been made on 10 December, by Riffautville, Bernard, & Cie,[130] a group whose chief chemist, Bernard, could claim as earlier industrial experience the mounting of several chlorine bleaching works, including one at Merville, near Lille, where before the war he had extracted 'soda' from his residues (see p. 211). Wishing to establish a soda works, and finding the Saint-Denis plant lying neglected, this company had asked if they might buy it at a favourable price, proposing greatly to expand it and adapt it to a different process. Fourcroy now referred their request to the *Commission d'agriculture et des arts* for a report, and the matter was there examined by Berthollet, one of the commissioners, who proceeded to check the company's credentials and to make inquiries of Leblanc and Dizé. Leblanc's initial response included a rather vague suggestion that he might be able to buy the works himself, if the Government were prepared to compromise over the surety and could assure supplies of raw materials, but when pressed for more precise proposals he replied on 6 March that he had been unable to complete arrangements with his backers and would have to give up this hope. Berthollet then presented his report, declaring himself satisfied that the prospective purchasers were competent and genuinely intended to make soda if they acquired the works. He pointed out, however, that Leblanc's deed of partnership showed that the surviving partners did have interests to be liquidated, and that these would be prejudiced if the works were sold at a price especially favourable to the purchaser. He therefore recommended public sale of the works to the highest bidder (with the condition that soda manufacture must be resumed), and liquidation of the existing partnership through the normal channels. On 21 March the Committee of Public Safety duly submitted a proposal to the Finance Committee for the sale of the works in this way, but in the event it was to remain unsold, for what reason we do not know.

There appears to have been no further action after this. The Committee of Public Safety seems now to have arrived at the conclusion that the factory was in any case unworkable in the prevailing circumstances since there was no practicable means of supplying it with sodium sulphate. Fourcroy, in some notes on this subject,[131] observed that even if the State itself were to acquire the works, by liquidating the interests of the partners, it would still be very difficult to make use of it because of the impossibility of obtaining sulphuric acid. Although it had been suggested that natural sulphate could be extracted from salt-works residues in the department of the Meurthe, 'the circumstances and the disposition of the rivers afford no means of bringing it to Franciade'. (We shall see later that in any case no great quantity of sulphate had in practice been extracted there.) There was also the possibility of producing the sulphate by means of pyrites, of which there were deposits along the banks of the Somme, but then the production of the sulphate would require a new establishment additional to that existing at Saint-Denis, and it would be simpler to create a new soda works altogether in a place where all the necessary materials were available. Fourcroy concluded that 'As long as peace is not made, it would appear that the idea of making soda at Franciade must be put aside, for want of sulphuric acid.'

There the affair seems to have rested. In June 1795 Leblanc left Paris for the far away departments of the Tarn and the Aveyron, where he was to spend the next year on a Government mission, investigating the region's mineral resources and in particular reviving the alum works at Saint-Georges.[132] On 26 October the Committee of Public Safety itself ceased to exist.

Gillispie, in his polemic against the romantic traditional portrayal of Leblanc and his misfortunes, has sought to blame the failure to revive manufacture at Saint-Denis on Leblanc himself, who by his dilatory, impractical, and obstructive behaviour, we are told, 'frustrated every attempt to reopen his factory'. This appears excessive, however. As far as Leblanc's early relations with the Committee are concerned—up to July 1794—there is no evidence that he was unco-operative and Gillispie's specific allegations are here quite ill-founded.[133] After the summer, it is true, Leblanc's behaviour might justly be described as dilatory and impractical, but it is doubtful whether this decisively affected the outcome. Probably the most that can be said is that Leblanc's delays might have contributed to the failure of the plans early in 1795 for the sale of the works to a third party.

It is evident from the account we have given that fundamentally the revival of the works was frustrated by two main obstacles. In the first place there was the awkward legal position which placed ownership jointly in the hands of the partners and the State, a position which became clear only after some time and which then proved irresolvable, since the partners lacked the resources to acquire full ownership for themselves and the Committee did not wish for it to be acquired by the State, while plans to liquidate the venture by sale to a third party fell through. Secondly, there was the basic problem, which again only gradually became fully clear, that there was in any case no reasonably practicable means by which the works could be supplied with the necessary sulphate. No doubt the Committee could have overcome these obstacles and roused Leblanc into action, if the need had appeared sufficient, but soda manufacture could not rank with munitions as a matter of imperious national importance. The emergency saltpetre programme itself, moreover, which soda manufacture had been primarily intended to assist, was perceptibly slackening by the end of 1794. The popular workshops were then beginning to abandon their operations, as revolutionary ardour faded, as the materials exploited became exhausted, and as the churches in which many workshops had been housed were restored for worship. By April 1795 the programme of 'Revolutionary' saltpetre extraction was effectively over.

Other soda works projected, 1794–1795

Turning now to consider the Committee's wider efforts, we have seen that by an *arrêté* of 2 July it sought to promote the mounting of soda ventures by offering help to intending producers and by ordering publication of the Darcet commission's findings as a guide. The *arrêté* ordered that the Darcet report itself, and an abridgement which had been prepared, were to be printed and sent out for distribution to all district administrations. The abridgement was then put out as a brochure within the month.[134] Publication of the full report was a little delayed, but in due course it appeared in an edition of 3000 copies at the beginning of November,[135] in the form of an eighty-page brochure entitled *Description de divers procédés pour extraire la soude du sel marin*. The full text of the report was here illustrated by comprehensive plates of Leblanc's plant,[136] and was accompanied by a short supplement detailing some minor processes that had been received after the main report was completed, and also by a description from the pen of Loysel of workshops for the refining of crude soda (probably those of the Saint-Gobain glass-works). Besides these brochure publications, there also appeared in the December issue of the *Journal des mines* an account of the Darcet report by Coquebert, with additional material by Coquebert himself.[137]

The Committee's measures brought little response. We know of eleven approaches made either to the Committee or to the *Commission d'agriculture et des arts* between June and December 1794. None is known to have resulted in a works, and most of the would-be manufacturers probably never got beyond the stage of thinking about it. The earliest was Giroud, the mining engineer mentioned above as having assisted the Darcet commission in its investigations. Giroud informed the Committee on 19 June that he had engaged the *négociant* Claude Périer of Grenoble, who had suitable buildings available, to back him in a venture there which would employ local pyrites. The outcome is not known.[138] There was also some early interest shown in the nearby department of Mont-Blanc, at least on the part of its administrative authorities. Wishing to initiate soda production with a view to reviving the glass-works at Thorens, then idle for want of potash, the department's administrators between August and October sent repeated requests to Paris for the Darcet report, and duly received copies when at last it appeared in November. Two new chemists in Paris to display an interest in the soda question were N. Deyeux and A. A. Parmentier, who on 29 October obtained 6000 *livres* from the *Commission d'agriculture* to mount a workshop at

Montdidier in the Somme for the trial of a process of their own devising. Their occupations in Paris prevented them from realizing their plans, however, and a year later they paid the money back. In Nantes, Pierre Athénas seems to have been inspired with new hopes of establishing a works, despite having lost (he said) over 30 000 *livres* in his previous venture. He and three associates requested copies of the Darcet report in December, complaining at the imperfect distribution that the report and its earlier abstract had received. At Troyes the question of soda production attracted the attention of Joseph Bosc, in a project which, almost alone among the sketchily recorded schemes of these months, is at least known to have resulted in some practical endeavours. Bosc had been engaged for several years in the commercial production of chemical salts, and was director at Troyes of a saltpetre works and of a chlorine bleachery. It was this latter concern which motivated his interest in soda when he first wrote to the Committee on 17 July, his primary intention then being to exploit the sodium sulphate in his distillation residues. By November, with his bleaching operations at a virtual standstill for want of sulphuric acid, he was intending instead to make soda from salt, and after having searched in vain for local pyrites and unsuccessfully attempted the lime process, he is last heard of at the end of November proposing to decompose salt by means of pyrolignate of baryta. The plans of four other individuals can be dismissed very briefly as appearing of scant promise. A pharmacist called Alyon, in asking to be released from detention, announced in July that he had persuaded a salt-works proprietor near Nantes to undertake the manufacture. In the following month interest in production was tepidly expressed by a certain Cormeray, a former friend of Daguin (recently killed by counter-revolutionaries), who was operating a works near Saint-Nazaire in Brittany for the extraction by Daguin's methods of sodium sulphate from peat ashes. Interest was also expressed by a Breton brandy distiller called Landry, of Bourg-des-Comptes, who plainly knew little about the matter, as also did Collignon, a potash producer at Dieuze in the east, who addressed an illiterate inquiry to Paris with the vague idea of using his boilers to make some soda as well.[139]

Amid this generally dismal response, there did arise two schemes which briefly appeared of some greater promise, and of these we might say a little more. The first was by Carny,[140] who approached the Committee in the month of vendémiaire (September/October 1794) expressing a wish to establish a soda plant as an adjunct to his hydrochloric acid works in the rue de Harlay. His intention was to increase his hydrochloric acid output and produce soda from the residual sulphate (he was evidently intending, therefore, as his primary plan, to work with sulphuric acid, though he later also spoke of a soda process using copperas). To enable him to mount a new workshop he requested a loan of 200 000 *livres* and the lease of a national property in the rue neuve Saint-Gilles, adjacent to his existing works. If granted these requests he undertook to produce 200 000 pounds a year of a white soda three times as rich in alkali as soda of Alicante. Berthollet and Chaptal visited his works on behalf of the *Commission d'agriculture et des arts*, and then in a report of frimaire (November/December) recommended to the Committee that Carny's requests be granted:

No doubt it would be preferable to undertake the manufacture of soda in a place abundant in pyrites, in coal, in marine salt; but would one find in the place combining these advantages buildings established and workers trained in this kind of industry? What sums would one not have to sacrifice to form an establishment of this kind at a moment when labour and materials are so costly? . . . The Committee of Public Safety has published several processes for the extraction of soda, but no-one has realised them in manufacture: it seems important to us that an example be given.

They remarked that the shortage of soda was currently such that its price had risen from half a *livre* to three *livres* a pound. Before the Committee could complete its consultations, however, Carny began to change his mind. On 8 January he announced that he had revised his plans: because of the rising cost of equipment and of construction work he had decided that he could not afford to expand on the scale at first intended, and so he now proposed to make only 50 000 pounds of soda a year, for which he requested a loan of 60 000 *livres*. By the end of the month he had given up the idea of making soda altogether, chiefly perhaps on account of the continuing rapid rise in costs. Prices in general had been rising progressively through the second half of 1794, as the end of the Terror brought open flouting of the Law of the Maximum designed to control them, and this rise steepened sharply after 24 December when the law was finally abolished. A further factor may have been the unhelpful response his various requests encountered in the administrative bodies the Committee consulted. Thus, the *Commission des armes* in a report of 21 January showed itself unwilling to supply him with potash (he was planning to operate by double decomposition of sodium sulphate with potash, proposing that his by-product potassium sulphate then be used in place of potash itself in the manufacture of saltpetre); nor could the Commission provide the transport he would need. It is open to doubt, too, whether the loan he sought would have been forthcoming, for another report a week later—probably from the *Agence des arts et manufactures*—commented unfavourably on his proposals, observing that 60 000 *livres* seemed a lot of money and that Carny offered insufficient guarantee that it would in fact be used for the manufacture of soda.[141] Moreover, even if Carny had begun soda production, it is in any case further questionable to what extent he would have been able to pursue it, in view of the problems in sulphuric acid supply: on 13 January he informed the Committee that lack of this acid was threatening the continued operation of his existing works, and he asked the State to sell him a quantity of sulphur and saltpetre (again almost certainly in vain).[142]

More ambitious and better founded than Carny's project was that of J. P. Champy, who approached the Committee with outline proposals on 2

December 1794.[143] Champy was an old friend of Guyton, the member of the Committee to whom his petition passed. In 1778, as *commissaire des poudres et salpêtres* for Burgundy, he had partnered Guyton in mounting his artificial saltpetre plantation, and in the 1780s, as a member of the Dijon Academy, he had been one of Guyton's scientific collaborators.[144] A long career in the service of the *Régie des poudres* had brought him during the Revolution to the top of his profession,[145] when in 1792 he became one of the *Régie*'s three presiding *régisseurs*. We may note in passing that he would thus for a time have worked as a colleague with Leblanc, who joined the *Régie* at the beginning of 1794. When the gunpowder administration was reorganized in July 1794—with the creation of a new *Agence des salpêtres et poudres* in place of the old *Régie* and *Administration révolutionnaire*—Champy became one of the three directors of the new body, in company with Chaptal and Bonjour. He would clearly appear a man who by his chemical experience and demonstrated administrative skills would have been admirably qualified to undertake a soda venture.

The plan with which Champy came forward was for the establishment of a works to exploit the pyritous coals of the Aisne and the Oise. Guyton responded to his approach by granting him a month's leave from his post so that he could visit these departments, examine their resources, and look out for any national properties which might be suitable for his projected factory. This Champy did, and on 18 January he then reported that he proposed to make his works at la Fère, where the best and most abundant deposits were to be found, and from where he would be able to supply the nearby Saint-Gobain glass-works, a particular sufferer from the soda shortage. Besides soda he also planned to produce copperas and alum. He did not need any funds from the Government but merely asked to purchase a number of national properties at their estimated value. He had already bought lead for his boilers and had also bought some salt supplies at Marennes, but he had been unable to obtain transport for his salt since all the barges were under requisition, and so he asked, in addition, to purchase 200 000 pounds of the impure salt which resulted as a by-product from saltpetre refining at the Arsenal. He stressed the importance of a prompt decision in view of the continuously rising prices of materials.

Aware of the need for a lead to be given, and with Carny having now abandoned his plans, Guyton particularly welcomed Champy's proposals and submitted the request for the purchase of national properties to the Finance Committee, urging a prompt and favourable reply. Unfortunately no such reply was forthcoming. Two months later Champy wrote again to the Committee of Public Safety, complaining of the bureaucratic delays caused by the financial authorities' interminable consultations:

Such a long delay is all the more prejudicial to me in that it has given rise to an enormous increase in the price of materials and utensils necessary to the manufacture of soda. I cannot . . . remain any longer in the uncertainty in which I find myself and if you persist in believing this establishment useful I beg you to secure the reply of the *Commission des revenus nationaux* [which the Finance Committee had consulted], and the decree which would allow me to realise my project.

Fourcroy tried to speed things up by writing to the Commission, but only on 1 July did the Finance Committee finally reach a decision, and then the decision was in the negative. By that time Champy had probably long since given up in despair anyway.

One further project—by Schemel in March 1795—need be recorded here only in passing, since although Schemel again had no immediate success his perseverance did eventually result in a works some years later, and we shall therefore have occasion to return to him below.

From the problems encountered by these various schemes, the Committee's failure to arouse any great interest in soda manufacture can readily be understood. The circumstances were manifestly unfavourable. The elevated price of soda could offer some inducement, it is true, but the general economic instability, the spiralling cost of materials and equipment, the continued pre-emption of goods and services for war work, the unavailability of sulphuric acid, such factors were more than a sufficient disincentive and obstacle, particularly since this was a new industry, without a single working factory to serve as an example.

Further ventures and the first successful factories, 1796–1806

The decade following the failure of the Revolutionary soda programme saw the mounting of ventures by a number of individuals, culminating at last, at the end of the period, in the first notable successes, when the rising price of imported soda as a consequence of the war was beginning to make its manufacture an ever more viable proposition (see Fig. 4.4).

One figure whom success tragically eluded, however, was Leblanc himself, who during these years was engaged in a protracted struggle, first to regain and then to revive his old works at Saint-Denis. We cannot here go into this sad and complex affair in detail.[146] Suffice it to say that after four years of bureaucratic delays and legal wranglings between Leblanc (supported by the Ministry of the Interior) and the Ministry of Finance, the works was eventually returned to Leblanc and his partners on 7 May 1800, though only on a provisional basis, pending a definitive settlement of their claims by the *Tribunal du commerce*.[147] Dizé and Shée were both now successfully established in new careers. Since 1795 Dizé had been chief pharmacist at the head of the army's central medical stores in Paris, and since April 1798 he had also been *affineur national* at the Paris Mint. Shée was now distinguishing himself in the civil administration in France's newly acquired eastern departments, and in 1801 was to succeed Lakanal as prefect of the Lower Rhine. Neither wished to resume his interest in the enterprise and so the old partnership was dissolved, and Leblanc, now fifty-seven, embarked on an unequal struggle to resurrect the works alone. The task must have been a daunting one in view of the total neglect to which the property had been abandoned for six years (already in December 1794 it was described as becoming overgrown with weeds). Nevertheless, by the end of August 1801

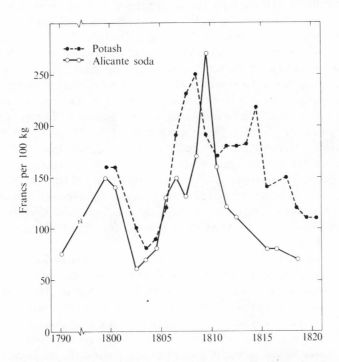

Fig. 4.4. Prices of natural alkali in Paris, 1790–1820. The soda price shown for 1790 represents the typical price prior to the Revolutionary Wars (AN, F¹²2245, [9]; Shée (1797), 292–3). According to Loysel in 1794 (Darcet *et al.* (an III [1794]), 68), the price of Alicante soda in Paris during the previous twenty years had varied between 15 and 90 *l.* per 100 pounds, most commonly standing at 30 *l.* (=61 fr. per 100 kg). The remainder of the graph is based on price series in Chabert (1945), 212, 215. No attempt has been made to show the price variations during the Revolutionary Wars since detailed information is lacking.

Leblanc could report that he had made several furnace loads ('fournées') of good soda, and had sent samples to the prefect of the Seine.[148] In the following years he evidently carried on production to some extent, though only in a small way and probably somewhat irregularly. How far this production was by the Leblanc process is uncertain. He had been experimenting with other methods since the later 1790s, and he seems now to have operated for at least a part of the time simply by double decomposition of sodium sulphate with potash (perhaps because this could be conducted with smaller resources).[149] He lacked the capital ever to turn the works into a going concern. Entirely without any personal fortune, he seems to have depended on loans to finance such limited production as he did undertake. The uncertainty of his position pending the definitive settlement of his claim was naturally a deterrent to prospective backers. Leblanc's hope was that the definitive settlement would give him the capital he needed, but for years that settlement was not forthcoming. In January 1804 the *Tribunal du commerce* declared itself unable to reach a decision, and only on 8 November 1805, after the matter had been sent before a special panel of arbitrators, did Leblanc and his former partners finally obtain a ruling on their claim. Besides the factory, this awarded them a sum of 116 607 fr. 89 centimes, of which Leblanc's share would have been about 50 000 francs.[150] The award would appear not unfair but it was very much less than the 2 243 000 francs Leblanc had claimed (largely on the basis of hypothetical lost profits). The disappointment of this ruling, coming after many years of poverty, frustration, and misfortune, is said to have shattered Leblanc's spirit, and on 16 January 1806 he committed suicide. Notoriously, even the sum awarded by the arbitrators was never in fact paid, despite continued efforts by Dizé—on behalf of himself and of Leblanc's family—for almost fifteen years after Leblanc's death.

Of the successful soda producers to appear in this period, one of the earliest was Payen, in whose chemical works at Grenelle soda manufacture was evidently under way by 1797. Towards the end of that year Carny remarked that the works currently made sal ammoniac, sodium sulphate, and soda, by processes of his own devising, but Carny does not seem any longer to have been involved in the enterprise and in reality soda production there was now almost certainly by the Leblanc process (Payen later wrote, in 1804, that this was the process by which he had been operating for nearly ten years).[151] Payen thus seems to have been the first manufacturer successfully to exploit the Leblanc process in regular production, though he conducted it only as a subsidiary interest, using by-product sodium sulphate from his manufacture of sal ammoniac. We may note in passing that it was among the products of soda manufacture at Grenelle that sodium thiosulphate was first discovered, by Chaussier in 1798.[152]

Another to take up soda manufacture was Nicolas Schemel, of Metz, who in 1795 decided to re-establish the chemical works he had briefly operated at Metz early in the Revolution, with the intention now of adding soda production as a new interest.[153] He planned to work by the recently published processes and may have had the Leblanc process particularly in mind, for when he approached the Committee of Public Safety in March 1795 he asked to visit the Saint-Denis factory on the mistaken presumption that it was working. He also requested assistance in obtaining equipment and materials. Schemel's requests were favourably received but he was to encounter many

Fig. 4.5. France: Places mentioned in connection with soda manufacture.

difficulties when he came to implement his plans. An initial set-back was his discovery, on visiting the salt-works of the Meurthe to arrange sodium sulphate supplies, that in fact there was very little sulphate ready available there.[154] During the winter of 1795–6 he therefore himself undertook to activate sulphate extraction in the salt-works, only to be frustrated by the uncommon mildness of the winter, which limited production to some 20 000 pounds (the extraction depended on fractional crystallization by cooling). The following winter, confident now of supplies, Schemel prepared to establish his works. To house it he bought from the State a property about 2 leagues (5½ miles) from Metz, consisting of the presbytery of Saint-Agnan, with its dependent church and cemetery. Here he encountered a new obstacle, however, in the hostility of the local inhabitants, who objected to the loss of the church and for some time prevented Schemel from taking possession of his property. By the spring of 1798, after legal proceedings, he had succeeded in establishing himself in the presbytery but was still having a difficult time with the local people, who continued to use the church for worship. Schemel's life was threatened, and on one occasion part of a Sunday congregation rioted in the cloisters and set fire to the presbytery. In January 1799 the Ministry of the Interior took action designed to see that his ownership was respected and his property protected. Despite all this Schemel does seem to have eventually succeeded in taking up soda manufacture, for in 1802 he sent a sample of sodium carbonate to the industrial exhibition in Paris.[155] His works was probably small and we have no information on its subsequent history.

Two other ventures might be noted briefly, both in the department of Mont-Blanc, where again there were inland salt-works and where we have seen some interest to have been stirred in 1794. In a book published in 1803–4, Herbin de Halle mentioned a soda factory at Carouge, of which no more is known. And in 1806 the mining engineer Lelivec wrote that a soda works had been established some years ago at Chambéry, but was no longer active. This had exploited natural sulphate from the Moûtiers salt-works.[156]

The most important early enterprises, however, and those considered by contemporaries to have been the principal pioneers in establishing soda manufacture in France, were the works severally created in the opening years of the new century by Carny, by Pelletan, and by Darcet *fils*, and these will now be considered in more detail.

Carny at Dieuze

In 1796 Carny's hydrochloric acid works in Paris ran into difficulties.[157] With the lease of his premises in the rue de Harlay shortly due to expire, Carny decided that year to move into the adjacent property in the rue neuve Saint-Gilles, where in addition he would have more room. In July he arranged the purchase of this property from the State, and paid for his acquisition in *mandats*, but to his embarrassment a law was introduced on 31 July requiring that in such purchases a quarter of the price was to be paid in specie. Unable to raise the specie payment within the prescribed time limit, Carny in October appealed to the Ministry of the Interior for help. The Ministry was sympathetic but largely powerless, and on 27 June 1797 Carny's acquisition was declared null, for want of his having completed payment in due time. This 'disastrous event' was said in September to have resulted in the destruction of his works. In the final months of the year Carny made endeavours to recover the property in order to re-establish his factory, but the eventual outcome of these further moves is not known.

This blow to his existing enterprise was no doubt an important factor motivating Carny in the new soda project which he conceived in Year VI (1797–8). With the help of Duquesnoy (one of the *fermiers des salines nationales*), he now arranged a scheme for the establishment of a soda plant in the great salt-works at Dieuze, one of the largest salt-works in Europe.[158] The intention was to utilize the largely unexploited sulphate-bearing residues which resulted as a by-product of salt manufacture. Carny was first to prove his processes by large-scale trials, and if these proved successful the company was then to finance him in the mounting of a soda works, for the direction of which he was promised a salary of 8000 francs a year, together with a 25 per cent share in the profits. In the event serious illness prevented Carny from realizing these initial plans, but after his recovery he made fresh arrangements with the salt-works administrators, and then went out to Dieuze, in eastern France, at the end of 1801. In 1802, after trials had been duly conducted, gardens and stockpiles of firewood were cleared from a site in the middle of the salt-works, and a soda plant was there established under Carny's direction.

Carny's first years at Dieuze were neither happy nor successful. His soda plant ran at a constant loss from the moment of its creation until the end of 1806, despite receiving its raw material (the sulphate-bearing residues) free of charge. Carny attributed the disappointing results to lack of co-operation and inadequate equipment. His friend Foucques (a minor chemical manufacturer in Paris) recalled in 1808 that Carny 'for the first two years was unceasingly thwarted by the ignorance and opposition of all the petty interests which were afraid of finding themselves injured'. Carny complained that little interest had been taken in the success of his plant, and that the directors of the salt-works had not even bothered to collect for him the residues he needed. 'While keeping my factory in a shameful mediocrity for want of furnaces, they wrote to me: Manufacture, produce profits for us and you shall have funds.' It seems likely that production was somewhat irregular at first, for Carny later wished to have power to run his factory 'without it being susceptible to stoppage, as it has been for four years through ill-will or thoughtlessness'. No doubt Carny's bitter complaints were in part well-founded, but one may question whether even with better facilities he would have been able at that time to compete with the prevailing price of natural soda, and it is hardly surprising that the company was unwilling to invest further in a steadily losing venture.

The turning point came in 1806. On 1 May that year the salt-works was taken over by a new company which showed decidedly more interest in the soda plant. Carny, moreover, got on very much better with the new director, Dufays, 'in whom I am immensely fortunate and to whom I owe all the moments of satisfaction that I have had here'. The new company's interest can readily be explained by the rising price of imported soda, as a consequence of the war (see Fig. 4.4), for this was now beginning to make soda manufacture an attractive proposition at last. A subsidiary factor, probably, was the re-introduction of a salt tax in the spring of 1806, which reduced other outlets for the company's saline residues and for sodium sulphate extracted from them, since the new tax was levied not only on salt itself but also on the sale of these various other materials: sales of the residues to glassmakers consequently fell, as too did sales of sulphate to pharmacists and to manufacturers of glass and faience.[159] Thus encouraged, the new company in 1806 spent 11 000 francs on the enlargement of its soda plant, enabling Carny to increase his production from an annual figure of some 50–60 000 kg (of various grades) in 1805–6,[160] to an output of 79 000 kg in 1807. And in 1807 the plant for the first time made a profit, of over 36 000 francs.

By now Carny's soda was acquiring a reputation for quality and was being well received in commerce. Samples analysed by the *Chambre consultative* at Nancy were found to be over twice as rich in alkali as soda of Alicante, and to give better results in use than ordinary soda, besides working out cheaper. Carny sent samples to the industrial exhibition in Paris in September 1806, and received an honourable mention from the jury.[161] Soda of various grades from Carny's works was now being sold in the important textiles centre of Rouen, where Descroizilles appended a eulogistic footnote on the new products to his 'Notices sur les alcalis du commerce' (1806): 'I have pleasure in publishing this new obligation of our country towards one of its most estimable citizens, who has not as yet received a recompense proportional to his merit'. Descroizilles reported that experiments had begun at Rouen, under the direction of Vitalis, on the use of Carny's soda for Turkey-red dyeing (one of the most exacting of uses), and that these were promising very good results: already a saving of 70 per cent had been found compared with the best Spanish sodas. By the autumn of 1807 Carny's products were in regular use in different parts of France, for glassmaking and for more delicate tasks such as the bleaching of cottons, the laundering of linen, and dyeing.[162]

Of the various kinds of residue which resulted as by-products from the salt-works operations, that which Carny chiefly employed was the material known as *schlot*, a mixture of the sulphates and chlorides of soda and lime, precipitated during the boiling down of the saline waters.[163] Carny extracted the sodium sulphate from this by means of lixiviation followed by crystallization. Two methods of crystallization were employed: 'par le chaud', which presumably means by evaporating the solution, and 'par le froid', which involved fractional crystallization by cooling. Crystallization in the cold way was the better, since apart from avoiding the expense of fuel, it gave a purer sulphate, from which better-quality sodas could be made. This method could only be practised during the winter, however, and although as much sulphate as possible was then made, the facilities of the factory allowed only a little over a third of the total sulphate production to be carried out by this method in 1807. Carny was thus obliged by his materials to manufacture low grade as well as high grade sodas. The records for 1807 show that twelve grades were then produced, ranging in alkaline strength, as measured by Descroizilles's alkalimeter, from 18° to 60° (equivalent to an Na_2CO_3 content of about 18–60 per cent). This compares with a typical strength for Alicante soda of some 20–33°. Besides soda Carny's plant also produced a material sold as *fondant vitreux* (presumably to glassmakers or potters), into the composition of which soda entered. There was also produced a small amount of sodium sulphate for direct sale, and there emerged a little salt as a by-product.

We do not know what process Carny was using for the conversion of the sulphate to soda. In 1810 he wrote that he had established the works following processes for which he had earlier obtained an exclusive privilege and a patent,[164] processes which we have seen above to have been lead oxide and pyrolignate methods (p. 208). He did not say whether he was still using the same methods in 1810, however. From later remarks by his son it appears that towards 1830 Carny had long been employing a method which was a variant on the Leblanc process, using lime in place of chalk, since the latter was locally unavailable.[165] This use of lime resulted in a crude soda containing much sodium sulphide, and so Carny then desulphurized the crude product by treating it in solution with lead oxide (now employing as a method of purification the kind of lead oxide reaction which he had originally developed for manufacture).

Encouraged by his growing success in 1806 and 1807, Carny wanted to expand production to the limit by utilizing all the suitable residues not only of the Dieuze works but also of the neighbouring and associated salt-works at Château-Salins and Moyenvic. The *Administrateurs des salines de l'est* were reluctant to invest further, however, feeling uncertain of the market:

The Company is aware that it owes the favourable returns of 1807 only to the war, and that from the moment foreign sodas are able to arrive as formerly (even only as they entered in Year XII and Year XIII) its factory will be burdensome.[166]

At the instigation of Carny's friend Foucques, the Ministry of the Interior looked into the matter early in 1808, desirous of seeing soda manufacture in France grow, and having ascertained that Carny's production could be increased fivefold by suitably expanding his plant, the Minister wrote an exhortatory letter to the company at the end of March. This had the desired effect, for a few days later the company unanimously adopted measures implementing all the Minister's suggestions. Arrangements were to be made for residues to be sent to Dieuze from Château-Salins and Moyenvic; the Dieuze director was to consult with Carny on the constructions needed to allow the greatest possible increase in production that year, and on further additions for subsequent years, with preference to be given to facilities for the crystallization of the sulphate 'par le froid'; and Carny was in future to be rewarded by a premium dependent on output. In provoking this decision to expand, the Minister's exhortations were reinforced by the continued improvement in the market: at the end of March it was reported that Carny was sending out soda as quickly as it could be made, and was unable to meet all the demands that were arriving, and indeed a proposal had been received from the Saint-Quirin glass-works to buy all the soda the plant could produce.

Additions were now begun which more than doubled the extent of Carny's factory. Thirty thousand francs were spent on the expansion in 1808, and a further 20 000 francs in 1809, this comparing with a total investment of about 50 000 francs up to 1807. Building was hampered by heavy rain, and the utilization of residues from the neighbouring works was delayed for some months because of difficulties raised regarding the transport of these dutiable materials, but Carny was able to increase his production from 79 000 kg in 1807 to 135 000 kg in 1808, and then to 421 000 kg in 1809. Profits, too, rose sharply, from 37 000 fr. in 1807 to 51 000 fr. in 1808, and 97 000 fr. in 1809. In 1810 production was expected to reach 600 000 kg. It had then reached a limit, however, with all the available *schlot* being fully utilized. Further increase in production would only have been possible by manufacturing sulphate from salt, to supplement the natural supply. Carny pressed for this, but the company would not agree to the heavy additional investment required for the erection of lead chambers.

In the mid-1820s the works was reported by Dupin to be producing a million kilogrammes of soda a year, still using only natural sulphate. Expansion to exploit salt was then projected, and was duly implemented towards 1830. The works was directed by Carny until the later 1820s, and then by his son until 1843.[167] The history of chemical manufacture at Dieuze continues to the present day, when the works belongs to the *Société des usines chimiques Ugine Kuhlmann*.

Pelletan at Rouen

The second of the pioneer manufacturers was Pierre Pelletan,[168] son of Philippe Pelletan, one of France's most distinguished surgeons. Pierre had studied at the *École polytechnique* from 1795 until 1797,[169] after which he had served as laboratory assistant to the physicist Charles, taught chemistry in Paris, and practised as a surgeon. He was a man with wide-ranging interests and abilities in medicine, science, and technology.

In 1804 Pelletan established a small soda works in Paris, almost certainly using the Leblanc process. One of the principal markets he had in mind was evidently the Rouen dyeing industry and in January 1805 he sent to the Rouen Academy a sample of soda crystals of his manufacture. The sample was examined by a commission including Vitalis and Descroizilles, which found the product to be very pure and to give good results in Turkey-red dyeing (a finding confirmed in trials by a number of local dyers). The product had the significant advantage, moreover, of being constant in its properties, in contrast with the notorious variability of natural sodas. The commission presented a highly favourable report on Pelletan's product on 12 June 1805, concluding that 'the Academy cannot too strongly advise and recommend its use'. The mayor of Rouen and the prefect of the Seine-inférieure agreed that it merited the fullest support. Pelletan later wrote that the Academy's report had been of great service to him in propagating the use of his product.[170]

At first Pelletan made his soda in Paris and sold it in Rouen through a local businessman, but at the end of 1805 he moved his works to Rouen in the hope of thereby increasing his sales and of manufacturing more economically. He was now in company with two businessmen, Haag and Muller (from Lübeck), who presumably financed the enterprise and who may already have been associated with him in Paris. The Rouen works was established at 12 rue Tous Vents, in the faubourg Saint-Sever, and it exploited as its raw material sodium sulphate residues from chlorine bleaching. These perhaps derived from the bleaching works of Descroizilles, who is described by one author[171] as having been an associate and sometime mentor of Pelletan. In the course of 1806 Pelletan's product came into extensive use at Rouen, and in September Pelletan, like Carny, received an honourable mention at the Paris exhibition.[172] By mid-1807 he was producing 200 kg a day (equivalent to some 60 000 kg a year) of a purified soda twice as rich in alkali as soda of Alicante.[173] His product seems to have been bought mainly by local dyers, who appreciated its high quality, finding it preferable to the best Alicante soda and capable even of replacing potash. They only wished that it were cheaper.

Pelletan's works was of some significance in helping to set the lead in this new branch of industry, and particularly in introducing the manufacture at Rouen, but it was not itself to develop into a major concern. Its output was small, limited by the supply of by-product sulphate. In the latter half of 1807 Pelletan petitioned the Government for exemption from the recently introduced salt duty, with a view to increasing his production by exploiting salt, and at about this same time the firm acquired a small lead-chamber plant nearby. As we shall see below, however, soda manufacturers were not granted exemption from salt duty until mid-1809, and it seems to have been only then that the works expanded to work salt. Pelletan left the works in 1809 to join Jean Holker in a much larger venture, and the old concern was then continued by Haag, now in partnership with Archambault. It folded in 1811 during the general collapse of the soda industry at Rouen. Pelletan's venture with Holker closed at the same time, and in 1813 Pelletan returned to Paris, where he was henceforth to pursue a career as a physician and teacher. In 1815 he became physician in ordinary to the King, a position in which he enjoyed great favour at Court. He seems to have retained some interest in chemical industry (see n. 61), but his plan to publish a *Traité de la fabrication des soudes*[174] does not appear to have been realized. He is incidentally familiar to historians of science as the Frenchman who in 1820 visited Dalton in Manchester and was astonished to find the great man employed in teaching arithmetic to a young boy with a slate.[175]

Darcet fils *and Anfrye, near Paris*

The most important of the pioneering enterprises was that of the chemists J. P. J. Darcet and J. J. J. Anfrye. The former was son of the Jean Darcet who had headed the commission on soda manufacture in 1794. Darcet *fils* had studied at the *Collège du Plessis* in Paris until its closure in 1789, and had then lived for several years with a family in Burgundy, where he was educated by the family tutor, principally in mathematics and the sciences. He returned to Paris in 1793, but did not enter the *École polytechnique* (established the following year) since his father intended him for the Mint. Instead he attended his father's chemistry lectures at the *Collège de France*, and then at the end of 1800 he was duly appointed assayer at the Paris Mint, where he was to work until his death: in 1805 he became *vérificateur général des essais* there, and he was subsequently *commissaire général des monnaies*. He was one of the leading technical chemists of his generation and a man noted for his modesty, his generosity, and the philanthropic character of much of his work.[176] Anfrye was a former student of the *École des mines* (in the later 1780s), and had subsequently served for a time as an *ingénieur des mines*. In the Napoleonic period he was a professional colleague of Darcet *fils* at the Mint, where he was *inspecteur général des essais*.[177]

The earliest indication that Darcet *fils* and Anfrye were giving their attention to the question of soda manufacture dates from 1802, when on 6 May they deposited with the First Class of the Institute sealed notes describing two general methods for the production of alkalis and alkaline earths.[178] One note concerned itself primarily with soda, although the method described was said to be equally applicable to the production of potash, baryta, and strontia. The note dismissed the familiar litharge process as uneconomical, and also the Leblanc process as being troublesome, in giving a crude product which needed refining. It then proceeded to describe a new method, based on reaction in the wet way between copper oxide and sodium sulphide (the latter made as usual by reduction of the sulphate); the reaction precipitated copper sulphide and left a solution of caustic soda.[179] The copper oxide was regenerated by simply heating the residual sulphide. The caustic soda could be carbonated by leaving the concentrated solution to stand (over sand), and the process was said to yield 150 kg of crystalline carbonate for 100 kg of sodium sulphate. The process described in the second note, although again a general method, was intended to be used in practice specifically for the carbonates of barium and strontium, for which alone it was profitable. The raw materials exploited were the naturally occurring sulphates, and the process consisted in reducing the sulphate to sulphide with charcoal, and then in precipitating the carbonate from a solution of the sulphide by passing in carbon dioxide gas. Anfrye and Darcet *fils* exhibited samples of their barium and strontium carbonates at the industrial exhibition in Paris in September 1802, and were awarded a gold medal for having devised processes which promised to furnish these materials at low cost for industrial use.[180]

By 1803, Anfrye and Darcet *fils* (to whom we shall hereafter refer simply as Darcet, his father now being dead), were engaged in mounting a works for soda manufacture at la Gare, just outside Paris. The process which it was intended to exploit was not, as one might have expected, the decomposition of sodium sulphide by copper oxide, but rather the decomposition of salt by baryta (barium hydroxide). The baryta itself was presumably made by the copper oxide method from natural barium sulphate. Such a plan may be presumed to have offered as a chief advantage its avoidance of the expense of sulphuric acid. Unfortunately, the plan proved to be thoroughly ill-founded, affording a notable example of misplaced faith in the received scientific knowledge of the day. For although the decomposition of salt by baryta was a reaction unquestioningly described by the generality of writers since Bergman, and although the reaction had long been regarded as promising an attractive manufacturing method if baryta could be inexpensively procured, the fact is that the reaction does not actually occur. (The belief that it did was based on such indirect indications as the production of an alkaline odour when the two solutions were mixed.) Anfrye and Darcet, having with remarkable incaution established plant for its exploitation, at first attributed their failure to themselves, and only after long and expensive endeavours did they finally come to recognize the long-standing chemical error by which they had been misled, as they confessed in a paper on the subject in the January 1804 issue of the *Annales de chimie*.

Undeterred by this embarrassing start, Anfrye and Darcet in September 1804 re-established their works on a new basis, at a cost it appears (up to 1807) of about 80 000 francs.[181] The process now employed is not definitely known but it seems likely that it was the Leblanc process. Their main early customer (from 1805) was the Saint-Gobain glass-works, which in 1806 bought three-quarters of their total output of 200 000 kg.[182] It was with this soda that the Saint-Gobain company made the glass for which it won a gold medal at the 1806 exhibition.[183] In May 1807 the partners moved their works from la Gare to Leblanc's old factory at Saint-Denis, which they rented from Dizé into whose hands it had passed after Leblanc's death. At Saint-Denis they installed plant for production on a larger scale, with the aim of supplying the Saint-Gobain glass-works with its total soda requirements, said to amount to 800 *milliers* (800 000 pounds) a year. Their product was said to be equal in strength to the best Alicante soda, and we can now be fairly certain that it was the Leblanc process they were using, selling their product in the crude state. The preliminary conversion of salt to sulphate seems to have been effected partly by an iron sulphate method and partly by sulphuric acid, which they bought in from outside.[184]

Anfrye and Darcet were partnered in their enterprise by two further colleagues from the Mint, Joseph Gautier and his brother-in-law J. B. L. L.

Barréra.[185] The partnership agreement,[186] of 2 March 1808, charged Anfrye specifically with the chemical direction of soda manufacture, while Darcet was to concern himself with a soap factory in which the firm also had an interest (discussed below). Gautier and Barréra were responsible for the firm's commercial direction and the firm went under the name Gautier, Barréra, & Cie. Already before this agreement, we find the enterprise referred to in August 1807 by the name 'Gauthier & Cie', and so it is possible that Gautier and Barréra had been involved with Darcet and Anfrye from an early stage.

As the rising price of imported soda made soda manufacture ever more attractive, the firm went from strength to strength. On 21 September 1808 it bought a six-acre site outside Paris at la Folie, near Nanterre,[187] and there established a subsidiary works for the initial production of the sulphate. The sulphate was then transported to Saint-Denis for conversion to soda. Complaints at Saint-Denis about the hydrochloric acid fumes evolved were the motivation for this transference of the offending first stage of manufacture to a more isolated spot.[188] The Saint-Denis and la Folie works together formed an impressive enterprise for the period. At the peak of its prosperity, in 1809 and early 1810 when the firm was sending large quantities of soda to Marseilles, the Saint-Denis factory had ten soda furnaces and employed a hundred workers, while the two plants together were said to employ up to three hundred. The firm's production capacity was given in a report of 20 August 1810 as 2 million kg of (crude) soda a year.[189]

In the latter part of 1810, Darcet and his partners established an additional works at Quessy (near la Fère, in the Aisne), for which preparations had begun towards the end of 1808.[190] This was intended to produce its sulphate by means of local pyritous materials instead of sulphuric acid. It was for the exploitation of this works that the partners formed a second soda company on 28 September 1810, with an initial capital of 200 000 fr., half of it provided by a sleeping partner, René Jacquemart. After a widening of the partnership six months later the new firm went under the name Auguste Jacquemart & Cie.[191] Auguste Jacquemart was a wallpaper manufacturer in Paris, a successor of Réveillon at the well-known factory in the rue de Montreuil, while his brother René was a principal partner in a Paris banking concern, the *Comptoir commercial Jabach*.[192] In the event the Quessy works seems to have made soda only briefly, if at all, but it was developed for the manufacture of copperas and alum, and was described in 1816 as one of the department's most important works in that line. It still retained a pre-eminent position in 1840,[193] and in 1853 continued to figure in the *Almanach du commerce* in the name of René Jacquemart's widow.

Besides making soda, the Saint-Denis company also had an associated interest in the manufacture of hard soap, a branch of industry which had formerly been virtually monopolized by Marseilles, but which became established on a significant scale in Paris with the growth of the soda industry there. In this development a central figure was J. G. Decroos, formerly a soap manufacturer in Calais. Decroos succeeded in producing a fine toilet soap similar to the highly reputed Windsor soap from England, a significant achievement since hitherto, although the Marseilles industry made common soap in large amount, France was said to have been dependent on England for her toilet soap. In 1804 (probably) Decroos established the manufacture of his toilet soap at Bagnolet, to the east of Paris, and in the middle of 1806 he set up a second works in the rue Culture Sainte-Catherine. He received an honourable mention at the exhibition in September of that year, and his products rapidly acquired popularity, enabling Cadet de Gassicourt to write in December that 'now, in Paris, people want no other luxury soap than that of M. Decroos'.[194] Besides toilet soap Decroos also made household soap, and his production in 1807—according to a Ministry of the Interior report of 24 November—amounted to 750 kg of soap a day.[195]

It was presumably to Decroos's works that the prefect of the Seine referred later, when he remarked that since 1806 Gautier, Barréra, & Cie had supplied two soap factories whose formation in Paris had been determined by the security of the soda supply their company offered. What part the soda company might have played in the establishment of the soap-works is not known, but from May 1807 it had a share in the soap firm in the name of Darcet. The Jacquemart brothers were also partners.[196] In mid-1809 the soap company moved its works into premises owned by René Jacquemart in the rue de Montreuil, with the aim of giving a greater extension to its operations, and when the works was visited some months later by members of the First Class of the Institute, they found it producing 7–8 *milliers* (7–8000 pounds) of soap a day, for which it consumed 4 *milliers* of artificial soda.[197] In November 1813, a report by Deyeux for the Paris *Conseil de salubrité* spoke glowingly of the factory's exemplary organization under the direction of Darcet: it was now able to produce the impressive quantity of some 30–4 *milliers* of soap a day, and was in addition making soda (for internal consumption), and also alum and copperas (from materials presumably brought from Quessy).[198] A biographer of Darcet later described the works as having been 'perhaps the largest soap-works which has existed'.[199] Darcet took a close personal interest in soap manufacture and was credited by the jury of the 1819 exhibition with having played a major part in the establishment of the manufacture in Paris since 1806:

There are here employed, to make the most sought-after soaps, materials which had previously been of little value. The processes are due to M. d'Arcet, who has carried them to a high degree of perfection.[200]

In 1815 he was commissioned by the First Class of the Institute to write a treatise on the manufacture of soap and soda,[201] intended as a continuation of the old Academy's *Descriptions des arts et métiers*, but the work does not seem to have appeared.

When in 1810 the First Class of the Institute came to award the Decennial Prize for industry—for the outstanding industrial establishment of the preceding ten years—Darcet and his Saint-Denis partners were among the seven firms short-listed for the award. The prize in fact went to Oberkampf, for his printed-cotton works at Jouy, but Darcet and company received an honourable mention for their role in the development of soda manufacture:

for the importance of the products, for the perseverence with which the proprietors have fought against the difficulties, finally for the chemical knowledge which has prepared their success, the establishment of M. Darcet and company has appeared to merit very honourable mention.[202]

The later history of the concern will be discussed below.

The salt duty, obstacle to expansion, 1806–1809

By an ironic coincidence, just as the rising price of imported soda began to make its manufacture a matter of increasing interest, the Government unwittingly put an obstacle in the way of the new industry's growth by in 1806 re-imposing an excise duty on salt. We have seen that the ancient *gabelle* had been abolished in 1790 as one of the most hated taxes of the *ancien régime*. It had been a major source of Government revenue, however, and as such it was missed in the years of budgetary difficulty which followed the Revolution. From the late 1790s it consequently began to be argued that the *gabelle* had been objectionable only because unfairly levied, and the possibility of instituting a new salt duty was mooted. Eventually, in the spring of 1806, such a duty was introduced, accompanied as a gesture of compensation by the abolition of tolls on main roads (a system which had proved very unpopular since its imposition in 1797). Only the curing of fish and the salting of provisions for the navy and colonies escaped the new duty.[203]

At 20 francs per 100 kg the new salt duty was quite a heavy one. It brought a three- or fourfold increase in the price of salt, from about 8–10 fr. to about 30–2 fr. per 100 kg (in Paris),[204] and was thus a matter of some concern to those chemical industries that employed salt as a raw material. Chaptal quickly sprang to the defence of its chief users, submitting a memoir to Napoleon in which he proposed exemption from the duty for salt employed in chlorine bleaching and in the manufacture of sal ammoniac. The matter was referred in June to the Ministry of the Interior for advice, and the Ministry's *Bureau consultatif des arts et manufactures* was asked for a report.[205] It was nearly a year before the *Bureau consultatif* reported, perhaps because of the extra burden of work on the administration at this time caused by Napoleon's absence. But when the report came, on 4 June 1807, it argued strongly and at length in favour of exemption, particularly for bleachers and manufacturers of sal ammoniac, who it was felt would both be severely affected otherwise. Exemption was advocated, too, for manufacturers of hydrochloric acid and tin chloride, and for those using salt in the production of glass and pottery, and in the preparation of leathers. Any danger of abuse might be prevented, the report proposed, by denaturing the salt to make it suitable only for industrial use, and by keeping a check on the salt consumption of factories granted exemption.[206] The Minister of the Interior transmitted the report to the Minister of Finance, who firmly opposed its recommendations, however, on the grounds that exemption might be abused.

In this report the manufacture of soda figured only rather incidentally, as a consumer of by-product sodium sulphate. Soon afterwards, though, it began to become clear that soda manufacture might be an important new factor in the exemption issue. By the late summer of 1807 the Minister of the Interior had received a number of petitions which suggested that the nascent soda industry was ready to develop into a major consumer of salt in its own right.[207] Thus, Darcet and company had announced that they were preparing to expand their works, but they threatened that they would on the contrary have to close it down if not granted exemption from the duty. Clément, Desormes, and Montgolfier, on 15 August, solicited exemption on their own behalf, announcing that they wished to undertake soda manufacture by utilizing the pyritous deposits from which they were already making alum and copperas in their important factory at Verberie, in the Oise. At the beginning of September Pelletan appealed for exemption (or alternatively a compensatory premium), so that he could increase his production at Rouen; his petition was strongly supported by the prefect of the Seine-inférieure, and was accompanied by certificates from eleven of the foremost dyers in the town, attesting to the quality of his product. And along with Pelletan's petition there also came one from Descroizilles, who declared that if the salt duty were removed he intended greatly to increase his manufacture of hydrochloric acid and tin chloride, and then to produce soda from the large quantities of sulphate which would result. These various petitioners drew attention to the major potential importance of the manufacture, arguing that if exemption were granted the industry would soon develop so as to break France's dependence on imports, to the great advantage of her industry and of her trade balance.

The Minister of the Interior responded on 21 September by ordering a special report on the subject. This was to indicate the current extent of soda manufacture compared with imports; it was to examine whether, with Government protection, the industry might one day put an end to importation; and it was to suggest means of preventing abuse if exemption were granted. The requested report was duly produced by the Ministry's *Bureau consultatif* on 24 November.[208] It found that already a significant amount of artificial soda was being produced, estimating the works of Carny, Pelletan, and Darcet to be together making the equivalent of some 646 000 kg of Alicante soda a year. This was compared with an average import figure of

4·42 million kg a year, for the Years X–XIV (1801–5), although it was admitted that this official figure was probably far below reality.[209] The report argued strongly that the industry should be exempted from the tax which was impeding its further development.

This kind of industry, altogether new for France, merits particular attention. The results already obtained are unequivocal, ... the manufacture of soda is a thing assured, and ... it needs only to be encouraged to acquire a very great extension.

Everything pointed to a prosperous future for the new industry. If exemption were granted, then even in peace-time artificial soda should be able to compete in price with soda from abroad, with at most only a small protective import duty. (Artificial soda currently sold at 90 fr. per 100 kg, and it was thought unlikely that peace would bring Alicante soda below its 1790 price of 70–80 fr. per 100 kg.) As for the prevention of abuse, the report here proposed that the best method would be to establish locked salt stores in the soda works under the key of a tax official, who would then deliver salt to the manufacturer in proportion to the soda made. One kilogramme of salt would be allowed for every two kilogrammes of soda produced, this being the quantity considered to be necessary for a product similar in strength to soda of Alicante. As further precautions the report again suggested the possibility of denaturing the salt, and of keeping a check on the factory's books. It was sure that by such methods abuse could be effectively prevented.

On the basis of this report the Ministry of the Interior renewed its endeavours to secure exemption for industrial users, and in particular now for manufacturers of soda. The need for soda manufacture was ever growing. After the prohibition of trade with Sicily early in 1808, and the insurrection in Spain in the middle of the same year, France's imports from her two main suppliers of soda were interrupted, causing the price to soar to very high levels. By early 1809, French industry—and in particular the soap industry of Marseilles—was threatened with serious soda shortage. The pressing solicitations of the Ministry of the Interior were for long opposed by the Ministry of Finance, which insisted that even if abuse could be prevented the necessary measures would involve unjustifiable expense. Eventually, however, on 9 May 1809, the *Conseil d'état* adopted a decision granting special exemption from duty for soda manufacturers alone (not for any other branch of industry). The decision received the approval of Napoleon, then at Ebersdorf, on 4 June.[210]

Before the decision could be put into effect a method of administering the exemption had to be worked out. This was in due course instituted by an imperial decree of 13 October.[211] The system adopted was essentially the same as that which the *Bureau consultatif* had proposed nearly two years earlier, and depended on a close quantitative check being kept on the salt by a customs or tax official (according to the location of the factory). The inspector checked the salt on its arrival at the works, and the manufacturer had to pay quadruple duty on any lost in transit. The salt was then put into the factory's store, which was locked with two keys, one of them kept by the manufacturer and the other by the inspector. In proportion as salt was needed for manufacture it was removed from store in the presence of the inspector. Records were kept of the quantity of salt in store and of the soda made and sold, so that it could be seen that the quantity of salt used corresponded with that of soda made: 50 kg of tax-free salt were allowed for the manufacture of 100 kg of soda. If the manufacturer used more salt than this he paid tax on the extra amount. To cover the administrative costs of exemption each manufacturer was required to pay 4000 francs a year for the privilege.

It will be seen that the exemption machinery involved a good deal of red-tape, and this brought some minor complaints from manufacturers about the time and expense it entailed. Manufacturers were naturally hardly pleased either at having to put up with the prying eyes of Government inspectors on their premises.[212] There may have been some cases, moreover, of inspectors interfering in the running of factories in an unauthorized and detrimental way: Darcet in 1814 complained of their preventing manufacturers from collecting and utilizing their hydrochloric acid (on the grounds that the exemption was solely for the production of soda).[213] Such cases were probably rare, however. On the whole the system seems to have worked quite satisfactorily, and it appears to have remained in force without essential change for the rest of the nineteenth century.

The general establishment of soda manufacture, 1809–1810

Already before exemption was granted the market had become so favourable as to persuade some new manufacturers to establish works, or to begin preparations, and when exemption finally came a very rapid expansion ensued. The remarkable proliferation of ventures which now occurred owed something to the fact that an attempt at soda manufacture did not necessarily call for a very large expenditure on plant. As Pelletan remarked in 1810, whereas the manufacture of sulphuric acid demanded a heavy capital investment in chambers, soda could be produced in sizeable quantity for a comparatively modest outlay on furnaces.[214] Of course, there arose many large, highly capitalized alkali concerns which besides making soda made their own sulphuric acid, too, but there were also at first many small ventures which did not. With the general establishment of manufacture the Leblanc process rapidly demonstrated its supremacy and became the basis of virtually all significant undertakings.

In the summer of 1810 the growth of the industry was promoted by further measures of Government encouragement and protection. In June the Minister of the Interior proposed to Napoleon that the charge made for the privilege of exemption should be reduced from 4000 fr. to 1500 fr. a year, and that the duty-free salt allowance should be raised from 50 to 67 kg for each 100 kg of soda made. Napoleon granted both requests by an imperial decree of 18 June, 'wishing to give to the manufacture of artificial soda in France new proofs of the interest we take in this kind of industry'.[215] More important than these detailed improvements in the conditions of exemption, however, was the action taken soon afterwards to protect the new industry against a threat from imports. There appears to have been some revival of imports early in 1810 through the granting of licences for the introduction of soda from Sicily (trade figures for 1810 record the entry of soda to the value of 9½ million francs—about two-thirds of it from 'Angleterre', i.e. Sicily—a figure which can be compared with total imports in 1808 evaluated at 4½ million francs).[216] The prefect at Rouen on 16 May informed the Minister of the Interior that the introduction of foreign soda was obliging local factories to work at only half their capacity, and a month later he transmitted a petition from the manufacturers demanding that imports be banned; the manufacturers had also made representations to Napoleon himself during his recent visit to the town.[217] At Saint-Denis Darcet's factory was severely affected at about this time and was said to be working at only one-tenth the level of a few months earlier.[218] The Minister responded by securing a decree on 11 July which imposed a total prohibition on soda imports, so allowing the industry henceforth to develop free of major competition from natural sodas.[219] Darcet, on 30 August, speaking as a member of the newly established *Conseil des fabriques et manufactures*, remarked that the Government had now done all that could be wished in the industry's favour.[220]

In order to gain an idea of the extension manufacture had acquired, the Minister of the Interior in April 1810 sent out a questionnaire to the prefects of the eighteen departments where he thought it most likely that the industry might have become established. The replies which came in during the following months showed that by mid-1810 manufacture was established in five of these departments, where there were already some 27 plants in activity and at least 3 more still in course of being installed, with a total intended production exceeding 24 million kg (crude) a year.[221] Since the returns were not always complete, and soda was also being made in some departments not covered by the survey, the total number of works engaged in soda production by mid-1810 must have been rather over 30. The Paris manufacturers, indeed, claimed that there were by then 70 works, capable of producing 30–40 million kg a year, but this was no doubt an exaggeration. The industry was continuing to grow, however.

The important centres of manufacture were the regions of Paris, Rouen, and Marseilles, and the departments of the Aisne and the Hérault.

Paris

In the Paris region Darcet's enterprise remained pre-eminent but soda also came to be made, for a time at least, by some half-dozen other producers.[222]

Payen, of course, was continuing to make soda in his sal ammoniac works at Grenelle, and he probably now greatly increased his production by working salt as well as by-product sulphate. In the Prefect's report of August 1810, his intended output was given as 600 000 kg a year, and twice that if foreign sodas were prohibited. The capital's other large sal ammoniac manufacturers, the Pluvinet brothers at Clichy, close associates of Payen, also developed an interest in soda. In 1809 they set up a small pilot plant near Belleville (to the east of Paris), but they very soon abandoned this when they wisely decided to erect their soda works near Marseilles instead. In their sal ammoniac factory at Clichy they continued to produce soda in a small way with a single furnace.

Soda manufacture was also taken up by the established chemical firm of Marc, Costel, & Cie, under the direction of Dizé, Leblanc's old partner. In its works at la Glacière this firm was occupied with the manufacture by August 1809, when the resulting fumes provoked complaint from the mayor of Gentilly. The fumes problem led the company two or three months later to remove the manufacture to an isolated spot near Belleville, where they evidently operated on quite a substantial scale: the Prefect in August reported their intended output to be 1 200 000 kg a year. They were soon again in trouble over their acid fumes, however. The plant was a cheap and hasty affair, its furnaces sheltered by a simple hangar, with no attempt made to condense the hydrochloric acid. In applying for authorization the company had claimed that the acid fumes would automatically be neutralized by ammoniacal vapours arising from a large sewage reservoir nearby! In reality the fumes travelled to Belleville and Pantin, and by October 1810 the Prefect of Police had ordered the firm to suspend its operations.

Three other ventures in the immediate vicinity of the capital were all rather smaller. In November 1809 Courcillon, Girard, & Cie sought authorization to start production at a works they had just erected on the plaine de Billancourt, near le Point-du-Jour. It had cost 20 000 francs and was said the following August to be producing up to 2–3 *milliers* of soda a day (equivalent to some 300–450 000 kg a year). By October 1810 soda was being made, too, by a pharmacist called Destouches, in a works at les carrières de Charenton for which he had requested authorization in January. This was continuing the following July, when Destouches was reported to have greatly improved his plant for the condensation of the hydrochloric acid. The third enterprise was at la Gare, where by April 1812 we encounter a certain Huskin making soda by a pyrolignate process.[223]

Finally, we may note a venture about twelve miles west of Paris at Saint-Germain-en-Laye. A certain Chauboy in February 1809 informed the Ministry of the Interior that he intended to form a works there, using 'new processes'. Trial samples which he submitted for the Ministry's opinion were reported by Vauquelin to be the worst he had ever seen, with only 13 per cent alkali and a great deal of sulphide, but this does not appear to have deterred

Chauboy from proceeding with his enterprise, for later in the year we find him sending samples of his soda (and also of soap made with it) to the *Société d'encouragement* in Paris.[224]

Rouen

At Rouen, the works established by Pelletan and Haag seems to have expanded to exploit salt instead of only by-product sulphate from about August 1809, and the second half of the year also saw the installation of soda furnaces by four sulphuric acid manufacturers in the Rouen region. Lefrançois began making soda in his works at Déville in July, and he was soon followed by Holker (now joined by Pelletan), by Le Bertre, and by Dubuc, who had all begun production in the faubourg Saint-Sever by the end of the year.

The new manufacture immediately provoked complaints because of the hydrochloric acid gas emitted, and in November this led Lefrançois to set up a separate plant for the production of his sulphate on the bruyères Saint-Julien, an isolated stretch of wasteland to the south of Rouen. The sulphate was then transported back to his main works for conversion to soda. The four manufacturers in the faubourg Saint-Sever tried at first to overcome the problem by condensing or absorbing the acid gas, but although various methods were tried none proved entirely satisfactory and complaints continued. In January 1810 their works were inspected by the Prefect, who by an *arrêté* of 15 January then ordered them to move the first stage of manufacture to the bruyères Saint-Julien, as Lefrançois had already done. They were allowed to continue operating in their main works until March, since the fumes would not damage crops in the winter, but they each then erected a subsidiary workshop for sulphate production on a specially designated area of the bruyères.[225]

The principal Rouen concern was that of Holker and Pelletan, which in February 1810 was making 3500 kg of crude soda a day, with 10 furnaces for sulphate production and 8 for the subsequent production of soda. About six months later Pelletan could claim their works to have 14 sulphate furnaces and 12–15 soda furnaces. In May 1810 the factory's intended production was reported by the Prefect to be a little over a million kilogrammes of crude soda a year, while that of the four other manufacturers was about ½ million kg each. The industry's total intended production was nearly 3 million kg, and the various works were said to be equipped to make double this if the market required.[226]

The Aisne

The first soda factory in the department of the Aisne was established by Pajot Descharmes in 1808. Descharmes had been interested in soda manufacture for a good many years. He later spoke of having experimented with the lead oxide method as early as 1784, and of having verified the suitability of artificial soda for the manufacture of glass, presumably at about that same time. Between 1779 and 1784 Descharmes had worked as an *aide de fabrique* at the Saint-Gobain glass company's Tourlaville works, and so his experiments were probably connected with the soda trials conducted by Hollenweger at Saint-Gobain and at Tourlaville in 1782–3 (see p. 202).[227] After the Revolution, Descharmes in 1798 and 1799 made large-scale trials of a process which used iron sulphate or alum to produce sodium sulphate from salt, perhaps then employing the Leblanc process to convert the sulphate to soda. These were at the Tourlaville works (of which in 1797 Descharmes became assistant director), and subsequently in Paris, in the quartier Saint-Antoine, no doubt in the Saint-Gobain company's workshops there.[228] Descharmes described his methods in a sealed note deposited with the First Class of the Institute in April 1803, but we do not have this since he withdrew it in 1832.[229] Between 1805 and 1807 Descharmes was director of the Saint-Gobain company's principal works at Saint-Gobain itself (in the Aisne),[230] and it was during this period that the works began using artificial soda from Darcet's factory near Paris. It was during this period, too, on 13 June 1806, that the company bought an old glass-works at Charles-Fontaine, nearby, with the aim of establishing a soda factory there to supply its own needs, although this plan was not to be realized until 1809–10.

In 1807 Descharmes left his post at Saint-Gobain, and on 21 November he petitioned the Minister of the Interior for exemption from salt duty for a soda works he was proposing to establish at la Fère, a short distance away. Although exemption was not forthcoming, he did establish a works in the summer of 1808, not in fact at la Fère but at Soissons in the same department. He there used local copperas or pyrites to convert salt to sodium sulphate, and then produced soda from this by the Leblanc process. The Saint-Gobain glass-works was one of his customers. By mid-1810 his works had become a considerable establishment, proposing to make 3 million (pounds presumably, i.e. 1½ million kg) of soda a year, and employing 80 workers; the firm was intending to increase its labour force to over 100 as it had begun making soap, too. In 1811 it was said to be employing 200 workers and to be producing 1½ million kg of soda a year.[231]

In October 1809 work was begun at a second factory in the department, at Chaillevet, near Laon. Louis Carpentier, a *négociant* and lace manufacturer in Paris, had in 1807 established a works at Chaillevet for the extraction of alum and copperas, and this now took up the production of soda as an additional interest. Local pyritous materials were used for the initial production of sodium sulphate, employing a method devised and introduced into the works by the mining engineer Lefroy. The works was proposing to make about ½ million kg of soda a year. According to the Prefect's report of September 1810 it had still not produced any soda, because the buildings were not yet ready, but it was engaged in the production of sodium sulphate in preparation. It was evidently not to pursue soda manufacture for very long, since in January 1816 it was reported to have abandoned the manufacture some years ago.[232]

In January 1810 work began in the soda factory established at Charles-Fontaine by the Saint-Gobain glass company.[233] This was an event of particular note, of course, in marking the first entry into the field of chemical manufacture of a company destined in our present century to become one of Europe's giant chemical concerns. The firm's intention in first taking up the manufacture, however, was simply to cater for its own needs. The Prefect's report in September 1810 indicated that the works had not yet produced any soda, but it was evidently making sulphate in readiness. There appear to have been a good many teething problems, with the furnaces initially erected—on the basis of the Darcet report—soon having to be torn down and rebuilt. Adenis-Colombeau *aîné*, the Saint-Gobain company's *contrôleur*, wryly remarked in 1816 that 'the founders of this works were far from having the talent of Amphion'.[234] He spoke of the enormous expense the works had entailed, and it can be seen from his figures that the establishment cost (including purchase of the property) had amounted to over 180 000 francs between 1809 and 30 June 1812. At first the works tried the production of sodium sulphate by means of pyritous materials, but it quickly adopted sulphuric acid instead and up to 30 June 1812 is recorded to have spent no less than 835 569 fr. on the purchase of acid. The high cost of transporting the acid to Charles-Fontaine, and the problem of breakages *en route*, led the firm before long to install a lead-chamber plant of its own, assisted by a certain Valette who joined the works from one near Rouen. Three chambers were built in the years 1811–13 and then a further two in 1816–17. The director at Charles-Fontaine in its early years was the chemist Tassaert, a former pupil of Vauquelin and later director of the Saint-Gobain glass-works.[235] It is perhaps of incidental interest to mention that it was Tassaert, in about 1813, who made the first observation of artificial ultramarine, as an accidental product of his soda furnaces at Charles-Fontaine.[236]

Besides the three enterprises we have described, the Prefect's report in September 1810 spoke of soda manufacture being projected by several others, too, including Dupuis (the bleacher and sulphuric acid manufacturer at Saint-Quentin), and Belly de Bussy & Cie (manufacturers of copperas and alum at Cuissy since about 1800).

The Hérault

In the department of the Hérault, in the south, the first to take up soda manufacture seems to have been the pharmacist Louis Audouard, who established a works at Béziers in May 1808 where he extracted potash from vegetable ashes as his main early activity but where he also made artificial soda (with a single furnace in December 1809). By July 1809 soda manufacture in a small way had also been begun by Bérard and Martin, in their old-established chemical works at Montpellier. Others followed in 1810. At the beginning of the year François Planche *aîné* in partnership with Martin *fils aîné* and Henry Reboul[237] began soda production at Pézenas. This was as a new addition to a works they had recently begun there for a range of chemical manufactures, including the distillation of marc (from pressed grapes), and the production of potash, lead acetate, and sulphuric acid. The works was highly spoken of by the mayor and employed 30 workers in mid-1810, when Planche *aîné* was counted among the department's leading industrialists. In April 1810 soda manufacture was taken up at Montpellier in the sulphuric acid works of Valedau & Laurent; and by May, Jessé & Cavaillé *frères* had formed a works at Béziers. According to the Prefect in June, these various concerns were proposing to make a total of 2 360 000 kg a year, the largest producer listed being Audouard, with a production of 40 000 kg a month and an intended output of 800 000 kg a year.[238]

The region of Marseilles

By far the largest centre of manufacture to arise was in Marseilles and its surrounding department of the Bouches-du-Rhône. The reasons for this are easily seen. In the first place the Marseilles soap industry was France's biggest single consumer of soda. Before the Revolution the industry had been producing some 20 million kg of soap a year (and on occasion perhaps as much as 40 million kg), consuming for this purpose roughly 10 (–20) million kg of natural soda of all kinds.[239] Although under the Empire soap production was rather diminished, through loss of markets, nevertheless towards 1810 it probably still amounted to some 10–20 million kg a year, requiring about 5–10 million kg of soda. In the second place the region of Marseilles was excellently provided with materials for the manufacture of soda: there were numerous salt-works and plentiful chalk deposits in the region, coal was abundantly available from the nearby department of the Gard, and the port of Marseilles received sulphur direct from Sicily. Despite these favourable circumstances, interest in soda manufacture at Marseilles was in fact late to develop. The early efforts at manufacture in France—from the abortive schemes of the 1780s to the first successful ventures two decades later—had been almost entirely the work of chemists, chemical manufacturers, and glass manufacturers in the north. It was only in 1809, when the Marseilles soap industry was threatened by a dearth of natural soda following the interruption of supplies from Sicily and Spain, that artificial soda began to attract attention there. In the course of 1809 Marseilles soap makers began to buy soda from Darcet's works near Paris. And at the same time soda works began to be thrown up in rapid succession in the region of Marseilles itself.

We cannot here attempt the detailed unravelling of the industry's early growth at Marseilles, and will have to content ourselves with an outline.

The first soda works in the region seems to have been that of Vasse & Cie, established early in 1809 at the salt-works of the Étang de Rassuen, about 25 miles north-west of Marseilles.[240] This plant was installed within the confines of the salt-works and hoped thereby to escape liability for payment of salt duty. It was probably in production by mid-1809. The Vasse in question is presumably to be identified with the mathematician of that name who was an active member of the Marseilles Academy, and who in 1806 had there spoken of the opportunities the region offered for soda manufacture.[241]

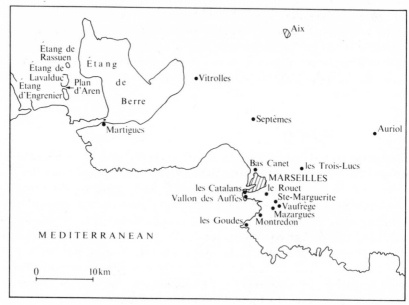

Fig. 4.6. The region of Marseilles. (Based on the map in Chanlaire (1815), vol. 1.)

In the later months of 1809 and the early part of 1810 works multiplied to an extraordinary degree. The list which the Prefect sent to the Minister of the Interior on 14 July 1810 indicated no fewer than 14 factories in the department, of which 10 had already begun work.[242] The total number of ventures, moreover, was a good deal higher than this, for the draft of the Prefect's list included a further 11 enterprises, of which at least 3 had already been abandoned and 4 had not yet begun work.[243] These further ventures, all in Marseilles itself, would seem to have been only very small affairs, however, and some may never have got beyond the stage of projects. Some of them were no doubt among the small workshops which the Prefect mentioned as having been formed in the beginning to make soda by the agency of tartar, but which had subsequently been abandoned since they could not compete with the large factories using sulphuric acid (i.e. the Leblanc process). According to one observer of the Marseilles scene—F. E. Fodéré, who as a member of the *Société de médecine* was active in inquiries into the new industry's pollution problems—over 30 works arose at Marseilles between the summer of 1809 and the end of 1810, briefly operating by a variety of processes which appear to have included the decomposition of salt by potash, by tartar, and by litharge.[244] The 14 works included by the Prefect in his list of 14 July were declared to have an intended production totalling over 11 million kg a year, though in practice their production at first was no doubt considerably below this (a figure of 1 893 300 kg was later given for the industry's actual production in 1810).[245]

Four particularly large enterprises figured in the list, each intending an annual production of $1\frac{1}{2}$ million kg or more, and it is noteworthy that three of these were created by chemical manufacturers from Paris. One was a factory established by the sons of Chaptal and Berthollet on the Plan d'Aren, not far from the Étang de Rassuen mentioned above. The beginnings of this enterprise were tragically clouded by the suicide of Berthollet *fils*, on 14 March 1810, partly precipitated perhaps by the strain of final preparations for the start of production. Despite this blow the works is recorded to have begun production on 15 March, and in July was said to be proposing to make 1·65 million kg a year, and even 2·5 million if there was the market. The factory was situated next to the salt-works of the Plan d'Aren, and was presumably formed in association with the salt-works owner, A. M. Bodin (a banker and *négociant* at Lyons), who had informed the Minister of the Interior at the beginning of 1809 that he was thinking of undertaking soda manufacture since reduced sales had made his salt-works no longer profitable.[246] The second major works to be formed in the region was mounted by the Pluvinet brothers (the sal ammoniac manufacturers from Paris), financed by a Marseilles soap manufacturer, Jean Bérard. It was located just outside the salt-works of the Étang de Rassuen, and was hoping to begin production in June, with an intended output of 1·8 million kg a year. The third concern was at Montredon, in the commune of Marseilles, and was established by Michel Gautier, younger brother of the Joseph Gautier whom we have met as a partner in Darcet's works near Paris. Michel had been actively involved in the Paris enterprise, but the Marseilles venture appears to have been an independent concern of his own rather than a direct subsidiary of that in Paris. It began production in April 1810 and was proposing to make 1·5 million kg a year. We later (by 1812) find it under the name of Gautier, Rabinel, & Cie. Finally, work began in May 1810 at a factory established in Marseilles itself, in the rue Saint-Maurice, by J. B. Vidal *fils*, a prominent banker and owner of a soap-works in the town; this again intended to produce 1·5 million kg a year. Besides soda these four leading concerns all made sulphuric acid too. Their importance may be judged from the fact that they all figured high among the fifteen names which the Prefect forwarded to the Minister of the Interior in mid-1810, when the Minister requested details of the department's principal industrialists. Bérard's works was then said to employ 120 workers, and the others 100 each.[247] Later figures show Chaptal *fils* to have been employing nearly 300 people in the spring of 1812, in the combined operations of his soda factory and the associated salt-works. Vidal *fils* was said at the beginning of 1813 normally to employ 63 workers, and Gautier 50–60.[248]

The remaining works on the Prefect's list of July 1810 nearly all had an intended production less than half that of the major concerns just described, but they were still by no means negligible. Soda was made at Vitrolles by Jacques Ricaud; at Aix by Olive (or Holive) *frères*, Berthe & Girard; and at les Trois-Lucs by Jules Baux (financed by Martin *fils d'*André, a *censeur* of the *Banque de France* and vice-president of the *Conseil général du commerce* in Paris). At Septèmes[249] there were three works: one formed by Blaise Rougier in company with a soap manufacturer called Mercoeur; another by Jean Toussaine Louis (a Marseilles pharmacist) and company; and a third begun (but since suspended) by Charpentier, another soap manufacturer. Another three manufacturers had formed works in the commune of Marseilles itself: Jean Baptiste Esprit Olive near the vallon des Auffes; Antoine Girard in the rue Saint-Suffren. with sulphate furnaces at Vaufrège and Montredon; and J. B. Michel (whom we have earlier encountered as a sulphuric acid manufacturer) in the rue Perrier, with sulphate furnaces at Vaufrège. Still more undertakings subsequently appeared. By the end of 1812 we find another two listed in the commune of Marseilles: Pierre Covello at Mazargues and Jean Baptiste Pontier at Sainte-Marguerite.[250] At la Bourine, near Auriol, a works was formed by Armand, Gazzino, and Deschamps, following prefectoral authorization of 25 April 1811.[251] And two more works were mounted at Septèmes, by Mallet and 'Durbec' (Dubuc?), both, it is interesting to note, from Rouen. It was perhaps to them that a writer in Rouen later referred when he recalled that some of the manufacturers in Rouen and Paris, on finding the market insufficient in the north, had formed works at Marseilles.[252] A further example of a chemical manufacturer from elsewhere being drawn to Marseilles by the opportunities of the new industry there was Charles Kestner of Alsace, whom we find listed at the beginning of 1811 as a producer of sulphuric acid at le Rouet (just outside the town) since 1809, and who also had a works at Aix.[253]

The development of the soda industry attracted the interest and support of the Marseilles Academy, which in its 1809–10 season devoted a considerable amount of attention to the subject.[254] A commission composed of Vasse, Laurens, Besson, Lautard, and Robert was named to occupy itself with the industry's improvement. Vasse read a memoir countering the popular prejudices which had arisen against soap made with artificial soda, and giving an account of some comparative experiments he had conducted on artificial and natural sodas. The pharmacist Laurens was working on techniques of estimating the strength of alkalis. Besson read a memoir discussing the legislative measures required by the establishment of objectionable factories in the heart of towns. Fontanier, a member of the *Société des amis des sciences d'Aix*, presented a memoir on the theory of soda manufacture. And Descroizilles, while in Marseilles on business, spoke to the Academy about his alkalimeter at several of its meetings, and read 'a memoir relating to the evaporation which artificial sodas undergo through the sole influence of the atmosphere, to the sulphides which are there formed, and to the admixtures which contaminate them'. In the 1810–11 season Laurens presented observations on the alkaline sulphides contained in artificial sodas, on means of detecting and estimating them, and on means of removing sulphides from lyes made with such sodas.[255] At its public meeting of 6 May 1810 the Academy offered a prize of 600 francs for the best memoir on the subject of soda manufacture, observing that 'Each of the soda factories newly established has thought it necessary to surround its operations with a secrecy harmful to their progress', and that in consequence 'several manufacturers, for want of knowledge, whose sources they did not know how to discover, have fallen into serious errors which have occasioned them considerable losses'. It was hoped that the competition would produce 'a good manual which can guide all the numerous soda manufacturers of the Midi in a sure manner'. The prize was not awarded in April 1811, as intended, but on being re-offered for 1812 it went to the manufacturer Rougier, whose piece was then published in the Academy's *Mémoires*.

It was only with some reluctance that the Marseilles soap industry adopted the use of artificial soda. A memoir addressed to the local chamber of commerce on 20 July 1810—provoked by the prohibition of soda imports and signed by a large majority of the town's soap manufacturers—complained particularly of the odour (hydrogen sulphide) which artificial soda produced in use, and of the smell and causticity of the resulting soap. Showing a somewhat shaky understanding of chemistry the authors ascribed the fault to the presence of sulphuric acid:

The soda alkali contained in marine salt being capable of development only with the aid of sulphuric acid, the artificial soda which results retains this acid in the amalgam of various materials which compose it. This acid is incorporated with the oil in manufacture; it manifests itself by the most foul odour in the operation of *empâtage* [saponification] and of boiling; the vapour evolved at the end of the *empâtage* goes to the head; it singularly affects the vision, and consequently cannot but be very harmful to the health of the workers ... The soap which results from these sodas remains impregnated with sulphuric acid: one recognises it in the odour and still more in the use of this soap, daily experience of which attests to the property it has of destroying linen by its causticity.[256]

Such complaints were no doubt exaggerated but not entirely without foundation. We can be sure that in the early days of manufacture some very bad soda was produced, and even well made soda would require a little getting used to since it differed somewhat in its properties from the natural product. Nevertheless, even the soap-makers themselves had to admit in their memoir that they were almost all in fact employing artificial soda. They adopted it for the simple reason that they had little option, given the scarcity of the natural article. A note by an official at the Ministry of the Interior, probably of about August 1810, recorded that in the past eighteen months the soap-makers of Marseilles had used 120 000 *quintaux* of artificial soda (5 or 6 million kg), drawn from Paris and Rouen as well as from the factories around Marseilles itself.[257]

Elsewhere

A number of works elsewhere can be mentioned briefly. Between 1806 and 1808 soda began to be made at Pellerey in the Côte-d'Or, at the chemical works of J. B. Mollerat (see p. 311). This works had as its basic concern the distillation of wood to obtain pyroligneous acid, and Mollerat made his soda by a pyrolignate process patented in 1806, apparently exploiting natural sodium sulphate from the salt-works at Montmorot.[258] His soda was well reputed for its high quality but in view of the process used is unlikely to have been made in very large amount. The other factories were probably all Leblanc works. By mid-1810 there was a works producing sulphuric acid and soda at Couternon, near Dijon, and another at Carpentras (see pp. 39, 41). There may also have been a works at Toulouse.[259] By April 1811 a works had arisen at Couëron, near Nantes, financed by a 'rich capitalist' called Butch and directed by Rivet, a former *négociant* in Nantes. It made soda and soap and was reported to be prospering, employing 40 workers and turning out 10 000 kg of soap a month, which was said to have largely replaced in the department that from Marseilles. The works was presumably of recent creation since a year earlier the Prefect had reported that the department had no soda works.[260] Later in 1811, a decree of 12 September authorized the establishment of a works by Richer *frères* on the Isle of Noirmoutier (off the Loire coast); this was in existence by April 1812, using pyrites instead of sulphuric acid.[261] In the east of France, soda manufacture on a modest scale had been taken up by May 1811 in the two chemical works at Strasbourg (p. 39). There was probably manufacture in other places too.

The final years of the Empire, 1811–1814

The final years of the Empire were a period of adjustment for the new industry. The excited expansion in 1809 and 1810 rapidly resulted in over-production, and this was aggravated by the depression of 1811. We have seen that by mid-1810 the intended production of the works in the five principal centres of manufacture totalled some 24 million kg a year, which must already have been considerably in excess of France's consumption at that time. In January 1811 Darcet told the *Conseil des fabriques et manufactures* that soda works had now multiplied to such an extent as to be capable of supplying ten times the amount required for the country's consumption (a figure later much quoted but no doubt much exaggerated).[262] Permission to export soda was granted by a decree of 11 February,[263] and it was hoped that manufacturers would be able to find new markets in Germany and Italy, but this hope was not, so far as we know, realized.

One result of this situation, of course, was a rapid fall in price. Already at the end of August 1810, the Minister of the Interior remarked that whereas

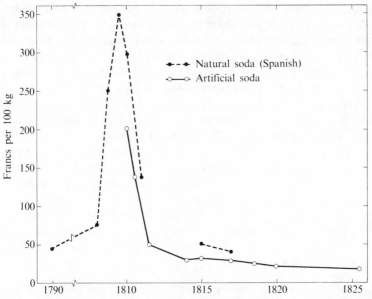

Fig. 4.7.   Soda prices at Marseilles, 1790–1825. The figure for 1790 represents what seems to have been the typical price prior to the Revolutionary Wars. The graph is based on data in: Julliany (1842–3), vol. 3, 306; *Mémoire* [1819], 5, 10, 25; *Arch. parl.*, 2nd series, vol. 13, 600–2; AN, $F^{12*}195$, 23 Jan. 1817; $F^{12}1966^L$, [7]; $F^{12}2245$, [4]; $F^{12}7589$, [1]; ADBR, $M^{14}4$; $M^{14}12$; $M^{14}226$.

prices would normally be expected to rise following a prohibition of imports, the price of artificial soda was on the contrary falling. Darcet confirmed the fact and explained that it was due to competition between the large number of factories established.[264] By the beginning of 1811 artificial soda was selling in Paris below the peace-time price of imported soda, and Darcet predicted that it would fall a further thirty-three per cent in the coming months.[265] In Marseilles, too, the price soon dropped sharply, to about the level at which Spanish soda had sold there before the Revolution (see Fig. 4.7). Such a fall in prices was obviously very disturbing for manufacturers, robbing them of their anticipated profits, but on the other hand we may presume it to have been not entirely without some beneficial effect for the industry. In the first place it must have presented manufacturers with a healthy stimulus to efficient and economical production; and in the second place it must have helped ease the acceptance of artificial soda by consumers, in the face of continuing prejudice in favour of the natural product.

Most manufacturers had to limit or cut their production to considerably less than originally intended, and some were soon obliged to abandon the manufacture altogether. The position probably varied a good deal from one place to another, but detailed information is generally lacking. The industry at Rouen is known to have been particularly badly hit, virtually collapsing in 1811 after its brief spell of prosperity. All the manufacturers there then abandoned their auxiliary plants on the bruyères Saint-Julien. Haag and Holker shut down their main works, too, wound up their businesses and left the town (p. 47). Lefrançois continued to make soda in his main works but only on a greatly reduced scale (in 1811 he produced 66 790 kg,[266] compared with his originally intended output of 432 000 kg a year).

In Marseilles and its department, where manufacturers had the advantage of the important local market in the soap industry, soda production is recorded to have grown from 1 893 300 kg in 1810 to 5 356 300 kg in 1812.[267] This was still only half the intended production declared in mid-1810, however, and the industry went through hard times. Between the second and fourth quarters of 1812, for example, the production of the seven manufacturers in the environs of Marseilles fell by half as a result of recession in the soap industry, and at the beginning of 1813 three of the works were idle.[268] The difficulties experienced even by so capable a manufacturer as Chaptal *fils* are reflected in the consoling letters written to him by his father.[269] Thus, in January 1811:

In the moment of crisis in which commerce finds itself, one must consider oneself fortunate not to lose ... The time will come when large establishments like yours, conducted with intelligence and economy, must necessarily prosper ... One should at the moment think oneself fortunate not to have made use of credit which one cannot support.

And again in May 1812:

I am sorry to see that your factories do not give you the profits you had a right to hope for; but where is the branch of commerce that gives profits today; those who subsist deem themselves fortunate ... On your return [from the Plan d'Aren] you must occupy yourself with research, it is the only means of advancing.

Apart, however, from the small ephemeral ventures that briefly proliferated in the industry's early months, most of the works begun in the department do seem to have survived, and probably over a dozen continued into the Restoration.[270]

The growth of soda manufacture under the Restoration

When, with the fall of the Empire, France's trade policies came under review, various interests petitioned for foreign sodas to be re-admitted, and there were those who would not have been sorry to see the demise of the soda industry now that it had served its purpose in supplying the country's needs during the war. The Marseilles Chamber of Commerce, which had been unfriendly from the outset, warned darkly that continued exclusion of natural soda would be seriously prejudicial to the town's soap industry, and would undermine France's trade with Spain and Italy. In 1814–16 considerable hostility to the soda industry developed in Marseilles, fuelled by the resentment of commercial interests injured by the diminution of trade in the natural product, by jealousy at the enormous profits which soda manufacturers were considered (erroneously) to be making, by annoyance at the acid fumes which emanated from the factories, and by a certain continuing prejudice in favour of what one petitioner called 'real sodas', as opposed to the new artificial concoction. There is said, too, to have been a degree of royalist animosity towards an industry which was considered a product of the Revolution.[271]

The Restoration Government, however, appreciated the value of the industry and continued to support it. Thus, the special exemption from the salt tax was maintained. And although foreign sodas were re-admitted by the tariff of 17 December 1814, a substantial import duty of 15 fr. per 100 kg was imposed as a protection.[272] This was necessary not because artificial soda was expensive—on the contrary, it was easily competitive in price with natural soda—but because it needed an appreciable price advantage to outweigh continuing prejudice in favour of the natural product. In Marseilles the question of tariff protection for the industry was complicated by association with another issue, that of the town's *franchise de port*. Before the Revolution Marseilles had enjoyed the privileged status of a free port, an arrangement which, to facilitate trade, had allowed foreign goods to enter the port and town without paying import duty. This important privilege having been lost during the Revolution, the commercial interests of Marseilles in 1814 pressed strongly for its restoration. The soda manufacturers objected on the grounds that such a franchise would allow the soap-makers to obtain foreign sodas free of duty. The fact that the conflict between Marseilles commercial interests and the soda manufacturers thus involved the important issue of the *franchise de port* was no doubt an aggravating factor in the hostility we have mentioned. The outcome was essentially a victory for the soda manufacturers, since although Marseilles did regain her free-port status (by a law of 16 December 1814), it was stipulated in the detailed dispositions that when soap made there passed into the interior of France, entry duty had to be paid on any foreign soda used in its manufacture, no duty being charged on soap made with artificial soda.[273] Since it was in France that by far the greater part of the soap made was sold, this meant that artificial soda did enjoy practically the same protection at Marseilles as in the rest of the country.

There was also action from the Government to promote the use of artificial soda. In response to a protest from glass-makers at Givors that only natural soda was suitable for their industry, the *Direction générale du commerce* published a detailed set of instructions by Pajot Descharmes on the manner of employing the new product. This was inserted in the widely circulated *Bulletin* of the *Société d'encouragement* early in 1815, with supporting pieces by Darcet, d'Artigues, and the Government's *Comité consultatif des arts et manufactures* (of which Darcet was a member). The collective work was also issued as a brochure, of which 600 copies were officially distributed in the departments. Demand exceeded supply and so two or three months later the work was additionally inserted in the *Journal de physique*, and a further 400 copies of the brochure were printed and distributed. The instruction proved of value not only to glass-makers, for whom it was primarily intended, but also to dyers and other users of soda.[274]

Under the Restoration the manufacture grew rapidly. This was no doubt partly due to a revival of consuming industries after the depression of the final years of the Empire. Another factor was the increasing production of soda in refined as well as crude form. The industry which sprang up so quickly in 1810 had at first marketed its product almost entirely in the crude state—particularly at Marseilles—but towards 1820 it also began to sell refined soda in growing quantity, partly in crystalline form but to a much larger extent as dry soda-ash. This purified product then began to displace potash from many of its traditional uses, so greatly extending the soda market. The Marseilles soda makers Daniel *frères* & Cie, in a memoir probably dating from 1823, tell us that in the past six years the production of soda ash in the Marseilles region had grown from 600 000 kg to 4 million kg a year.[275] In Paris the substitution of soda for potash was promoted by the cunning of certain chemical producers who by suitably doctoring the refined soda adjusted it to the strength and appearance of red American potash (the most esteemed variety on the market) and then sold it under that name. In 1824 there were five establishments in the capital producing 324 000 kg a year of this artificial 'potash', for which laundry houses happily paid twice the price of the soda ash that went into its confection.[276] With the Restoration the soda industry also began to export to some extent. A memoir by four of the Marseilles manufacturers claimed in mid-1819 that in the past two years $2\frac{1}{2}$ million kg of soda had been exported from Marseilles to England, for the soap-works of London, Bristol, and Liverpool, and a further 1 million kg to Trieste.[277] The writers hoped that with experience of the product the foreign market might grow considerably. This was not to be, however, and in the early 1820s Marseilles exported only a very minor part of her production, in the form of occasional sales to Italy and the United States. It was said that sales to England had been ended by 'un bill prohibitif' of the British Government at the beginning of 1819.[278]

Records for the industry's tax-free salt consumption for the years 1816–18 afford a reliable guide to the state and distribution of the manufacture in France after the Empire (see Table 4.1). The country's total soda production in those years, as estimated from the salt consumed, is shown in Table 4.2. For comparison, this also presents details of France's foreign trade in soda. The amount of natural soda produced in France was probably very small, and so from the data here assembled we can surmise that perhaps some 80 per cent of the soda now used by French industry was artificial.

TABLE 4.1
Duty-free salt consumption of French soda industry, 1816–1818 (kg)

	1816	1817	1818
Factories under surveillance of the Direction des contributions indirectes			
Tassaert at Charles-Fontaine (Aisne)	279 707	369 200	366 800
Korn at Soissons (Aisne)	678 900	372 000	18 900
Dubruel at le Picquenard, nr. Poissy (Seine-et-Oise)	522 000	415 100	381 000
Dizé at Saint-Denis (Seine)	—	—	59 800
Bonnaire at Vaugirard (Seine)	155 970	113 273	128 957
Holker at la Folie (Seine)	194 200	333 000	479 100
Chervaux at Couternon (Côte-d'Or)	98 800	92 600	184 800
Cluchier at Lambesc (Bouches-du-Rhône)	136 422	7 075	—
Chavagnac at Lambesc	—	—	42 511
Armand at Auriol (Bouches-du-Rhône)	611 600	534 800	577 500
Castinel at Auriol	—	—	280 600
Michel at Cavaillon (Vaucluse)	48 500	—	—
Daumar at Cavaillon	—	27 537	—
Factories under surveillance of the Douanes			
Rouen	—	—	70 000
Marseilles	6 801 502	6 704 477	7 817 157
Toulon	—	351 400	746 850
Strasbourg	30 800	26 930	—
Paris	41 900	45 600	—
Total	9 600 301	9 392 992	11 153 975

Note: Based on figures given in AN, F[12] 2245, [10]. The picture of the soda industry presented by this table is not quite complete. It does not include Carny's factory at Dieuze, for instance, since this used natural sulphate instead of salt. Nor does it include Payen's works at Grenelle, since this was officially regarded as a sal ammoniac works rather than a soda factory and so was not allowed exemption from salt duty (AN, F[12] 2242, [3]).

It will be seen from Table 4.1 that soda manufacture was now very largely concentrated in the region of Marseilles, and that many of the works elsewhere which originally took up the manufacture in about 1810 had by now apparently abandoned it. Three-quarters of all the soda made in France was now made in Marseilles and its department, which since 1810 had suddenly acquired a major new industry, becoming the largest centre of chemical manufacture in the country. This was a position which it was

TABLE 4.2

French soda production and alkali imports, 1816–1818 (kg)

		1816	1817	1818
Estimated production of soda industry (crude)		15 400 000	15 000 000	17 800 000
Natural soda imports	(a)	1 620 000	2 344 000	1 679 800
	(b)	1 244 551	2 254 462	1 679 776
Soda exports	(b)	422 102	1 044 653	c. 3 000 000
Potash imports	(a)	5 070 000	4 462 000	3 057 000
	(b)	4 023 242	4 717 285	3 051 096
Potash exports	(b)	118 610	69 911	

Note: We have estimated soda production from salt consumption, assuming a yield of 160 parts of crude soda from 100 of salt (the yield commonly obtained was 150–170). Trade figures are from: (a) AN, F^{12}7591; (b) 'Tableaux' (1818) and Admin. Douanes (1818). Alkali exports in 1818 were recorded as 3 236 299 kg, under the head 'Alkalis; soudes de toutes sortes'. This may have included a small amount of potash but was presumably very largely soda.

subsequently to consolidate as the manufacture there continued to grow steadily (see Fig. 4.8). In the later 1820s two-thirds of the soda made in the department was consumed by the Marseilles soap industry, the remaining third being sent into the interior in refined form. A tiny proportion of less than 2 per cent was exported.[279]

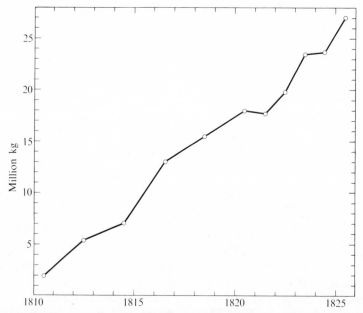

Fig. 4.8. Soda production in the region of Marseilles, 1810–25. Figures are for the department of the Bouches-du-Rhône and are expressed in terms of crude soda. Sources: Julliany (1842–3), vol. 3, 308; Villeneuve (1821–34), vol. 4, 787, 794.

For many years the largest works in the region was that established by Chaptal *fils* on the Plan d'Aren. In second place there came the nearby Rassuen works, created by the Pluvinet brothers. The concern of Chaptal *fils* · in 1819 became a *société anonyme* (a joint-stock company), under the name *Compagnie des salines et produits chimiques du Plan d'Aren*. The works was at that date reckoned to be worth 600 000 fr., with materials in store to the further value of 200 000 fr. It was said to be making about 1·6 million kg of crude soda a year. Following the formation of the *société anonyme* it was expanded considerably: in 1823 it made 2·95 million kg of sulphuric acid (reckoned as concentrated), this then being largely applied to the production of 5·66 million kg of crude soda, of which 3·61 million kg were refined to yield 1·26 million kg of soda ash.[280] In 1821 the firm was reported to be employing 250 workers in chemical manufacture, and a further 150 in the associated salt-works (though the numbers may have been rather variable); in 1825 the figures were 360 in chemical manufacture and 100 in the salt-works. The Rassuen works in the same period employed about 150–250.[281] The Plan d'Aren works was still about the largest in the region towards 1840, when it was said to be making 4 million kg of soda a year. It was continuing to operate in 1860, though now rather diminished, with 91 workers.[282] Today it can be seen in ruins beside the Étang de Lavalduc.

Outside the region of Marseilles, we find Darcet's Paris enterprise still prominent under the Restoration. After relinquishing its Saint-Denis plant to Dizé in 1813, Darcet's company concentrated its production at la Folie, where by about that time it had been joined by Jean Holker from Rouen. At the end of July 1816 Darcet and Holker formed an association with Chaptal *fils*,[283] and the la Folie works was then run in conjunction with the latter's factory at les Ternes. Holker was charged particularly with the direction at la Folie, and Darcet with that at les Ternes. Darcet and Holker were also at this time preparing to mount a works at Liège, but the venture there seems to have been short-lived.[284] The two Paris works for some time prospered, together forming what must have been the largest chemical concern in the north of France. That at la Folie was the larger of the two, where most of the heavy chemical manufacture was conducted. The joint concern won a gold medal at the 1819 exhibition, and was then said to be employing 150 workers and producing 1 200 000 kg of crude soda a year (of which a sizeable part was then refined), together with large quantities of sulphuric acid (360 000–400 000 kg), hydrochloric acid (525 000–600 000 kg), and alum (600 000 kg), and smaller amounts of copperas, nitric acid, oxalic acid, tin chloride, and bleaching powder.[285] This important concern collapsed, however, following the economic crisis of 1825, when Chaptal *fils* was ruined by an unfortunate commercial speculation.[286] The works at les Ternes had closed down by November 1825 and seems never to have reopened. The works at la Folie was continued by Poisat *oncle* & Cie, under whose name it figures in the *Almanach du commerce* until at least 1853. Darcet and Holker evidently retained an interest in the works in association with Poisat.[287] Chaptal *fils*, after his ruin, went to Mexico in 1831 and there died two years later.

As for the historic factory at Saint-Denis, after being returned to Dizé this stood idle from about 1814 until 1818, when Dizé resumed production. In 1823 Dizé moved to Belgium, to mount a works at Brussels, and the Saint-Denis works was then continued by his son Émile in company with Joseph Ador *jeune*. From 1827 until the 1830s it figured in the *Almanach du commerce* under the name of Anselme Payen, and thereafter, until at least 1853, under the name of Arnould & Bertrand. By the time Anastasi wrote in 1884 it had disappeared.[288]

Of all the manufacturers we have discussed, the one destined to acquire the greatest importance in the long term was the Saint-Gobain glass company, but under the Restoration this firm as yet gave no hint of its future eminence as a chemical producer. Its soda plant at Charles-Fontaine was continuing to manufacture simply for internal use. Figures for the plant's salt consumption in the years 1816–18 (Table 4.1) show it to have been making some half a million kilogrammes of crude soda a year, about equal to the requirements of the Saint-Gobain glass-works.[289] From a description of about 1817–22 we learn that the plant, directed now by Valette, possessed four furnaces for the preliminary production of sulphate (of which three worked, while the fourth stood in reserve), and two soda furnaces (worked one at a time), besides the five sulphuric acid chambers which we have mentioned earlier. The manufacture of both sulphuric acid and soda proceeded without interruption day and night, employing a dozen workers divided into two shifts. In addition about 15 men and half a dozen boys were employed in the associated refinery.[290] In 1822–4 the company moved the works to Chauny where it would be better situated, and it was there that it was to begin developing in a more important way, profiting from the guidance of Clément and Gay-Lussac, who both had close associations with the company.

Early understanding of the Leblanc process

To complete our account of the early soda industry, we shall consider briefly some aspects of the technical state and development of Leblanc soda manufacture in the period of our study, beginning with the question of how far the process was then understood.

Writers have commonly represented the discovery of the Leblanc process, and its subsequent early history, as a more or less totally empirical affair. Dumas in 1856 wrote of the discovery that 'artificial soda, like so many other inventions, was to be born of stubborn trials and efforts, whose results theory was unable to anticipate'. A generation later Lunge, in his great treatise on soda manufacture, went further, asserting that the first explanation of the Leblanc process had been that given by Dumas in 1830. More recently, Gillispie, in his discussion of the origins and early history of the process, has alleged that not only did Leblanc not discover his process through any theoretical insight, 'but after he had worked it out, neither he nor any of the other artisans interested in alkali production made any attempt to investigate or explain the reactions involved'. Of the various soda processes then known, the Leblanc process (Gillispie writes) was that with the least understood mechanism, inevitably so, since the reaction is 'an exceedingly complicated one', only properly described towards the end of the nineteenth century.[291]

It is true that in the mid-nineteenth century the reactions of the Leblanc process came to be much discussed, and that the modern theory was only satisfactorily established in the later part of the century. This was not so much due to any exceeding complexity in the reactions, however, as to the difficulty of experimentally establishing which of the various suggested mechanisms actually occurred. As Lunge explains, this was difficult because

... the processes going on in the white hot furnace admit of observation only by means of some secondary signs, such as the colour of the flame, the appearance of the melting mass, or the analysis of the fire-gas, and because our analyses of black ash [crude soda] are necessarily limited to the cooled down product, in which already on cooling, and still more during the subsequent treatment with water, changes can and must take place.[292]

The various theories proposed did not involve particularly complicated chemistry, and that ultimately accepted as the modern explanation is remarkable for its simplicity and its obviousness. It explains the reaction as occurring in two stages, in which the sodium sulphate is first reduced to sulphide by the coal, and the sulphide then reacts with the chalk by double decomposition:

$$Na_2SO_4 + 2C = Na_2S + 2CO_2$$
$$Na_2S + CaCO_3 = Na_2CO_3 + CaS$$

This interpretation, in essence, was quite within the capabilities of late eighteenth- and early nineteenth-century chemistry.

The literature of that period does, admittedly, make very little reference to the reactions of the process, and it is disappointing to find that Leblanc in particular offers no explanation in either his published or unpublished writings. Early chemists and manufacturers, nevertheless, were not entirely unmindful of the chemistry of the process, and the picture of naïve empiricism which has been presented has been rather overdrawn. That Leblanc himself did have some early rationale we can perhaps infer from a remark in 1791 by Darcet. After visiting the Saint-Denis works with other commissioners in connection with Leblanc's patent application, Darcet described how Leblanc had shown them round the plant he had installed 'for the large-scale application of his means and processes, of which he both showed us the products and unfolded the chemical theory, which we found to be based on the best principles'.[293] From the brief interpretative comments that one does encounter in the subsequent literature it is possible to discern something of the explanations that came to be entertained between the 1790s and the 1820s. These invariably saw the process as a two-stage reaction beginning with reduction of the sulphate to sulphide, as in the modern theory. To explain how sodium carbonate was then formed from the sulphide two main views developed.

The earliest interpretation saw the sulphide of soda as being converted to carbonate by carbonic acid from the chalk, while the sulphur of the sulphide was in considerable part supposed to be evolved in gaseous form. Such a view would seem to have been held by Dizé, for example, to judge from a remark made in 1794 by Coquebert, when he spoke of the method developed by Leblanc and Dizé

to break the union which is formed between the sulphur and the alkali when the sulphuric acid of the sulphate of soda has been converted to sulphur. The auxiliary substance which they employ is carbonate of lime (chalk), whose effect would appear to be primarily to neutralise the alkali by saturating it in part with carbonic acid. Such at least is the opinion of citizen Dizé, who thinks that at a high temperature the carbonic acid acquires more affinity for the soda than for the lime.

Although we are not told what Dizé believed the fate of the sulphur to be, he probably agreed with the account given in the Darcet report, which pictured the sulphur as being evolved (to a considerable extent) in the form of sulphuretted hydrogen. It was the formation of this gas which was there considered to be the cause of the candle-like flames appearing on the surface of the mass (in fact they are flames of carbon monoxide), and it was to promote the disengagement of the sulphur in this manner that the molten mass was directed to be vigorously stirred. The commission presumably supposed the sulphuretted hydrogen to be formed by reaction of the sulphur with hydrogen in the charcoal (the distillation of sulphur with charcoal was one early method of preparation). The commission's analyses of the crude soda showed there to be also a portion of sulphur retained, in combination with the chalk. The crude Leblanc product was thus considered to be essentially a mixture of carbonate of soda and chalk, the latter in a partly sulphuretted state. A more explicit explanation of the process, broadly similar to that of the Darcet report, was given by Fourcroy in his chemistry text of 1800:

The carbon, by decomposing the sulphuric acid of the sulphate of soda, disengages the sulphur, which unites with the lime of the carbonate of lime, and which is in part volatilised, while a portion of the carbonic acid combines with the soda; so that the product is a mixture of carbonate of soda, lime and charcoal, similar to the soda of commerce.

Explanations along these lines were evidently regarded by Rougier in 1812 as having a certain currency, when he offered a variant theory of his own. This differed in supposing the carbonic acid responsible for decomposing the sulphide to derive not from the chalk but from the combustion of the charcoal (for he had observed that lime could serve in the process instead of chalk); the displaced sulphur was then supposed to be evolved, either by simply burning off or by combining with hydrogen.[294]

In course of time it came to be recognized that the sulphur was not in fact evolved in the process to any significant degree, but remained combined in the crude product. The reaction then came to be interpreted, as in the modern theory, as a simple double decomposition, giving a product consisting essentially of a mixture of carbonate of soda and insoluble sulphide of lime. Such a view may have been in the mind of Vauquelin already in 1802, when he commented that in the Leblanc process a sufficiently large quantity of chalk must be used 'to saturate the totality of the sulphur resulting from the decomposition of the sulphate of soda by the charcoal'. A view of this kind was also probably held by Welter and Gay-Lussac in 1820, when they clearly stated that 'Crude soda is a mixture of sulphide of lime, of very small solubility, and carbonate of soda'. The earliest explicit statement of this explanation that we have found, however, is in a chemistry textbook published in 1823 by the Marseilles chemist J. C. E. Péclet:

The coal acts first on the sulphate and transforms it into sulphide, and by a reciprocal decomposition of the sulphide of soda and the sub-carbonate of lime [calcium carbonate], there is formed sub-carbonate of soda [sodium carbonate] and sulphide of lime.

The same explanation is also found in a work published by Pelletan at about the same time, and in 1828 in a work by Poutet. Clément in 1825 gave a variant interpretation, which resembled that of Rougier in supposing the decomposition of the sodium sulphide to be by carbon dioxide from the charcoal, but which now recognized the displaced sulphur to be retained as sulphide of lime.[295]

Explanations of the Leblanc process thus certainly existed from the 1790s, and by the 1820s an understanding essentially similar to the modern theory seems to have become quite common. It is true, however, that these early explanations were merely interpretations based on casual observation rather than elaborated theories built on detailed study. Dumas, in 1830, seems to have been the first to give closer scientific thought to the matter, and it is ironic that he was thereby led to depart from the simple and essentially correct interpretation then commonly assumed, introducing an erroneous elaboration that was to be the subject of much of the discussion of the mid-nineteenth century.

The development of plant and technique

The general establishment of Leblanc soda manufacture brought no dramatic change to the basic manufacturing technique developed by Leblanc himself in the early 1790s, and indeed the process was to remain fundamentally unchanged throughout its history. The industry's technical development in the course of the nineteenth century was to be a matter of progressive amelioration of plant and methods in the directions of greater handling efficiency and convenience, improved yields and fuel economy, reduced environmental pollution, and increased utilization of the by-products. Detailed improvements of this kind were beginning already in the Napoleonic period. Thus, Darcet and Payen separately introduced the use of soda furnaces with large oval beds, instead of the smaller rectangular bed which Leblanc himself had employed. The new bed ensured that the reaction mixture was more evenly heated, resulting in a crude product containing less sulphate and sulphide impurity, and with a correspondingly higher yield in sodium carbonate. It was a significant improvement which was subsequently widely adopted.[296] How Payen and Darcet each came to make the change we do not know, but it is no less plausible to imagine it to have arisen from a scientific diagnosis of the shortcomings of the earlier furnace, based on recognition of sulphate and sulphide in the product as a sign of uneven reaction in the mixture, than to suppose it to have resulted from mere empiricism.

Another innovation of the period which was to become commonly adopted was the twin-bedded furnace, in which a single fire heated two working beds used for different stages of manufacture. This type of furnace seems to have been first described in 1810 by Pelletan, who employed one bed for the preliminary reaction of the salt with sulphuric acid, and the other for higher temperature calcination to complete the conversion to sulphate (see Fig. 4.9). Subsequently, at Marseilles, there came into use similar twin-bedded furnaces in which the production of sulphate on one bed was coupled with soda production on the other. The chief advantage of such double furnaces was obviously the saving in fuel they offered. Another improvement in the direction of fuel economy was the use of coal in the reaction mixture, instead of the charcoal employed by Leblanc himself. This was an innovation which Rougier in 1812 attributed to Darcet at la Folie, and again it was to find general adoption, although charcoal continued to be employed by some manufacturers since it was considered to give a better quality product.

A technical development relating rather to the soda trade than to soda manufacture itself, but which we might nevertheless appropriately mention here, was Descroizilles's introduction in 1806 of his alkalimeter.[297] We have already discussed the berthollimeter, a graduated measuring cylinder whose use Descroizilles had earlier popularized for determining the strength of bleaching liquors, by titrating them against a standard dye solution. His alkalimeter was a somewhat similar instrument—in fact an early form of burette—which he devised for the estimation of alkalis, by titration with a standard solution of sulphuric acid. It was the first simple means to become available for measuring the strength of commercial alkalis. The instrument was marketed from 1806, and soon began to be taken up by alkali dealers and consumers, and then, with the growth of the Leblanc industry, by the new soda manufacturers, too. It seems to have early become common for the value of artificial sodas to be adjudged by use of the alkalimeter, and Descroizilles remarked in 1818 that the practice had now for some years been established at Marseilles of quoting the alkalimetric strength of artificial sodas in the *prix courants*, along with their price. By providing an objective means by which the real alkali content of the new products could be assessed, the alkalimeter may have been of some significance in helping to facilitate their adoption.

The aspect of manufacture which above all attracted early attention was the problem posed by the hydrochloric acid by-product. The first stage of the Leblanc process gave rise to a ton of hydrochloric acid gas for every three or four tons of crude soda made. From the production figures for the early factories it can thus be seen that in the largest of them the gas was issuing from the sulphate furnaces at the rate of something like a ton a day. With the rapid expansion of the industry in 1809 and 1810, the vast quantities of acid fumes released very soon proved a serious nuisance, causing damage, discomfort, disquiet, and consequently complaint. This resulted in efforts by manufacturers to absorb the offending gas, and also in legislative action by the Government, which it will be convenient to consider first.

Legislation on offensive factories

The pungent emanations which announced the birth of the new industry speedily came to the attention of local authorities. In Rouen, for example, where the problem was particularly felt because of the proximity of the chemical works to houses and other properties, the neighbouring inhabitants and property owners protested to the Prefect. They complained that the fumes caused choking, headache, and nausea, that they damaged vegetation, clothing, and metallic objects, and that they diminished the value of their properties. The concern caused by the real effects of the fumes was exaggerated in the popular mind, moreover, by other more or less imaginary fears regarding their possible effects on health. A Rouen physician by the name of Henri Pillore, in a pamphlet full of purple prose and curious chemistry, went so far as to claim (among other things) that the fumes would cause denaturation of mothers' milk, with permanent effects on their offspring, and that they would interfere with the onset of puberty in girls, with 'terrible consequences'. The manufacturers disingenuously replied that the fumes from their factories, far from being unhealthy, were a proven disinfectant, recommending the good doctor to send his patients for a stroll in their neighbourhood so as to benefit from the salubrious air.[298] When the Prefect's investigation of the early complaints showed that they were not unfounded, he responded with an *arrêté* of 10 October 1809,[299] which brought the development of the chemical industry under his own control. He ordered that in future no new chemical works could be established, nor could any addition be made to an existing works, without his prior authorization, and he appointed a special commission, headed by the local chemist Vitalis, to act as an advisory and supervisory body. In the region of Marseilles, similarly, the first soda factories gave rise to protests, leading the prefect there to order on 10 November 1809 that in future no factory for the manufacture of soda or other chemicals could be established in the department without his permission, which would be accorded on the advice of the appropriate mayor.[300] In the town of Marseilles itself, the industry became the subject of inquiries and reports by the public-health committee of the local *Société de médecine*, which served as adviser to the mayor on the matter.[301] In Paris, the industry came under the surveillance of the *Conseil de salubrité*, an advisory body which had been created in 1802 by the Prefect of Police, to concern itself with the capital's full range of public-health problems. The *Conseil* advised on the authorization of factories (required in Paris since 12 February 1806, for all workshops posing a health- or fire-risk), and then investigated subsequent complaints against them, with visits of inspection.[302] This kind of regulation at local level was very soon followed by national legislation.

The problem of air pollution was by no means a new one, of course. The fumes and odours emitted by a wide variety of arts and manufactures had for many years been increasingly a cause of sporadic complaint and controversy, and had already in 1804 attracted the attention of the Minister of the Interior, Champagny. The Minister had consulted the First Class of the Institute on the matter, but he was advised at that time—in a report of 17 December by Guyton and Chaptal—that the gaseous emanations from most branches of manufacture were at most only somewhat unpleasant, not truly harmful, and that the complaints of neighbours were generally unfounded.[303] With the commencement of large-scale soda manufacture, however, the problem began to take on altogether new dimensions. It was undeniable that the fumes emitted by the new industry did cause real damage, and the Minister of the Interior (Montalivet) now decided that national legislation was needed to control the siting of offensive factories. The First Class of the Institute was asked to prepare detailed proposals, and the resulting report—by Chaptal, Fourcroy, Vauquelin, Guyton, and Deyeux—then served as the basis for legislative measures introduced by an imperial decree of 15 October 1810.[304]

This important decree enumerated the various manufactures (forty-four in all) which involved the emission of objectionable fumes and odours, dividing them into three classes according to the degree of danger or nuisance they presented. A works in any class could henceforward be established only after authorization had been obtained, though the authorizing body was different for each class. The conversion of salt to sodium sulphate (the offensive first stage of soda manufacture) came in the first class, that of manufactures whose fumes constituted the greatest nuisance, and which required authorization by the central government. When a request was made to establish a works in this class, the request had to be publicly posted in the locality where the works was proposed, after which any objections were to be judged by the *conseil de préfecture*. Authorization was then to be accorded by a decree of the *Conseil d'état*, on the advice of the prefect and the report of the Minister of the Interior. The decree left it to local authorities to judge the suitability of the chosen locality for the manufacture in question, since although the Institute had attempted to define precise minimum distances from dwellings, it had decided in the end that it was impossible to lay down general rules on this point because of the great variation in local circumstances. Once a first class factory had been established, with authorization, anyone who subsequently built in its neighbourhood could not later justifiably demand its removal. While it was possible for factories to be suppressed, this was only to be in the case of grave inconvenience, and then only by special decree of the *Conseil d'état*, rendered after hearing the advice of the local authorities and the defence of the manufacturer.

The decree thus fulfilled two objectives. In the first place it afforded protection to property owners against the establishment of offensive factories near their property, and in the second place it afforded protection to manufacturers against demands for the removal of their works once they had been established with authorization. The detailed dispositions of the decree were subsequently to be modified, and the lists of the three classes of manufactures were to undergo successive amendment and augmentation, but this decree has remained the basis of French legislation on the siting of offensive factories ever since.[305]

Efforts to condense the hydrochloric acid gas

From the very beginnings of Leblanc soda manufacture, attention was paid to the collection of the hydrochloric acid in order to profit from it as a by-product. We have seen that Leblanc's own plant incorporated a lead chamber for this purpose, in which the acid could either be condensed as such or alternatively combined with ammonia gas to yield sal ammoniac. One cannot imagine such plant proving effective on any very large scale, however. Payen, the first manufacturer to work the Leblanc process regularly, is said by his son to have early absorbed the gas by means of a stream of water: the salt was decomposed in a large leaden pan, and the gas was passed into a long brick conduit down which a slow current of water passed. Payen's method of working had its drawbacks, though, and half the gas was still lost in the subsequent calcination stage.[306] With the growth in the scale of manufacture and the expansion of the industry in 1809–10, it became a major concern to find some means of absorbing the acid fumes efficiently, economically, and completely, not now primarily for reasons of profit but simply to prevent the nuisance of their release. A great deal of inventive effort was expended to this end.

In Rouen a number of different kinds of absorption system were proposed and tried, as appendages to reverberatory furnaces for the production of the sulphate.[307] Haag sought to absorb the fumes in water, by passing them through a series of leaden reservoirs terminating in a chimney to provide the draught for the furnace. Pelletan, too, at Holker's works, at first attempted to absorb the gas in water, and built quite a sophisticated plant designed to secure a large surface contact between the water and the gas. This consisted in a stepwise conduit, with a leaden reservoir on each step; water from a tank at the top passed down the series of reservoirs, and the arrangement was such that each reservoir, in overflowing to the one below, produced a curtain or shower of water through which the fumes had to pass. This plant was worked for a time until it became apparent that it was incapable of achieving complete condensation. Other systems depended on reaction of the acid gas with chalk. Descroizilles, in the December 1809 issue of the *Annales de chimie*, proposed the passage of the fumes up a leaden tower, packed with chalk, which was to be wetted with a water spray. Le Bertre and Dubuc, by January 1810, had both adopted a method which seems to have been prescribed to them by the Prefect's commission, and which consisted in passing the fumes through an underground channel containing lime and quarry-stone, and supplied with a constant current of water, the channel terminating in a reservoir surmounted by a thirty-foot chimney. Pelletan, when his water system proved inadequate, adopted a similar arrangement, in which the fumes met with a water spray and then with calcareous stone in traversing a lengthy conduit terminating in a chimney. On this system, Holker commented in January 1810:

The vapours leaving my workshops, after having passed through a suspension of chalk, after having received an abundant and continual rain of cold pure water for the space of sixty feet, after having travelled through underground channels filled with calcareous stone for the space of nearly 200 feet, these vapours I say, do they retain sufficient acidity to be essentially harmful to vegetable or animal life?[308]

Pelletan also devised a somewhat simpler system, consisting of an underground channel, sixty feet long, packed with chalk and now divided by walls in such a way as to cause the fumes to follow a circuitous course (see Fig. 4.9). He described this plant in a memoir read to the First Class of the Institute in March 1810, and shortly afterwards published.

In the Paris region there were similar endeavours. Destouches used a Woulfe's apparatus; Courcillon, leaden reservoirs of water; Payen, a chalk-filled channel. Darcet's firm found a more spectacular solution. Even after moving their sulphate production to the comparative isolation of la Folie, where they were said to be two miles from the nearest dwelling and over a mile from the nearest tree or garden, in the midst of a sandy region where only root crops were grown, this firm was still sorely pestered by claims for

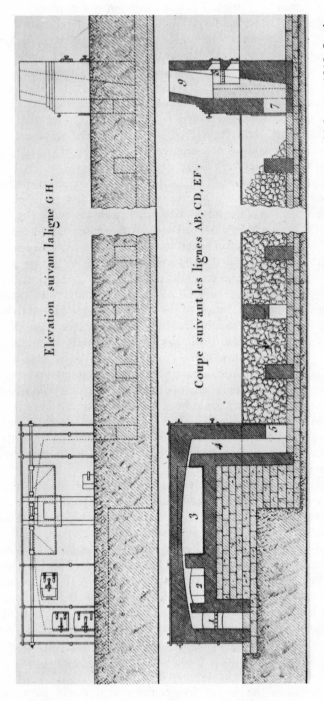

Elévation suivant la ligne G H.

Coupe suivant les lignes AB, CD, EF.

Fig. 4.9. Pelletan's plant for production of the sulphate with absorption of the hydrochloric acid fumes, 1810. In the twin-bedded furnace on the left, the initial reaction of the salt with sulphuric acid was carried out in the leaden pan marked 3, whose product was then calcined to complete the reaction on the second bed, marked 2. 1 is the fire-grate. The flue gases passed into a channel packed with calcareous stone and divided by walls at 3-foot intervals so as to force them to follow a circuitous course. The channel terminated in a 30-foot chimney, with a fire at the base to provide the draught for the furnace. (Pelletan (1810).)

damages. As a result they in 1810 adopted an absorption system in which the fumes were first passed through a large vaulted chamber, supplied with a continuous rain of water, and then underground into an old quarry, packed with calcareous stone and carefully sealed over, with a chimney at the end to produce the necessary draught.[309] At Marseilles, too, the commencement of soda manufacture is said to have been accompanied by efforts to absorb the acid fumes,[310] though we do not have detailed information on these. The Marseilles Academy made the suggestion of plant for this purpose part of the subject of its soda prize, and Rougier, the prize-winner, proposed an arrangement in which the salt was decomposed in a covered pan and the gas then condensed in a large brickwork chamber.

Unfortunately, the readiness with which the gas could be absorbed (or condensed) proved to depend in an inconvenient way on the kind of plant employed for the production of the sulphate. If production was carried out in closed pans, then the gas could be absorbed with relative ease; but the use of closed pans was slow and troublesome, and the cast iron of which they were generally made was quickly corroded. Production was quicker, cheaper, and easier in a reverberatory furnace, as generally practised; but then absorption of the gas became very much more difficult because it was hotter and mixed with fire gases, and because the design of the absorption system had to be such as not to interfere with the draught necessary for the furnace. And while it was technically possible to absorb the gas, at least in part, by suitable systems of chalk or water, such systems obviously involved trouble and expense. Because of these difficulties a good many manufacturers under the Empire seem to have opted for the expedient of carrying out their production in some isolated spot, where they could happily disgorge their fumes with little or no thought for their absorption. This practice seems to have been common in the region of Marseilles, and we have seen that it was also the ultimate outcome at Rouen, where manufacturers moved to the bruyères Saint-Julien in the spring of 1810.

Apart from the difficulty and expense of absorbing the gas, another disincentive was the absence of any ready market for the hydrochloric acid or calcium chloride collected. Demand for the acid was small, and that for the chloride non-existent. Recognizing this problem, the *Société d'encouragement* in August 1810 offered a prize of 2000 francs for suggestions of major new uses for these materials.[311] The soda prize offered by the Marseilles Academy in May asked for similar suggestions. A few proposals resulted. It was suggested that calcium chloride, in view of its hygroscopic properties, might be employed agriculturally to improve the soil, but this suggestion does not seem to have found any adoption and the chloride remained a useless by-product. A number of applications for hydrochloric acid were proposed, but no genuinely novel idea of any value emerged and for some years the market for the acid remained small.

Under the Restoration the use of hydrochloric acid began to grow, however, partly through its application by Darcet to the extraction of gelatine from bones. It had long been known that a gelatinous foodstuff could be extracted from bones by boiling them in water, or by treating them in a Papin's digester, but this extraction was little practised. Darcet, as a result of researches begun in 1810, developed a technique for the extraction of the gelatine by digesting the bones in dilute hydrochloric acid. This had the advantage of giving a much greater yield, and so of being more economical than previous methods. The gelatinous material obtained could be used as the basis for a foodstuff, or for the production of glue. In December 1813 Darcet was rewarded for his work with the award of a free patent by the Government, and at about that time he established his process on an industrial scale in the factory of Robert, near Paris, which was to win a silver medal at the 1819 exhibition for its alimentary preparations and glues. Darcet's work aroused a great deal of interest, promising as it did to provide almost unlimited quantities of cheap food for the poor, and for the army and navy, from materials formerly unused; but there was controversy over the nutritional value of the product and it never acquired the major and lasting importance in this respect for which Darcet had hoped. His process did find use for the production of glues, however, and in the mid-nineteenth century large quantities of hydrochloric acid were said to be employed for this purpose.[312]

By far the more important use of the acid to develop under the Restoration was in chlorine bleaching. Chlorine bleaching itself was by then long established, of course, but bleachers had hitherto been in the habit of preparing their chlorine with sulphuric acid (and salt), rather than with hydrochloric acid, since the former was cheaper. Under the Restoration cheap hydrochloric acid came to find adoption instead, and at the same time bleaching powder began to be manufactured to some extent. Soda manufacturers in the north of France now increasingly collected their hydrochloric acid for use. In 1819 the works at la Folie and les Ternes were producing 525 000–600 000 kg of hydrochloric acid a year, and some bleaching powder, too, from which we can see that the greater part of their acid was now being collected. The higher price of raw materials in the north of France meant that manufacturers there needed to turn their acid by-product to profit in order to compete with soda producers at Marseilles (the latter for the most part did not collect their acid since the local market for bleaching was small).[313] There is said to have been a good deal of experimenting with plant, but that generally adopted did not differ remarkably from the kind used under the Empire. The sulphate was made in a gallery of cast iron boilers or cylinders (the latter a new refinement), or alternatively in a reverberatory furnace, the acid being collected in each case by passing it through a series of stoneware bottles (see Fig. 4.10).[314] This was the kind of plant which remained in general use in France into the second half of the nineteenth century.

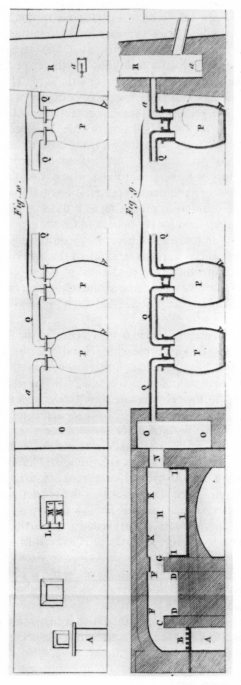

Fig. 4.10. Plant for sulphate production with collection of the hydrochloric acid, c. 1830. The salt was treated with sulphuric acid in the leaden pan I, and the resulting sulphate was then calcined on the bed D. The furnace gases were cooled somewhat by passage through a small brick chamber O, and the hydrochloric acid was then condensed in a series of 20–5 large bottles, P, terminating in a chimney R, here shown also connected to the flue of a soda furnace to provide the necessary draught. (*Dict. tech.*, vol. 19 (1831), 434–5, and *Dict. tech. Atlas*, vol. 1 (1835), 'Arts chimiques', Pl. 75.)

As soda production grew under the Restoration, manufacturers not wishing to collect their acid for sale nevertheless increasingly took precautions to prevent its escape, under pressure of complaints and proceedings for damages. In the region of Marseilles, manufacturers at Septèmes, for example, had to give serious attention to this problem in the late 1810s and early 1820s, when it became apparent that even though their works were reasonably isolated the fumes still caused damage because they travelled considerable distances. In the summer of 1816 the works at Septèmes were threatened with mob violence, obliging the Prefect to issue a firm warning, on instructions from the Minister of the Interior, that they would be protected by force if necessary.[315] The manufacturers at Septèmes, and elsewhere, then adopted the use of absorption channels packed with chalk (up to some 500 metres long), and systems of this kind seem to have remained in use in France for many years. In 1823 Péclet at Marseilles suggested the use of Clément's *cascade chimique* (pp. 87, 166), to absorb the gases in water.[316] This, of course, was essentially the same as the absorption tower later to be developed in England by William Gossage, and in due course adopted as standard in British plants in the mid-century. Such absorption towers, however, never acquired the importance in France that they acquired in Britain.

NOTES

1. On the natural alkalis: Chaptal (1807), vol. 2, 92–144; id. (1798); Coquebert (1794); Descroizilles (1806); Julia (1804).
2. The terms 'potash' and 'soda', confusingly, were used in several different senses in the late eighteenth century. Whereas in commerce they signified the crude materials employed industrially, in scientific writings they signified the pure alkaline carbonates, or alternatively, in the new nomenclature introduced by Lavoisier and his colleagues in 1787, these alkalis free of their carbonic acid.
3. Coquebert (1794), 83; Julia (1804); Ribero.
4. Scoville (1950), 50, 55. On the history of varech production see Sauvageau (1920), 49–66.
5. Coquebert, 35–6, 88 (describing a quarter of the imports from Spain as going to the north of France and three-quarters to the south); Magnien and Deu (1809), vol. 2, part 2, p. 550; AN, $F^{12}2244$, [4]. With regard to potash, detailed figures for 1788 show net imports into the free ports of 3 041 898 pounds (1·49 million kg, value 866 225*l*.), but the country's total imports cannot be ascertained because figures are mixed with those for soda (AN, $F^{12}1835$). After the Empire potash imports averaged 3·83 million kg for the years 1816–20 ('Tableaux' (1818); Admin. douanes). Import *values* for alkali in the years 1787–9, in millions of *livres*, were as follows (AN, $F^{12}513$, [1]):

	1787	1788	1789
Cendres diverses	1·375	2·412	1·026
Potasse, védasses et gravelées	1·728	1·214	2·178
Soude	2·682	1·831	5·676

6. Duhamel (1739).

7. The reactions are:

$$2NaCl + H_2SO_4 = Na_2SO_4 + 2HCl$$
$$Na_2SO_4 + 2C = Na_2S + 2CO_2$$
$$Na_2S + 2CH_3COOH = 2CH_3COONa + H_2S$$
$$2CH_3COONa = Na_2CO_3 + (CH_3)_2CO$$

(Partington, vol. 3, 70).

8. On early British developments see: Padley (1951–2); Clow (1952),Ch. 4; Gittins (1966); Moilliet (1966); Musson and Robinson (1969), Ch. 10.

9. AN, $F^{12*}166$, p. 203.

10. Previous accounts of the various early soda ventures in France are in: Baud (1933a, b), (1934); Gillispie (1957a).

11. This account is based on: Nouv. biog. gén., 'Malherbe'; Biog. univ. Supp., 'Athénas'; J. Darcet et al. (1797a), 81–7, 93–4 (for process details); AAS, [4]; CNAMA, [2]; AN, $F^{12}1505$, [1], [5]; $F^{12}1507$, [3]; $F^{12}1508$, [11]; $F^{12*}30$, pp. 89–91; $F^{12*}106$, pp. 86–7; $F^{12*}166$, p. 240.

12. See e.g. Macquer (1766), vol. 2, 380–1.

13. AN, $F^{12}1507$, [2].

14. See Lunge (1879–80), vol. 2, 340–5.

15. Yearly figures 1776–88 in Payan (1934), 157.

16. Multhauf (1971), 177–8.

17. Scheler ((1961), 260–4) gives the text. See also the preliminary draft in Lavoisier (1862–93), vol. 6, 16–19.

18. AAS, [3]. The entries are anonymous. Three survive from 1783. They were numbered for identification, with brief ranking comments as follows: No. 2, 'le 4e en mérite' (4 pp., in Latin and dated from Erfurt); No. 3, 'le 3e en mérite' (11 pp., from an author in Nantes, describing the formation of soda as an efflorescence when iron, copper, or zinc were exposed to brine and the air, a method he reiterated in 1787); No. 5, 'le meilleur après le no. 1' (13 pp., describing, among other methods: double decomposition of salt and potash; efflorescence of soda from salt and lime; trituration of litharge with brine; fusion of sodium sulphate with charcoal; and calcination of sodium acetate, made from lead acetate and salt). The 1786 submission (129 pp., mostly tangential) described experiments closely similar to those subsequently published in November 1786 by the Italian chemist Lorgna, in French translation by Champy (see below, p. 211); the resemblance is far too close for coincidence.

19. AAS, [4] ('La proclamation . . . a écarté les capitalistes sans lesquels on ne peut faire d'entreprise de conséquence'); AN, $F^{12*}166$, p. 240.

20. AN, $F^{12}2242$, [1].

21. Dict. sci. biog.; Bouchard (1938).

22. The following account is mainly from AN, $F^{12}1507$, [2].

23. Acad. Dijon (1819), 133–5; AN, $F^{12}997$, [1]; Darcet et al. (1797a), 103–4. W. V. and K. R. Farrar have convincingly argued that the chemistry of such a process depended on decomposition not of salt but of sodium sulphate also present in sea-water:

$$Na_2SO_4 + Ca(OH)_2 = CaSO_4 + 2NaOH$$

Absorption of CO_2 from the air would then give sodium carbonate. (See Musson and Robinson (1969), 369–71.)

24. Scheele (1931), 200–2.

25. Tolozan offered in March 1785 to seek exemption from duty on Guyton's behalf if he would draw up a request, but Guyton does not seem to have responded. The duty was only very small, amounting to 12 s. per *quintal* (Magnien (1786), vol. 3, 87).

26. Darcet *et al.* (1797a), 103–4.

27. See primarily Pris (1975), 569–70, 1223; AN, F^{12}2242, [1]; also F^{12*}167, p. 1.

28. Scoville (1950), 30, 51–2.

29. Some indication of his methods is provided by a list of materials he employed in experiments at Tourlaville in 1782–3: 'Marine salt, saltpetre, sulphur, vitriolic acid, vinegar, Glauber's salt, acids, litharge, sal ammoniac, Tripoli' (AN, 26 AQ 1, [2]).

30. Its existence in Oct. 1786 is referred to by Chassagne (1971), 285.

31. AN, F^{12}1505, [4].

32. AN, F^{12}680. The same remark was applied to Guyton's works.

33. AN, F^{12}1505, [4]; F^{12}1508, [19]; F^{12}2242, [4].

34. Author of several papers on technical chemistry and in 1789 director of the *Société royale d'agriculture*. The following account is mainly from: AN, F^{12}2242, [2]; F^{12}1507, [2].

35. AN, F^{12}997, [1].

36. *Dict. biog. fr.*; *Carny* [c. 1850?]; AN, F^{12}2244, [4].

37. AN, F^{12}997, [1].

38. Bouchard (1938), 104.

39. The Assembly referred the matter in December to its *Comité d'agriculture*, which nine months later passed on the pieces to the *Société d'agriculture* for an opinion (Gerbaux and Schmidt, vol. 1, 691, 722, and vol. 2, 405–6).

40. Thiéry (1787), vol. 2, 643; Musson and Robinson (1969), 274; English patent no. 1677, March 1789, in name of Boneuil; Darcet *et al.* (1797a), 77–82; *J. de Normandie*, 7 Jan. 1789; AN, F^{12}652; F^{12}1508, [6].

41. Partington, vol. 3, 218–19; Darcet *et al.* (1797a), 96–102; Chaptal (1791b), vol. 1, 261, and vol. 2, 305; id. (1807), vol. 2, 144–7; AN, F^{12}1508, [30]. The reaction is:

$$5PbO + H_2O + 2NaCl \rightleftharpoons 2NaOH + PbCl_2.4PbO$$

42. Bonnassieux and Lelong (1900), 481; AN, F^{12}1505, [2]; F^{12}2242, [1]; F^{12*}108, p. 51.

43. Gerbaux and Schmidt, vol. 1, 740–1; Hassenfratz (1792).

44. AN, F^{12}2244, [4].

45. Barral (1873), 67–8; Lambeau (1912), 68, 72–4, 514. The property was a former hunting lodge called the Maison Blanche, situated at the head of the present rue de Javel.

46. INPI, [2]; AN, F^{12}997, [1].

47. AN, F^{12}2244, [4].

48. Darcet *et al.* (1797a), 79.

49. The basic biography is by Leblanc's grandson (Anastasi (1884)), and is a work of family piety, to be used with caution. Other accounts have generally derived heavily from this, perhaps the best such being Oesper (1942–3). The only important critical study is Gillispie (1957a), which makes extensive use of the archive sources and to which we are deeply indebted for our own introduction to the subject, but which is rather unsympathetic towards Leblanc, and not

without its errors. Some earlier use of the archive sources was made by Baud (1933*a*, *b*).

50. Then Louis Philippe, father of Leblanc's future partner, Louis Philippe Joseph, who acceded to the dukedom in 1785.

51. Anastasi, 9.

52. Leblanc (1800), 467–8; he gives the date 1784 in several other places, too.

53. Delamétherie (1789), 44.

54. AN, F^{12}2244, [2].

55. Darcet *et al.* (1797*a*), 113–15.

56. Experiments by Mactear in the 1870s, using closed crucibles, showed that up to 27 per cent of the sulphate employed—and up to 35 per cent of that actually decomposed—was converted to carbonate (Mactear (1878), 481).

57. Leblanc (1800), 468; cf. id. (1803).

58. Darcet *et al.* ('1794' [1797*b*]), 118; Delamétherie (1809), 425.

59. Dizé (1810); Boudet (1852). See also the outline by Dizé in a letter of 1843, published by Pillas and Balland (1906), 20–2.

60. AN, F^{12}997, [1].

61. Its large-scale exploitation was to be attempted by many in the nineteenth century—among the first by Pelletan in the 1820s—but without success. (Dumas (1856), 556; Lunge (1879–80), vol. 2, 345–52.)

62. e.g. Suliac (1950).

63. Dumas (1856), 572. The jury included Chaptal, Berthollet, and Darcet *fils*. It is possible that Dizé's publication of his story in the spring of 1810 was partly motivated by the fact that the Academy of Sciences was that summer to make the award of the Decennial Prize for industry; we are told that he did communicate his story to the Academy in 1810, but received no reply (AAS, [1]).

64. Dumas (1856), 576–8; Pillas and Balland (1906), 123.

65. Gillispie (1957*a*), 169. Gillispie seems to have been incompletely informed on Dizé, citing neither Boudet nor Pillas and Balland.

66. Dizé elsewhere stated that the discovery was made on 14 July 1789 (Pillas and Balland (1906), 21), which also seems wrong. The chronology would make good sense, however, if one supposed the first trials to have begun in mid-July 1789, and the new experiments in August, this putting the discovery itself early in November, which would agree with the formation of the partnership in the new year.

67. Dizé (1803).

68. There is a copy of the first eight articles in AAS, [1]. The full agreement is said to have run to some thirty articles.

69. Dizé (1810), 296–7.

70. Shée was born of an expatriate Irish family (O'Shee), and was an uncle of Henry Clarke, who as the duc de Feltre became Napoleon's War Minister. See *Biog. univ. Supp.*, and Hayes (1949), 260.

71. MC, Étude XXIII, [4]; published in Dumas (1856), 559–62. Brichard's inventory records an earlier deposition of a packet by Leblanc and Dizé on 26 January (MC, Étude XXIII), but this item cannot now be found. It might have related to the association which the two are said by Boudet ((1852), 106) to have formed together under the auspices of Darcet, or it might just be an inventory error for the deposition of 27 March, which is not itself recorded.

72. There are documents on the purchase in AN, R⁴901.

73. MC, Étude XXIII, [3].

74. This specified that profits from soda and sal ammoniac were to go three-fifths to Leblanc and two-fifths to Dizé, while 'white lead' profits were to be divided in the reverse fashion. It was essentially a reaffirmation of an earlier agreement of 1 April 1790. (MC, Étude XXIII, [1], [2].)

75. INPI, [1]; Dizé (1810), 296; Pillas and Balland (1906), 22.

76. Gerbaux and Schmidt, vol. 2, 230, 364, 400–1; INPI, [1].

77. Leblanc's description (of 19 Sept.) and the commissioners' report are in AN, F¹²997, [1]. The description incorporated in his patent (INPI, [1]), and that later published (*Description*, vol. 1, (1811), 170–7), are essentially the same as the description he deposited. In January 1792 Carny secured the secrecy of his own patent on the basis of Leblanc's precedent.

78. The reaction occurs in two stages, the second of which requires a higher temperature than the first:

$$NaCl + H_2SO_4 = NaHSO_4 + HCl$$
$$NaHSO_4 + NaCl = Na_2SO_4 + HCl$$

The stages are not very sharply separated and in practice about two-thirds of the HCl is evolved at a relatively low temperature.

79. Scheurer-Kestner (1885), 393.

80. Shée (1797), 290–1.

81. Anastasi, 19; Darcet *et al.* (1797*a*), 60–76; AN, F¹²2244, [2]; F¹²1508, [9].

82. According to some costings given by Shée in 1794, based on pre-war (1791) prices, the acid, bought at 8 *s.* per pound, formed 62 per cent of total production costs, including materials, fuel, labour and interest on capital. (Shée (1797), 291–3).

83. The following technical description is from Darcet *et al.* (1797*a*), 60–76.

84. See next note. This is consistent with the description by Darcet *et al.* of two 250-pound trial batches being worked in 7 hours.

85. e.g. Shée (1797), 291; Anastasi, 116; Dizé in AN, F¹²1508, [13]. Varying data is found in different places. Leblanc himself in July 1794 gave figures indicating a daily production capacity of 750 pounds—from six charges in 24 hours—equivalent to only 225 000 pounds a year, though his note is hasty and he might have intended a figure of 1500 (AN, F¹²1508, [9]). An anonymous note in F¹²1508, [13] reckons on the basis of six 300-pound charges per 24 hours, and a year of 324 days, which would give 330 000 pounds of crude soda a year.

86. AN, F¹²2243, [1]; INPI, [1].

87. INPI, [1].

88. Calvet (1933).

89. INPI, [1]; AN, F¹²2243, [1]; F¹²2244, [2].

90. Darcet *et al.* (1797*a*), 60, 73–4.

91. AN, F¹²2244, [2].

92. Anastasi, 125.

93. Shée (1797), 297.

94. Leblanc (1800), 468; Dizé (1803).

95. Richard (1922); Bouchard (1946); Arbelet (1918); Multhauf (1971).

96. Richard, 538, 541; Payan (1934), 157.

97. Richard, 490–1, 542–53.
98. Richard, 547.
99. Richard, 552; Caron (1925), 139. Chaptal replied that it was impossible (AN, F^{12}1508, [2]).
100. Not found but referred to in Darcet *et al.* (an III [1794]), 2, and in the Committee's *arrêté* of 27 Jan.
101. AN, F^{12}1508, [8]; Aulard, vol. 18, 626.
102. Pillas and Balland (1906), 85.
103. AN, F^{12}1508, [8].
104. Dizé (an X [1802]), 32.
105. For publication details see p. 240.
106. From Malherbe (1795) we learn that the Munzthal works, whose lease Jourdan acquired in 1791 (Marcus (1887), 119), was idle from Oct. 1793 until the following autumn, as a result first of enemy invasion and then of French requisitioning of its potash. He does not mention soda manufacture. Elsewhere he speaks of a method of his devising for the production of sodium sulphate by the agency of gypsum, which had been successfully tried at Munzthal by Jourdan's associate d'Artigues, but whether this was with a view to soda manufacture he does not say (*J. des mines*, No. 4 (Jan. 1795), 46).
107. Dumas (1856), 566–7.
108. With the rank of principal assistant medical officer. He had been commissioned in Sept. 1792 as an apprentice pharmacist in the hospitals of the military camp outside Paris, with pay of 100*l.* a month, and he had then been posted to the Saint-Denis hospital on 28 Jan. 1793. (Pillas and Balland (1915), 111–13; Rouquette (1964–5). For Dizé's career as a military pharmacist see also Blaessinger (1952).)
109. Richard, 428–9; Anastasi, 33.
110. AN, F^{12}1508, [19], [8].
111. When the administrators of the Loire-inférieure asked Paris what they should do with this works, the *Commission d'agriculture* on 2 July simply advised them to seek a purchaser who would revive it (AN, F^{12}1505, [4]; see also F^{12}1508, [19]).
112. Darcet *et al.* (1797*a*), 61.
113. Id. (an III [1794]), 10.
114. Id. (1797*a*), 140–1.
115. *Moniteur*, 16 July 1794.
116. AN, F^{12}1508, [13]. Another anonymous piece of about the same time spoke of a proposal to establish an *agence temporaire* to direct manufacture of soda and soap, with workshops to be installed in the abbey at Saint-Denis (F^{12}1508, [4]).
117. AN, F^{12}1508, [19].
118. AN, F^{12}1508, [13]. Notes headed 'Généralités sur la manufacture de soude', signed by Dizé 'in the name of my co-associates'.
119. Pillas and Balland (1906), 74.
120. AN, F^{12}1508, [9].
121. Leblanc had written to the Committee on 10 June, speaking of his poor circumstances (he was 7000–8000*l.* in debt), and suggesting that if the Darcet commission should find his process of value he might be eligible for an award. He was granted 4000*l.* (AN, F^{12}1508, [13]).

122. Prieur left the Committee on 6 Oct. 1794, but remained for a time as an adviser in its offices. Fourcroy was a member of the Committee from 1 Sept. 1794 to 3 June 1795 (with a break in January), and Guyton from 6 Oct. 1794 to 3 Feb. 1795.

123. AN, F^{12}1508, [13].

124. Gerbaux and Schmidt, vol. 3, 318.

125. *Arrêté* by Fourcroy in Aulard, vol. 18, 11. A later *arrêté*, of 30 Jan. 1795, ordered the inspector of mines J. H. Hassenfratz to go immediately to the eastern salt-works to activate sulphate extraction there, but on 10 March he was dismissed and recalled from his mission, perhaps for political reasons (Aulard, vol. 19, 763, and vol. 20, 773).

126. AN, F^{12}1508, [10], [26].

127. See Leblanc to Dizé, 29 Nov. 1794, in Pillas and Balland (1906), 74–5.

128. *Bull. lois*, no. 97 (an III), piece no. 497.

129. AN, F^{12}1508, [13].

130. AN, F^{12}1508, [13]; F^{12}2244, [2]. The group was not, as Gillispie would have it, 'a well-known chemical firm'.

131. AN, F^{12}1508, [13]. Notes unsigned and undated but in Fourcroy's hand and probably *c.* April-May 1795.

132. Anastasi, 38 ff.

133. Gillispie alleges that Leblanc's reply of 14 July 1794 to the Committee's questions about the works came only after a delay of some four months, and was then unrealistic in presenting grossly exaggerated figures on production capacity. In reality the Committee's questions had almost certainly followed the Darcet report (20 June), and Leblanc's figures are actually slightly lower than those of Gillispie's own estimate. Gillispie appears to have been making erroneous inferences from some (unsigned) notes by Guyton in AN, F^{12}1508, [13].

134. Acknowledgements dated 19 July–1 Aug. in AN, F^{12}1508, [19].

135. AN, F^{12}2243, [8].

136. The original water-coloured drawings for these are now in the museum of the *Conservatoire des arts et métiers* (reserve collection, No. 13 571–342).

137. Later, in 1797, edited versions of the report were also to appear in the *Ann. chim.* and *J. de phys.*, when these journals resumed publication after the Revolution; the relevant issue of the *J. de phys.* bears the date Aug. 1794, but in fact it appeared in Aug. 1797 (see Lalande (1803), 641). In 1798 there also appeared a shorter abridgement by Bouillon-Lagrange in the *J. Soc. pharmaciens*, and the full report was reprinted that year in the collected works of Pelletier.

138. Périer was one of the region's prominent industrialists, with banking and textile manufacturing interests. He was later to be a principal founder of the *Banque de France*, and is remembered as the father of a remarkable family including Casimir Périer (the political figure). The soda project is not mentioned in the standard biographical studies (e.g. Barral (1964)).

139. This paragraph is from: AN, F^{12}1508, [4], [7], [18], [19], [21], [23], [24], [32]; F^{12}2242, [4]; F^{12}2243, [8]; F^{12}2244, [3].

140. AN, F^{12}2244, [4]; F^{12}1508, [13], [33].

141. AN, F^{12}2244, [4]. Gillispie is entirely mistaken when he writes ((1957a), 165) that with the aid of a government loan Carny established a soda works which proved a failure, forcing him to seek government assistance. Carny did not

obtain a loan, did not establish a soda works, and his petitions of 1796–7 were not concerned with soda manufacture (see p. 248).

142. If granted 20 000 pounds of sulphur and 3000 pounds of saltpetre he offered to supply 70 000 pounds of sodium sulphate for hospital use (having thus evidently already given up his plan to make soda). The *Commission de santé* and *Commission des armes* recommended that his request be rejected since enough medicinal sulphate could be obtained from other sources (AN, F^{12}1508, [16]). Carny had been having difficulty in obtaining sulphuric acid for some time, and on 15 Nov. 1794 had been authorized to import 20 000 pounds from Brussels (AN, F^{12}1508, [15]).

143. AN, F^{12}1508, [14].

144. Toraude (1921), 27–8, 135; Bouchard (1938), 102, 120.

145. Smeaton (1962), 42; Richard, 423, 646–7.

146. See particularly: Anastasi; AN, F^{12}2243.

147. Anastasi, 95–6.

148. Anastasi, 100.

149. In January 1805 Leblanc presented to the First Class a note on saltpetre manufacture in which he proposed the substitution of potassium sulphate for the potash generally employed. Deyeux and Vauquelin, reporting on his note, remarked that to obtain the potassium sulphate 'it would be sufficient . . . to put into activity the manufacture of artificial soda, in conformity . . . with the process which he has indicated and followed with success for a certain time, but which special circumstances have forced him to abandon' (Acad. sci. (1910–22), vol.3, 184). Leblanc had described the potash process in a paper of 1803, which concluded with a characteristically mysterious remark that he also had 'means more simple, more certain and more economical than all those used up to the present' (see also Leblanc (1800), 468–71).

150. Anastasi, 121–30.

151. AN, F^{12}2244, [4]; Anastasi, 120. See also: A. Payen (1866), 967; AN, F^{12}2242, [3].

152. Partington, vol. 3, 95; Vauquelin (1799).

153. AN, F^{12}1508, [1], [25], [29]. On Schemel see also Bouvet (1928).

154. During the winter of 1794–5 the promotional efforts of the Committee of Public Safety had resulted in the rather modest production of some 30 000–40 000 pounds of sulphate in the great salt-works at Dieuze (in the Meurthe), together with lesser quantities elsewhere in the eastern departments ('Avis' (1795)). Much of this may have been absorbed by other uses.

155. *Troisième exposition* [1802], 29.

156. Herbin de Halle (1803–4), *Atlas*, 29; Lelivec (1806), 492.

157. AN, F^{12}2244, [4].

158. The following account derives primarily from the detailed documentation in AN, F^{12}2244, [5], [6]. Carny was also active in 1797–8 at Saint-Quentin, being described in September 1797 as 'fabricant de produits chymiques, à St Quentin'; his activities there may have involved sulphuric acid manufacture, since he was at that time seeking (unsuccessfully) to obtain from the State a quantity of lead, sulphur, and saltpetre (AN, F^{12}2244, [4]).

159. AN, F^{12}2234, [5].

160. *Notices* (1806), 163; Descroizilles (1806*b*), 21.

161. Chaptal (1807), vol. 2, 152; *Notices* (1806), 163; *Exposition* (1806), 174.

162. AN, F¹²2245, [9].

163. For a detailed account of the salt-works and its operations see Nicolas (1797).

164. *J. du commerce*, 30 July 1810.

165. AAS, [1].

166. Meunier, 30 Jan. 1808, in AN, F¹²2244, [6].

167. Dupin (1827), vol. 1, 218; Ancelon (1879), 176–7; AAS, [1].

168. *Nouv. biog. gén.*; Sonolet and Poulet (1972).

169. Fourcy (1828), 399.

170. ADSM, 5 MP 1253, [6]; AN, F¹²2245, [12]; *Précis*, année 1805 (1807), 66–7; Descroizilles (1806*b*), 19–20; [Pelletan (1810)].

171. *Nouv. biog. gén.*, 'Pelletan'.

172. *Exposition* (1806), 174. Soda was also exhibited by a certain Savary of Rouen, of whom no more is known.

173. AN, F¹²2245, [9].

174. Pelletan (1822–4), article 'Soude d'Alicante'.

175. I assume the Pelletan in question to have been Pierre rather than his aged father or younger brother.

176. *Dict. biog. fr.*; Dumas [1844?]; Perraud de Thoury (1853); Paillet.

177. Taton (1964), 401; Richard, 218, 420.

178. AAS, [5]; *Mém. soc. sav.* **2** (1802), 378–9.

179. The use of copper oxide was perhaps connected with Anfrye's interest in the exploitation of the dross of bell-metal, for which he had recently formed a works with Lecour (*Ann. chim.* **41** (1802), 167–76).

180. *Exposition* [1802], 53–4.

181. AN, F¹²2245, [1], [3].

182. AN, F¹²2245, [9].

183. *Bull. Soc. enc.* **6** (1807), 113.

184. [Pelletan (1810)].

185. Gautier and Barréra had since 23 Nov. 1802 been partners in the joint exercise of a position as *affineur national*, in succession to Gautier *père* (MC, Étude VI, [1]).

186. AS, D³¹U³, année 1808, no. 54.

187. MC, Étude X, [1].

188. PP, [1], [2].

189. AN, F¹²2245, [1].

190. On this works see: AN, F⁸94, [1]; F¹⁴4234, [1]. It was on the banks of the canal from Saint-Quentin to Chauny.

191. AS, D³¹U³, année 1810, no. 194; MC, Étude XV, [1].

192. On René's banking firm see Bergeron (1975), 283–93. On the wallpaper concern see Clouzot and Follot (1935), Ch. 9.

193. Desmier de Saint-Simon (1843), Table D.

194. See Cadet de Gassicourt (1806); also AN, F¹²1007, [2].

195. AN, F¹²2245, [9].

196. A partnership was formed on 2 May 1807 between Decroos, Darcet, and Leullier. From a later act dissolving the partnership—on 23 March 1811—we learn that under Darcet's name there were also included Anfrye, Gautier, and

Barréra, and that Leullier represented the Jacquemart brothers (AS, $D^{31}U^3$, année 1811, no. 61). The Saint-Denis company's share was said in September 1809 to be nine twenty-fourths (AS, $D^{31}U^3$, année 1809, no. 164). Following the dissolution of the 1807 partnership, a new partnership was formed on 25 March 1811, this now also embracing the Quessy soda works, with a combined capital of 420 000 fr.; Decroos no longer figured (MC, Étude XV, [1]).

197. Institut (1810), 122–3.
198. PP, [4], no. 256.
199. Perraud de Thoury (1853), 7.
200. L. Costaz (1819), 284; cf. Chaptal (1819), vol. 2, 80.
201. AAS, [2].
202. Institut (1810), 123.
203. Cochois (1902), 105–15; decrees of 16 and 17 March, and law of 24 April, in *Bull. lois*, 4th series, vol. 4 (1806), pp. 328, 347, 443; Becquey, pp. xiv–xvi.
204. Price series 1799–1820 in Chabert (1945), 222. I have assumed that Chabert's figures are actually double the true values, as would appear from other evidence to be obviously the case (he has probably confused pounds with kilogrammes). Payen and Pluvinet *frères*, in a petition of 1 March 1807, wrote that throughout the period when salt was duty-free it had cost only 3*l*. 10 *s*. per *quintal* (7 fr. per 100 kg), but that its price had now quadrupled (AN, $F^{12}2234$, [4]). The duty was raised to 40 fr. per 100 kg on 11 Nov. 1813, and was then lowered to 30 fr., probably at the end of 1814 (*Arch. parl.*, 2nd series, vol. 12, 703). It still stood at 30 fr. in 1819 (AN, $F^{12}2245$, [10]).
205. For the salt-tax discussions, 1806–10, see esp. AN, $F^{12}2245$, [9]; also $F^{12}2456$, [1].
206. Report by Bardel, Montgolfier, Molard, Gay-Lussac, and A. Ampère, in AN, $F^{12}2245$, [9].
207. AN, $F^{12}2245$, [3], [15], [11], [12].
208. Report by Gay-Lussac, Molard, and Bardel, in AN, $F^{12}2245$, [9].
209. It was a calculated figure. Soda was subject only to a nominal import duty which was levied on a value basis, and so imports did not have to be declared by weight.
210. *Bull. lois*, 4th series, **10** (1809), 235–6.
211. *Bull. lois*, 4th series, **11** (1810), 155–7.
212. See e.g. AN, $F^{12}2245$, [1], [2], [7], [8], [13]; $F^{12}2244$, [9].
213. AN, $F^{12}2245$, [5]. Such a problem had been encountered by Payen & Pluvinet *frères* (AN, $F^{12}2242$, [3]).
214. [Pelletan (1810)], 5.
215. AN, $F^{12}2245$, [9].
216. AN, $F^{12}7591$.
217. AN, $F^{12}2245$, [1]; ADSM, 5 MP 1252, [1].
218. AN, $F^{12}2245$, [1].
219. *J. de l'Empire*, 20 July 1810.
220. AN, $F^{12*}194$, p. 2.
221. AN, $F^{12}2245$, [1].
222. The following is mainly from: AN, $F^{12}2245$, [1]; PP, [1], [2], [3].
223. Fresnel (1866–70), vol. 2, 816.
224. AN, $F^{12}2244$, [8]; *Bull. Soc. enc.* **8** (1809), 258.

225. ADSM, 5 MP 1048, [1]; 5 MP 1253, [5]; 5 MP 1252, [1]; *J. de Rouen*, 17 and 28 Jan., 3 March 1810.

226. ADSM, 5 MP 1335, [3]; [Pelletan (1810)], 4; AN, $F^{12}2245$, [1].

227. Pajot Descharmes (1798), 209–10; Pris (1975), 352, 1228.

228. AN, $F^{12}2245$, [14]; Pajot Descharmes (1826), 266–7; *J. de phys.* **52** (1801), 210–11. Cf. Pris (1975), 839.

229. Acad. sci. (1910–22), vol. 2, 641, and vol. 10, 113.

230. Cochin, 166.

231. AN, $F^{12}2245$, [14], [1]; $F^{12}2243$, [4]; $F^{12}1549$, doss. 'Soissons. Soude artificielle. 1811'. By the end of 1815 the works (together with its associated soap-works) had been taken over by Korn. Its production was then said to have diminished somewhat since the re-introduction of foreign soda at the end of the Empire, but it was nevertheless said to be employing 110 workers and to be producing 800 000 kg of soda a year, still using pyrites rather than sulphuric acid (AN, $F^{14}4234$, [1]). In 1819 the works was transferred to Rouen (Brayer (1824–5), vol. 2, 252).

232. AN, $F^{12}2245$, [1]; $F^{14}4234$, [1]; Lefroy (1810).

233. There is some account of the Charles-Fontaine works in Hennezel (1933), 238–40, and there are some fragmentary early records in SG, [1], and more particularly in AN, 26 AQ 1, [1], [2]. For a detailed description by Tassaert of manufacturing methods and plant, presumably as used at Charles-Fontaine, see *Enc. méth. Chymie*, vol. 6 (1815), 160–5. (*Note added in proof:* On the Saint-Gobain company and its early chemical interests see now above all the important but obscurely published study by Pris (1975), of which we have been able to make only late and limited use.)

234. SG, [1].

235. Tassaert was an occasional contributor to the *Ann. chim.*, and in 1798 was a member of the chemical group which controlled its publication (Court (1972), 122). In 1800 he was described by Fourcroy as a Prussian chemist in charge of the laboratory of the *École des mines* in Paris (Fourcroy (1800), vol. 5, 108–9). In 1807 he was the translator of the French edition of Klaproth's collected mineralogical writings. He was director at Saint-Gobain from 1816 to 1833 (Pris (1975), 573, 1231).

236. Vauquelin (1814).

237. H.P.I. Reboul, geologist and chemist, a correspondent of the First Class of the Institute (*Index biographique* (1954); *Nouv. biog. gén.*).

238. AN, $F^{12}2245$, [1], [18], [19], [20]; $F^{12}937$; ADH, 109 M 33, [1]; 109 M 139, [3]; 109 M 187, [1]; 3 S 5.

239. In a memoir of 1819, the Marseilles soda manufacturers wrote that prior to the Revolution Marseilles had imported 140 000 *q.m.* (14 million kg) of soda a year from Spain and Sicily, and received a further 3000–4000 *q.m.* from the Languedoc; about 100 000 *q.m.* (10 million kg) were used by the Marseilles soap industry and the remaining 40 000 *q.m.* passed into the interior (*Mémoire* [1819], 5). Masson, in his standard work on the industrial history of Marseilles, has written that in 1789 there were produced about 0·8–1·0 million *quintaux* of soap (local measure, i.e. about 32–40 million kg); since on average about one part of soda was required for two parts of soap, we can reckon that this would have called for some 16–20 million kg of soda (Masson (1926), 2, 33). Carrière, more

recently, has considered such a figure for soap production too high, however, and after considering various contemporary and modern estimates for the town's soap production in the later 1780s, ranging from 0·5 to 1·0 million *quintaux*, he himself has opted for a figure of 0·7 million *quintaux* (28 million kg). He indicates soda consumption to have been 250 000 *quintaux* (10 million kg) (Carrière [1973?], vol. 1, 212, 316–19). Masson remains a highly informative source on the history of Marseilles industry in general, and includes some account of the early soda industry there.

240. AN, F^{12}2245, [1]; F^{12}2244, [1].

241. *Mém. Acad. Marseille* **5** (1807), 122–7.

242. AN, F^{12}2245, [1].

243. ADBR, M^{14}2. Masson ((1926), 47) is wrong in giving the date of this piece as 1811. The same works figure in another list sent to the Ministry of the Interior by the mayor of Marseilles on 23 June [1810] (AN, F^{12}2245, [1]).

244. Fodéré (1813), vol. 6, 320–2.

245. Villeneuve (1821–34), vol. 4, 787.

246. AN, F^{12}2244, [1].

247. AN, F^{12}937.

248. AN, F^{20}289^{I}, [1]; AMM, 22 F 1, [1].

249. De Laget (1962) gives some very brief details.

250. AMM, 22 F 1, [1]. There was no doubt some connection here with the factory described in mid-1810 as being under construction at Mazargues by Flautard & Cie, and with the (possibly identical) works listed in February 1811 as operating at Sainte-Marguerite in the name of Guillaume Flautard *fils* (ADBR, M^{14}2 and M^{14}47; AN, F^{12}2245, [1]). A third new manufacturer encountered by 1812, André Gilly at les Catalans, was probably exploiting the works near the vallon des Auffes established in 1810 by Olive. Antoine Girard had by now evidently given up making soda but was continuing as a sulphuric acid manufacturer.

251. ADBR, K^{2}31; INPI, [7].

252. De Laget (1962), 75; *Annuaire* (1823), vol. 1, 224.

253. ADBR, M^{14}47. See also above, p. 39.

254. Summarized in *Mém. Acad. Marseille*, vol. 8 (1812), Année 1810, Première partie, 8–9.

255. *Mém. Acad. Marseille*, vol. 9 (1812), Année 1811, Première partie, 16; Laurens (1812).

256. CCM, [1]. There are 59 signatories.

257. AN, F^{12}2245, [1].

258. *Description*, vol. 9 (1824), 260; INPI, [3]; AN, F^{12}2245, [9].

259. In May 1810 the Gironde was reported to be drawing artificial soda from Marseilles, Montpellier, and Toulouse (AN, F^{12}2245, [6]).

260. AN, F^{12}2244, [7]; F^{12}2245, [6].

261. *J. mines*, **30** (1811), 396–7; Fresnel (1866–70), vol. 2, 816.

262. AN, F^{12*}194, p. 54.

263. Ibid.; *Tarif* (1853–4), vol. 2, 798–9.

264. AN, F^{12*}194, p. 2.

265. AN, F^{12*}194, p. 54.

266. ADSM, Series M (unclass.), [1].

267. Villeneuve (1821–34), vol. 4, 787.
268. The manufacturers in question were Gautier, Vidal, Michel, Gilly, Pontier, and Covello, all in the commune of Marseilles, and Baux at les Trois-Lucs (the last three idle). Production fell from 619 000 kg (with 195 workers) in the second quarter, to 300 000 kg (with 100) in the fourth. (ADBR, M^{14}4, M^{14}6; AMM, 22 F 1, [1]).
269. Pigeire (1932), 327–8.
270. There is some very fragmentary statistical data for the period 1810–25 in ADBR, M^{14}2 to M^{14}12, and M^{14}47. In 1817 six works were indicated in the commune of Marseilles, five at Septèmes, and one at Vitrolles; elsewhere in the department the factories of Chaptal *fils*, Pluvinet *frères*, and Armand are known to have continued.
271. See e.g.: *Mémoire* [1819]; C. A. Costaz (1843), vol. 2, 43–6; *Arch. parl.*, 2nd series, vol. 13, 600–3, 608–9; AN, F^{12}7589, [1]; F^{12}2243, [7]; F^{12}633–637, doss. nos. 137, 186; CCM, [1], [2], pp. 30–45.
272. *Tarif* (1853–4), vol. 2, 794–7. Reduced to 10 fr. on 27 March 1817.
273. Royal ordonnance of 20 Feb. 1815, in *Bull. lois*, 5th series, vol. 3, p. 717.
274. AN, F^{12}2245, [4]; F^{12}2243, [5], [6]; Pajot Descharmes [1815].
275. ADBR, M^{14}226.
276. A. Payen (1855), 237–8; id. (1866), 977; *Recherches*, vol. 3 (1826), Table 107. Payen ascribed the development of this product to the chemical manufacturer Ador.
277. ADBR, M^{14}226; cf. Mactear (1877), 25.
278. ADBR, M^{14}12; Julliany (1842–3), vol. 3, 316. We have not found any Act of Parliament specifically affecting French or artificial sodas, but an Act of 19 May 1819 introduced a new schedule of import duties which might have had the incidental effect of increasing the duty charged on the French product (59 Geo. III, c. 29).
279. Villeneuve (1821–34), vol. 4, 795–6.
280. For these and further details see: *Bull. lois*, 7th series, vol. 9, no. 299 (1819), piece 7144; *Moniteur*, 29 June 1824 (Supplement); Héricart de Thury (1819), 202; AN, F^{12}6728, [1]. The new company began with an issue of 120 shares of 10 000 fr. each; in 1824 a further issue was made to the same amount. Chaptal *fils*, Bodin, Darcet, and Holker were among the shareholders.
281. ADBR, M^{14}8; M^{14}12;116 E N1.
282. Julliany (1842–3), vol. 3, 310; Tinthoin (n.d.), 202.
283. AS, D^{31}U^3, année 1816, no. 186. The old partnership of Gautier, Barréra, & Cie had been dissolved on 20 May 1815 (AS, D^{31}U^3, année 1815, no. 134).
284. Briavoinne (1839), vol. 1, 390.
285. Héricart de Thury (1819), 201–2; cf. Pigeire (1932), 309–10.
286. Lévy-Leboyer (1964), 486; Pigeire (1931), 519–25; MC, Étude X, [1]. By 1825 Chaptal *fils* had acquired three-quarters ownership of the la Folie works, the other quarter belonging to Darcet. Holker withdrew from the 1816 partnership on 5 March 1824, and the remaining partnership between Chaptal *fils* and Darcet was dissolved in August 1826 (AS, D^{31}U^3, année 1824, no. 125, and année 1826, no. 655).
287. Laboulaye (1853–61), vol. 2, 'Sulfurique (Acide)'. On Poisat see Ure (1839), 1061.

288. AN, F^{12}2245, [10]; Briavoinne (1839), vol. 1, 388; AS, D^{31}U^3, année 1823, no. 521; Anastasi, 135.

289. In the years 1811–18 the annual soda consumption at Saint-Gobain—as indicated by the through-put of the refinery—ranged between 660 160 pounds (in 1814) and 1 419 470 pounds (in 1812), averaging about 1·1 million pounds (AN, 26 AQ 1, [1]).

290. AN, 26 AQ 1, [1], [2].

291. Dumas (1856), 557; Lunge (1879–80), vol. 2, 428; Gillispie (1957b), 401 (see also id. (1957a), 170, and (1965), 143).

292. Lunge (1891–6), vol. 2, 459; for the various theories see pp. 458–74.

293. AN, F^{12}997, [1].

294. Coquebert (1794), 69–70; Darcet et al. (1797a), 67–75; Fourcroy (1800), vol. 3, 186; Rougier (1812), 91–2, 96. See also Loysel (an VIII [1799–1800]), 248–9.

295. Vauquelin (1802), 192; Welter and Gay-Lussac (1820), 218; Péclet (1823), 439; Pelletan (1822–4), vol. 2, 'Soude factice'; Poutet (1828), 15; CNAMB, [1], vol. 2, cahier 1, pp. 86–7.

296. L. Costaz (1819), 275; A. Payen (1855), 212.

297. See: Descroizilles (1806a, b), (1818); Madsen (1958); Szabadvary (1966); Christophe (1971).

298. H. P. [1810], and L. [1810 or 1811], both in ADSM, 5 MP 1252, [1], where there are also various other petitions and pamphlets on the subject.

299. ADSM, 5 MP 1253, [1].

300. ADBR, K^{2}31, p. 117.

301. Fodéré (1813), vol. 6, 317–30.

302. Collection (1880), 158; PP, [1]–[5]. For general studies of the Conseil see: La Berge (1975), Weiner (1974).

303. Guyton and Chaptal (1805).

304. Acad. sci. (1910–22), vol. 4, 268–73; Conseil d'état (1810); Bull. lois, 4th series, vol. 13 (1811), No. 323, piece 6059.

305. For a historical sketch see Quevauviller (1955).

306. Dict. tech., vol. 2 (1822), 88–9.

307. J. de Rouen, 28 Jan. 1810; Pelletan (1810).

308. ADSM, 5 MP 1253, [1].

309. PP, [3], nos. 23, 79; Acad. sci. (1910–22), vol. 4, 342.

310. AN, F^{12}2245, [5].

311. Bull. Soc. enc. 9 (1810), Prize programme; ibid. 10 (1811), 215–16.

312. Darcet (1821); L. Costaz (1819), 290–1; Paillet, 5; Dumas [1844?], 8. On the nutrition controversy see Holmes (1974), 6–15.

313. According to a note of July 1820 (in AN, F^{12}2245, [1]) salt in Marseilles then cost 3 fr. 50 per 100 kg and sulphur 7 fr., while in Paris the prices were double this; coal, too, was much dearer in Paris.

314. Dict. tech., vol. 1 (1822), 'Acide hydrochlorique'; vol. 19 (1831), 'Soude'.

315. On the fumes problem at Marseilles after the Empire see e.g.: Villeneuve (1821–34), vol. 4, 785–95; Mémoire [1819]; Mémoire (1818); Rapport (1819); Rapport (1826).

316. Péclet (1823), 442.

5

CONCLUSION

By the Restoration France had acquired a large and diversified chemical industry. Sulphuric acid production was widely established in industrial centres throughout the country, generally accompanied in the same works by a range of dependent manufactures, such as nitric and hydrochloric acids, and chemical salts for dyers and others. In addition there were now the new soda factories, chiefly around Marseilles and more specialized in their activities, but usually, besides soda, making their own sulphuric acid, and sometimes collecting their hydrochloric acid as a further saleable product. Sulphuric acid and soda were now established in their classic position as the twin mainstays of the nineteenth-century heavy chemical industry. The pre-eminent importance of these particular manufactures, however, whose evolution it has been our primary concern to trace, should not lead us entirely to overlook the other branches of chemical production which saw significant development in our period. While we cannot examine these at any length, we should not conclude our study without briefly recording some of the more notable of them.

The most important of these further manufactures were those of alum and copperas, two materials which had long been among the leading chemicals in industrial use. Alum was principally employed as a mordant in the dyeing and printing of textiles, though it had quite a large number of lesser applications, too: in the preparation of leathers, for example, and in the manufacture of paper. Copperas (ferrous sulphate) was again chiefly used in dyeing, and was employed to a lesser extent in the manufacture of Prussian blue and the preparation of writing inks. In the mid-eighteenth century France's own production of these materials was negligible and they were therefore imported. Alum of the finest quality, for use in high grade dyeing, come from the famous papal works at Tolfa, near Rome, while poorer alum for common use was bought from Britain, Liège, Sweden, and the eastern Mediterranean. Copperas seems to have come largely from Britain. In all these places the materials were extracted from naturally occurring ores by traditional processes, involving lixiviation and crystallization, generally after weathering the ore for some months in heaps.

In the fifteen or twenty years prior to the Revolution, serious efforts began to be made to develop home production in France. No doubt a stimulus, as in the case of sulphuric acid manufacture, was a growing need for these materials in dyeing and textiles printing. A number of works were now established for the extraction of copperas from pyritous deposits, and by the outbreak of the Revolution these were already beginning to make a significant contribution to the country's consumption: there were works of growing importance at Saint-Julien-de-Valgagues in the Gard (privilege accorded on 25 October 1774), at Goincourt and Becquet near Beauvais in the Oise (letters patent of 15 May 1780 and 12 March 1781), and at Urcel in the Aisne (privilege of 30 May 1786, to Chamberlain).[1] The production of alum was also starting to be developed, although by the Revolution it was still of little importance: there was a works at Monchy-Humières near Compiègne in the Oise (privilege of 22 January 1785), and in the later 1780s exploitation was beginning in the Aveyron.[2]

At the same time as France's mineral deposits were thus beginning to be exploited, the country's new sulphuric acid manufacturers were turning their attention to the production of alum and copperas by chemical synthesis. It was alum which particularly attracted their interest, because its extraction from ores was scarcely beginning, and good ores were rare in France. The synthetic manufacture of alum derived ultimately from scientific studies by E. F. Geoffroy in the 1720s, which had shown that alum could be obtained by digesting roasted clay with sulphuric acid.[3] The pioneers in exploiting this knowledge industrially were Alban at Javel and Chaptal at Montpellier. The Javel manufacturers were occupied with the matter very soon after establishing their works, and on 5 May 1781 they obtained an exclusive privilege for the manufacture of 'artificial alum' in the *généralités* of Paris, Alençon, Caen, and Rouen. They subsequently made alum regularly at Javel in quite large quantity, as one of their principal activities: a capacity of 300 *milliers* (300 000 pounds) a year was claimed in 1780, though in 1794 Alban wrote, 'My alum works is mounted to make 2000 [pounds] of alum per *décade* [ten days], and as much again if the materials could become more abundant' (this amounting to some 70–140 000 pounds a year).[4] Their method was quite straightforward, involving digestion of calcined clay with sulphuric acid, followed by lixiviation, addition of alkali, and crystallization. Chaptal tried this method at Montpellier in the mid-1780s, but finding it uncompetitive with low-priced imported alums he devised a different method of his own, which instead of employing sulphuric acid exposed the clay for three or four weeks to the sulphurous (and nitrous) gases in a lead chamber.[5] He evidently worked his process for a time but it was the method of the Javel manufacturers that was to be of more lasting significance.

Production of alum and copperas, both synthetically and from ores, suffered as a result of the Revolutionary disruptions of 1793–4, and it was in the Napoleonic period that these manufactures grew in an important way. Synthetic alum continued to be made at Javel and at Montpellier, and it was also now made in Chaptal's new works at les Ternes, and by Bouvier and Curaudau, both in the capital. The methods now employed were all based on digestion of clay with sulphuric acid, though there were differences in detail, Bouvier, for example, using a process devised by Chaptal, in which the manufacture of alum was coupled with that of nitric acid by distilling together clay, sulphuric acid, and saltpetre.[6] A significant manufacturing advance made by Chaptal in the 1790s, and popularized from the turn of the century, was the discovery that potassium sulphate could be profitably substituted for the potash that was traditionally added to the lixivium—in both natural and synthetic processes—to induce crystallization of the alum. In close association with this advance in manufacture there also came a notable theoretical advance, with the first recognition of the true chemical constitution of alum as a double salt (a discovery made independently by Descroizilles, Chaptal, and Vauquelin).[7] The principal synthetic alums were submitted to exhaustive tests in 1806 by Thenard and Roard, for the *Société d'encouragement*, with the aim of convincing a sceptical public of their value. The products of Bouvier and Curaudau were proved to be of high quality, second only to Roman alum in purity, while that of Javel was fully equal to the common alums from abroad.[8] The manufacture of synthetic alum was regarded as one of the triumphs of chemical manufacture in the period, another illustration of the fruitful application of science to industry, and it was to remain a resource of some significance. The production of synthetic copperas (by simply dissolving iron in sulphuric acid) was also practised under the Empire, and continued to some extent thereafter.

Of greater economic importance than manufacture by synthesis, however, was the development of the extraction of these materials from mineral deposits. The extraction of copperas grew very rapidly from about 1804, especially in the departments of the Aisne and the Oise, rendering France virtually independent of imports after 1806. At the same time the extraction of alum also grew greatly, though more slowly, again particularly in the Aisne and the Oise, where alum came to be extracted from the hitherto rejected residues resulting from copperas production. The lead in this was given by Clément, Desormes, and J. M. Montgolfier, who established an important works for the purpose at Verberie in the Oise in 1804. The response of Clément and Desormes to the problem of consumer resistance against alum made in France was to doctor the product to give it the appearance of the esteemed Roman variety, and then to sell it under that name. By the beginning of 1806, when they disclosed their practice to the Institute, they had successfully sold 153 000 pounds of 'Roman' alum in eighteen months, and Clément later recalled that by this subterfuge, conducted in a plant near Jouy, they made profits of 100 000 francs in a year with a capital of 1000 *écus* (3000 francs), and could scarcely keep up with the demand.[9] By the final years of the Empire, France had not only become self-sufficient in copperas but also virtually so in alum as well.[10]

Apart from alum and copperas, another material of large consumption was white lead (basic lead carbonate), used as a pigment and in the manufacture of pottery. This was a material imported chiefly from Holland, where it was made by a traditional process involving slow reaction between lead, vinegar, and atmospheric carbon dioxide. France's net imports in 1787 amounted to 1 200 000 kg.[11] There were manufacturing projects and ventures in France from the 1780s, but production only became established there in an important way under the Empire. In 1802 the *Société d'encouragement* offered a prize on the subject, with the aim of encouraging the establishment and development of manufacture. The prize was awarded in 1809 to Brechoz and Leseur, who had just established a works at Pontoise, near Paris, using a process patented in 1808 and based on a reaction suggested in 1803 by Thenard: this involved the precipitation of white lead by passing carbon dioxide through a solution of lead oxide in vinegar or pyroligneous acid.[12] From 1809 a much larger works using the same process was established at Clichy-la-Garenne, by Roard, who had probably collaborated in the works at Pontoise. He was a former student of the *École polytechnique* and director of the Gobelins dyeworks under the Empire. The Clichy factory was said in 1819 to produce 350 000–400 000 kg of white lead a year, together with a slightly smaller amount of minium, and it employed fifty workers and a steam engine said to be equal to a further hundred. It was greatly praised by the *Société d'encouragement* for its chemical and mechanical perfection, and was considered another major development in France's chemical manufactures. The works was hindered in its progress, however, by the reluctance of the market to accept the new product, despite considerable support and publicity from the *Société d'encouragement* and the Government. Although scientifically neater, the new process was never in fact to displace the traditional method for the production of white lead pigment, because it gave a product of inferior covering power. Under the Restoration, France's imports of white lead still ran at about the same level as before the Revolution (a little over 1 million kg a year).[13]

A material employed in much smaller quantity than white lead, but still of some significance for its use in dyeing and in the cleaning of metal surfaces for plating, was sal ammoniac (ammonium chloride). The chemist Baumé mounted an early works for this product, at Gravelle (outside Paris) in 1766 or 1767, but Baumé himself withdrew from the enterprise after seven or eight years, and after languishing for some time it is said to have finally closed in 1787 at a heavy loss.[14] After the Revolution two important works arose near Paris. One was that of Payen, who was taking up the manufacture at Grenelle by the beginning of 1797. The other was that of the brothers Bernard Nicolas and Charles Pierre Pluvinet, who in 1797 established a works at Rouen (probably in association with Descroizilles),[15] and then in 1800 transferred it to Clichy-la-Garenne near Paris, where the requisite waste materials would be more abundant. The Pluvinet brothers presumably had some connection with the druggist of that name in the rue des Lombards, who had been a friend and supplier of Lavoisier.[16] After the removal of their works to Paris it was run in close association with that of Payen. The process employed by both consisted in the distillation of bones and rags in iron cylinders to obtain ammonium carbonate, which was then converted to the sulphate by reaction with calcium sulphate, and thence to the chloride by double decomposition with salt. In the period 1800–20 the combined output of the two factories was in the region of 25 000–80 000 kg a year, and they together employed 50–120 workers. Their works seem to have been the only undertakings of any significance in France at the time, and it was they who must have been chiefly responsible for the reduction in France's imports from 82 042 pounds (41 021 kg, mostly from Egypt) in 1788, to only 2420 kg in Year XIII (1804–5).[17] Both works seem to have continued into the mid-nineteenth century.[18]

One further branch of production that deserves mention was the distillation of wood, the outstanding manufacturer in this line being J. B. Mollerat in the Côte-d'Or.[19] Mollerat was a cousin of Prieur de la Côte-d'Or, whom he partnered for some years in a copper sulphate concern in Paris, until withdrawing early in 1804. Later that year he patented plant for the distillation of wood, and he then proceeded to establish works for the exploitation of his patent at Pellerey near Nuits, in his home department. His products were charcoal, tar, and pyroligneous acid, and he was principally renowned for his success in purifying the latter to obtain acetic acid, which he then applied to the production of a range of chemical salts—chiefly pigments and mordants—and to the confection of some controversial vinegars (wood vinegar was in recent times still known in the Côte-d'Or as 'vinaigre Mollerat'). We have seen that he also made soda by a pyrolignate process, and this indeed seems to have been a primary motivation in his first development of the enterprise, for according to his own account it was following his discovery of his soda process (towards 1800) that he was led to give his attention to the economical production of pyroligneous acid. By 1810 Mollerat was counted among the department's principal industrialists, then employing thirty workers at Pellerey and being engaged in establishing another works at Pouilly-sur-Saône, to which he was said to have devoted a capital of 500 000 francs. He won a gold medal at the 1819 exhibition.

While some of France's smaller chemical works no doubt still had more in common with an eighteenth-century workshop than a nineteenth-century factory, the larger works clearly belonged in character as well as in time to the new century, in the scale of their production and the scale and the nature of the plant employed: the developments in lead-chamber design, and in the design of furnaces and absorption systems for soda manufacture, are clearly a part of the early history of chemical engineering, and are far removed from the earthen retorts which were the chief apparatus of the distillers of aqua fortis. The industry might also be seen as presenting a modern appearance in its close relations with science. Not only did it have important roots in scientific discovery (in particular, of course, in Duhamel's discovery of the 'presence' of soda in salt), but its development was also, in part at least, a matter of scientific technology. It is remarkable how many of the early pioneers and manufacturers were instructed in chemistry, being arguably early 'manufacturing chemists' in the nineteenth-century mould. And their grounding in chemistry was by no means irrelevant for their work: though workshop experience and empiricism still counted for much, they were able to profit from their knowledge of chemical properties, and from their familiarity with analytical and other laboratory techniques, in the development of manufacturing methods, in the solution of problems that arose, and in the day-to-day running of the works. We have seen, moreover, that the degree of theoretical understanding was often greater than might commonly be imagined, and in the case of sulphuric acid manufacture, at least, this was able to assist in the improvement of manufacture. It was not without justification that the industry was regarded by contemporaries not only as one of the main fields of industrial development of the period, but also as an outstanding and inspiring example of the impact of science upon industry.

Chaptal, in 1819, in his detailed survey of the development of French industry over the past thirty years, considered that it was in the chemical arts, and particularly in the manufacture of chemical products, that the greatest progress had been made, and claimed that in this field France now knew no rival.[20] As regards the fundamental manufactures, sulphuric acid and soda, this was certainly true. In sulphuric acid manufacture France was now ahead of Britain in the volume of her production, and led the way in the improvement of manufacturing techniques. And above all in the manufacture of Leblanc soda France now enjoyed a sizeable lead. Although the production of soda from salt had been conducted in Britain by various processes since the 1780s or earlier, and particularly during the Napoleonic Wars, when the price of imported soda rose, nevertheless the scale of production was relatively small, and it was not until after the war that the Leblanc process was introduced in any significant way. It may be that the introduction of the process into Britain was delayed by that same wartime interruption of communications which, as is well known, delayed the passage of improvements in the mechanical industries in the reverse direction. After the fall of the Empire, one of the first to adopt it was Tennant at St. Rollox, who took it up in 1818 following correspondence on the subject with Darcet and Chaptal *fils*.[21] It was only in the mid-1820s, with the abolition of the British salt duty and a rapid growth in the alkali requirements of British industry, that Leblanc soda began to be made extensively and on a very large scale in Britain. And only in the course of the second quarter of the century did the British industry, favoured by a larger home market and better export facilities, overtake the French.[22]

NOTES

1. Ballot (1923), 545–6; Bardon (1896); AN, F^{14}4234, [1]; F^{14}4244, [3].
2. AN, F^{12}2255, [1]; F^{14}4236, [2]; Chaptal (1791a), 769.

3. *Mém. Acad.*, année 1724 (1726), 395–7, and année 1728 (1730), 303–4.

4. *Correspondance* [1780]; AN, F¹²2255, [1], [2]; F¹²1508, [20].

5. Chaptal (1791*a*).

6. Chaptal (1807), vol. 4, 71–6.

7. Chaptal (1797); Vauquelin (1797).

8. Thenard and Roard (1806).

9. Chaptal and Vauquelin (1806), 327–9; CNAMB, [1], vol. 2, cahier 1, pp. 148–9.

10. In 1787 France imported 19 426 *quintaux* (950 903 kg) of copperas and exported 1778 *quintaux* (87 033 kg) ('Aperçu' (1794), Table). Between Years V and XII (1796–1804) imports averaged 973 481 kg (annual figures in AN, F¹²1966ᴸ, [5]). In 1806 imports were 215 700 kg, while 169 100 kg of green, white, and blue vitriols were exported (Magnien and Deu (1809), vol. 1, 230). Alum imports were 28 102 *quintaux* (1 375 593 kg) in 1787, valued at 1 396 690*l*., and exports were 2951 *quintaux* (144 451 kg) ('Aperçu' (1794), Table). Between Years V and XIII imports averaged 1 365 344 fr. (annual figures in AN, F¹²7591). They were 736 643 kg in 1807, 250 501 kg in 1811, and 15 595 kg in 1813 (annual figures 1806–18 in AN, F¹²2512, [1]). There was a temporary resurgence of alum imports after the fall of the Empire, but in the early 1820s France was a net exporter of both alum and copperas (AN, F¹²251). On extraction in the Aisne and the Oise see: AN, F¹⁴4234, [1]; F¹⁴4244, [3]; F¹²633–637, doss. no. 137.

11. 'Aperçu' (1794), Table.

12. *Description*, vol. 10 (1825), 369–71; *Bull. Soc. enc.* **1** (1802–3), 99–100.

13. AN, F¹²7591. On the Pontoise and Clichy works see: *Bull. Soc. enc.* **8** (1809), 275–9; ib. **12** (1813), 128, 179–87; ib. **13** (1814), 79; Héricart de Thury (1819), 202–3; Chaptal (1819), vol. 2, 76–8.

14. R. Davy (1955); Multhauf (1965).

15. AN, F¹²1508, [31]. Of the two Pluvinet brothers the chemist was Bernard Nicolas, who had studied medicine in Paris in the 1780s, and who was presumably the Pluvinet—described as a *docteur régent* of the Paris Faculty of Medicine—whose death at the age of sixty-one was reported in the *Moniteur* on 5 May 1822, with the remark that 'He was one of the chemists who have most contributed to the progress of our manufactures'. (I owe the latter details to Mr John Perkins.) The name Pluvinet (Bernard Nicolas?) is encountered in the Revolutionary saltpetre programme as one of the teachers in the instruction courses held in Paris in Feb.–March 1794, and then from 9 March as one of the provincial inspectors, in succession to Descroizilles (Richard (1922), 474, 492). He is later (1794–5) found as 'conservateur de chimie' at the nascent *École polytechnique*. And in Jan. 1796 a Pluvinet (of Rouen) was a candidate for the position of non-resident associate in the chemistry section of the Institute.

16. McKie (1952), 298.

17. Magnien and Deu (1809), vol. 2, part 2, p. 502.

18. On these works see: *Dict. tech.*, vol. 19 (1831), 218–42; Héricart de Thury (1819), 204; AN, F¹²2234, [4]; F¹²2242, [3]; F¹²2245, [17]. At the beginning of the century (by 1801) Payen was in partnership at Grenelle with the chemist Nicolas Bourlier.

19. Bouchard (1946), 363, 385; Toraude (1921), 17; Fourcroy *et al.* (1808); *Notices* (1806), 325; L. Costaz (1819), 277–9; *Description*, vol. 7 (1824), 324–45; INPI, [3]; AN, F¹²937.

20. Chaptal (1819), vol. 1, p. xlv; vol. 2, pp. 41–2, 64.
21. Mactear (1877), 25.
22. Haber (1969), 10, 42.

BIBLIOGRAPHY

I: MANUSCRIPT SOURCES

To facilitate concise citation in footnotes, we have introduced arbitrary reference numbers (in square brackets) for the identification of individual dossiers or other items within cartons or archives.

AAS: ARCHIVES DE L'ACADÉMIE DES SCIENCES
[1] Doss. 'Leblanc (N.)'; [2] Doss. 'Darcet (J. P. J.)'; [3] Doss. 'Prix, 1783'; [4] Envelope '30 avril 1783'; [5] Pli cacheté no. 12, deposited 16 floréal an X (6 May 1802) by Anfrye and Darcet.

ADBR: ARCHIVES DÉPARTEMENTALES DES BOUCHES-DU-RHÔNE
116 E N1: Commune of Istres, 19th century
$K^2 31$: Register of prefectoral arrêtés, 1809–15
$M^{14}2$–12: Notes and reports on industry, an XIII [1805]–1825 (11 liasses)
$M^{14}47$: Établissements insalubres, 1806–71
$M^{14}51$: Commerce and industry, 1814–55
Doss: [1] 'Fabriques de soude . . . 1816'; [2] 'Soude . . . 1814–1816'.
$M^{14}226$: Industrial exhibitions in Paris, 1801–34

ADH: ARCHIVES DÉPARTEMENTALES DE L'HÉRAULT
C 2739: Iron and other works, 1788
C 2741: Chemical works, 1754–88
109 M 5: Établissements insalubres, an XII [1803–4]–1851
109 M 33: Établissements insalubres, Béziers
Doss: [1] 'Soude, 1810'.
109 M 139: Montpellier chemical works, 1807–94
Doss: [1] 'Laurent-Valedau . . . 1816'; [2] 'Jaumes . . . 1813'; [3] 'Alun, potasse, soude . . . 1811–1852'; [4] 'Delmas . . . 1807–1825'.
109 M 187: Établissements insalubres
Doss: [1] 'Soude artificielle, 1810' (at Pézenas).
3 S 5: Correspondence on industry, an XI [1802–3]–1859

ADR: ARCHIVES DÉPARTEMENTALES DU RHÔNE
1 L 524: Chemical and other industries, an II–V [1793–8]
1 Q 328: Property purchases during Revolution*
1 Q 724: Revolutionary sequestrations
2 Q 261: Revolutionary sequestrations*
96 Q 3: Tables of property purchases, 1783–93*
Series X, unclassified: liasse 'Ateliers incommodes. Produits chimiques liquides. 1813–1847' (formerly 7 M 14)
Doss: [1] '1813 . . . Srs Paris . . .'; [2] 'Perret . . . 1822'.

316   BIBLIOGRAPHY

ADSM: ARCHIVES DÉPARTEMENTALES DE LA SEINE-MARITIME
C 80: Industrial administration, 1775–88
Doss: [1] On Stourme, 1779.
3 J 168: Reports and memoirs of Société d'émulation de Rouen, 1793–1819
[1] 'Discours sur les établissements de blanchisserie Bertholienne', by Arvers, 9 June 1818.
5 MP 1048: Établissements insalubres, Déville, 1809–1935
Doss: [1] 'Acide sulfurique . . . Anfrye' (1809–14).
5 MP 1194: Établissements insalubres, Petit-Quevilly, 1805–1927
Doss: [1] 'Bougon' (1810); [2] 'Delacroix' (1809–10); [3] 'Germain' (1810); [4] 'Vallée' (1810); [5] 'Baril' (1809–14); [6] 'Delacroix et Malétra' (1805).
5 MP 1252: Établissements insalubres, Rouen, 1804–1934
Doss: [1] 'Correspondances' (an XII [1803–4]–1813); [2] 'Dubuc, Roulier et Bazire' (1810–15); [3] 'Dubuc & Delahalle' (1809–18); [4] 'Foucard et Baril & Dubuc' (1809–11); [5] 'Gessard' (1821).
5 MP 1253: Établissements insalubres, Rouen, 1805–1940
Doss: [1] 'Holker' (1809–11); [2] 'Dubuc' (1809); [3] 'Haag et Archambault' (1809–10); [4] 'Lebertre' (1809–10); [5] 'Lefrançois' (1809); [6] 'Pelletan' (1805–6).
5 MP 1335: Établissements insalubres, Rouen, 1807–1934
Doss: [1] '1807. Demande d'exemption de droits . . . sur le sel'; [2] On soda works, 1809–10; [3] '1809 . . . Visite des établissemens'.
Series M, unclassified: [1] *liasse 'Industrie et commerce. Statistique industrielle. Année 1811'*

AML: ARCHIVES MUNICIPALES DE LYON
'I 1 Guillotière' (liasse)
Doss: [1] 'Colin . . . 1816–1827'; [2] 'Estienne et Jalabert . . . 1838–1851'; [3] 'Perret. Successeur d'Alban . . . 1836'.

AMM; ARCHIVES MUNICIPALES DE MARSEILLE
10 F 1: 'Faillites', 1809–18
22 F 1: Industry, 1810–14
Doss: [1] '1813. Enquête sur le nombre d'ouvriers'; [2] '1813'.
22 F 2: Industry, 1815–20
23 F 1: Sulphuric acid works, 1815–44

AN: ARCHIVES NATIONALES
AF^{IV} 395 and 523: Imperial decrees, 5 Aug. 1809 and 8 Feb. 1811
F^7 3041: Fragmentary prices-current, an IV [1795–6]–1815
F^8 94: Établissements insalubres, 1791–1821
Doss: [1] 'Usine de Quessy' (1818–19).
F^{12*} 30: Intendance du commerce, Tolozan's department, decrees of Conseil d'état, 1777–87
F^{12*} 65: Procès-verbaux of Conseil de commerce, 1719
F^{12*} 106: Deliberations of Bureau du commerce, 1783–8
F^{12*} 107–108: Id., 1788–91
F^{12*} 166: Intendance du commerce, Tolozan's department, matters referred by Controller-General, 1777–83

$F^{12*}167$: Id., 1783–7

$F^{12*}172$: Inventory of Trudaine's affairs

$F^{12*}194$: Procès-verbaux of Conseil des fabriques et manufactures, 1810–17

$F^{12*}195$: Id., 1817–18

$F^{12*}242$–250: Foreign trade figures, 1775–80, 1782, 1787 (8 registers)

$F^{12*}251$: Printed tables of imports and exports, 1787–9, 1796–1826

$F^{12}513$: Miscellanea from Ministry of Interior

Doss: [1] 'Importations et exportations . . . 1787, 1788, 1789 et 1811–12'.

$F^{12}633$–637: Petitions regarding commerce, 1814–15

$F^{12}650$: Memorials by inspectors of manufactures, 1738–89

Doss: [1] '14—Rouen'.

$F^{12}652$: Petitions to Comités d'agriculture et du commerce, 1789– an II [1794]

$F^{12}658^A$: Industrial administration, 18th century

Doss: [1] 'Bureau d'encouragement de Rouen, 1788'.

$F^{12}680$: 'Usines à feu', 1788–9

$F^{12}724$: 'Avis' of deputies of commerce, 1788–91

Doss: [1] '1789'; [2] '1790–1791'.

$F^{12}740$: Inspectors of manufactures (nominations etc.), 1776–82

Doss: [1] '. . . notes sur MM. Holker . . . 1777'; [2] 'Arrêt du Conseil . . . Holker fils . . . 1777'; [3] 'Nomination du Sr Holker fils . . . 1777'.

$F^{12}822^C$: Loans to manufacturers, 1753–93

Doss: [1] 'Prêt . . . à Messrs Holker . . . 1770'.

$F^{12}871^A$: 'Faillites', early 19th century

Doss: [1] 'Rouen'.

$F^{12}879$: Disputes concerning manufacturers, 18th century

Doss: [1] 'Chatel & Comp^e . . . 1774'.

$F^{12}937$: Organisation of Conseil général des fabriques et manufactures (information gathered on prominent manufacturers), 1810

$F^{12}997$: Privileges and patents, 1788–93

[1] 'Dossier 3e'.

$F^{12}1007$: Patents, 1806

Doss: [1] 'Boullanger'; [2] 'Decroos'.

$F^{12}1017^A$: Patents, 1810

Doss: [1] 'Pelletan'.

$F^{12}1327$: Bleaching and laundry, 1690– an X [1801–2]

Doss: [1] 'Les curandiers blanchisseurs de . . . Caen . . . 1782'; [2] 'Pellerin . . . 1783'; [3] 'Brisson . . . 1779'.

$F^{12}1329$: Dyeing, 18th century

Doss: [1] 'Les Srs Fourcroy, Berthollet [etc.] . . . demandent la place . . . de Mr Macquer . . . 1784'; [2] 'Mr de Calonne à Mr Bertholet . . . 1784'; [3] '. . . traitement du Sr Bertholet . . . 1790'.

$F^{12}1505$: Soda, potash, soap, etc., 1701– an V [1796–7]

Doss: [1] 'Dom Malherbe' (1777); [2] 'Bourgogne et Beaudouin . . . 1788'; [3] 'Rifauville et compagnie' (1794–5); [4] 'Aulveger [Hollenweger] . . . An 2'; [5] 'Dominique Malherbe . . . 1778'.

$F^{12}1506$: Chemical products, 1724–81

Doss: [1] 'Vincent . . . 1779–1781'; [2] 'Chatel & Cie . . . 1768–1778'; [3] 'Delaferté, Bourboulon & Buffaut . . . 1778'; [4] 'Holker . . . 1764–1766'; [5] 'Péters et Alban . . .

1777'; [6] 'Holker . . . 1769'; [7] '. . . la manuf^re d'huile de vitriol établie à St-Sever-lès-Rouen. 1768–1771'; [8] 'Holker père & fils . . . 1768'; [9] 'Stourme . . . 1777–1779'; [10] 'Chatel & Comp^e . . . 1775'; [11] 'Anfry . . . 1778'; [12] 'Pitou, Caizac et Cie . . . 1781'; [13] 'Pitou, Caizac et Cie . . . 1780'.

F¹²1507: Chemical products, 1782–92
Doss: [1] 'Janvier & Cie . . . 1786–1787'; [2] 'Mr le Marquis de Bullion et Mr Guyton de Morveau . . . 1783–1788'; [3] 'Malherbe . . . 1783'; [4] 'Murry . . . 1782'; [5] 'Lapenne . . . 1782'.

F¹²1508: Chemical products, an II–X [1793–1802]
Doss: [1] 'Schemel . . . An 3'; [2] 'Le citoyen Chaptal . . . An 2'; [3] 'Chaptal . . . An 3'; [4] 'Giroud & Périer . . . An 2'; [5] 'Mathieu . . . An 7'; [6] 'Peeters et Alban . . . 1791'; [7] 'Alyon . . . An 2'; [8] 'Projets d'arrêtés . . . & pièces relatives à la soude & potasse. An 2'; [9] 'le Blanc & Dizé . . . An 2'; [10] 'Nicolas. An 2'; [11] 'Malherbe . . . an 3'; [12] 'Athénas . . . An 2'; [13] 'Observations . . . relatives à la manufacture de soude de Franciade . . . An 3'; [14] 'Champy . . . An 3'; [15] 'Carny, propriétaire d'une fabrique d'acide muriatique . . . An 3'; [16] 'Offre du C. Carny . . . An 3'; [17] 'Notice détaillée des produits et de la consommation de différentes fabriques de produits chimiques en activité à Rouen' (probably late 1794); [18] 'Bosc . . . An 2'; [19] 'Rapports & autres pièces relatives à la fabrication de la soude & du savon. Ans 2 & 3'; [20] 'Alban . . . alun. An 3'; [21] 'Bosc . . . An 3'; [22] 'Chamberlain . . . An 3'; [23] 'Athénas . . . An 3'; [24] 'Landry . . . An 3'; [25] 'Chemel . . . An 3'; [26] 'Notes du Cit. Nicolas . . . An 3'; [27] 'Chatel . . . An 5'; [28] 'Holker fils . . . An 4'; [29] 'Schemel . . . An 7'; [30] 'Chaptal & Bérard. An 2'; [31] 'Mallet . . . An 6'; [32] 'Broc [Bosc]. An 3'; [33] 'Charny [Carny] . . . An 3'; [34] 'La Salle . . . An 2'.

F¹²1549–1551: Statistical notes on industry, 1800–11 (3 cartons)

F¹²1552–1553: Statistical notes on industry and agriculture, 1810–11

F¹²1554: Prices-current, an XI [1802–3]–1813

F¹²1569: Statistical notes on industry
Doss: [1] 'Bas-Rhin'.

F¹²1835: Tables of imports and exports, 1782–8

F¹²1966ᴸ: Customs, chemical products, an IV [1795–6]–1814
Doss: [1] On request to import sulphur from Sicily, 1808; [2] On fraudulent importation of sulphuric acid in the Ourthe, an X; [3] On fraudulent importation of sulphuric acid in Belgian departments, an X; [4] 'Prohibition des acides minéraux . . . Honfleur' (an V–VI); [5] 'Proposition d'augmenter les droits d'entrée sur la couperose' (1806–7); [6] On proposal to prohibit importation of refined sulphur, July 1814; [7] On Spanish soda, 1808–9.

F¹²2234: Chemical products, an III [1794–5]–1855
Doss: [1] 'Acides minéraux. Renseignemens donnés au Conseil des 500 . . . An 6'; [2] 'Chamberlain . . . An 4–11'; [3] 'Chamberlain . . . Ans 2, 3 et 7'; [4] 'Payen & Pluvinet frères' (1807–8); [5] 'les Administrateurs des Salines de l'est . . . 1809'; [6] 'Mariette . . . An 11'; [7] 'Janvier & Caminet . . . An 10'; [8] 'Ribaucourt . . . An 9'.

F¹²2242: Chemical products
Doss: [1] 'Hollenweger . . . 1783'; [2] 'de Bullion, Guyton de Morveau, Fontmartin et Carny . . . 1788 et 1789'; [3] 'Payen et Pluvinet . . . 1810'; [4] 'Cormeray . . . An 2'.

F¹²2243: Soda, potash, etc., an II [1793–4]–1841
Doss: [1] Untitled dossier on Leblanc, 1797–9; [2] Untitled dossier on Leblanc, 1795–1806; [3] Dossier on claims made by Leblanc family in 1856; [4] 'Notice demandée à

Mr Pajot Descharmes . . . 1815'; [5] 'Pajot Descharmes à Paris . . . 1815'; [6] 'Mémoire de Mr Pajot Descharmes' (1814–15); [7] 'Soude naturelle et soude factice comparées' (1814–19); [8] 'Procédés pour extraire la soude . . . Ans 2 et 3'.

$F^{12}2244$: *Soda and potash, an II* [*1793–4*]–*1841*
Doss: [1] 'Bodin . . . 1809'; [2] 'Fabrique de soude . . . à Franciade . . . An 3'; [3] 'Deyeux et Parmentier . . . Ans 3 et 4'; [4] 'Carny à Paris . . . An 3, 5, 6, 9'; [5] 'Carny. Extraction de la soude' (an IX); [6] 'Carny à Dieuze . . . 1808, 1809, 1810'; [7] 'Rivet . . . 1811'; [8] 'Chauboy . . . 1809'; [9] 'Martin, fils d'André . . . 1810'; [10] 'Lefrançois . . . 1814'.

$F^{12}2245$: *Soda, etc., 19th century*
Doss: [1] 'Soude artificielle. Renseignements demandés . . . Réponses de Messrs les Préfets . . . 1810. Aisne, Calvados, Hérault, Bouches-du-Rhône, Seine, Seine-Inférieure'; [2] 'Les administrateurs de la Manufacture des Glaces . . . 1810'; [3] 'Anfrye & Darcet' (1807?); [4] '. . . mémoire de plusieurs verriers . . . 1814'; [5] 'Darcet . . . 1814'; [6] 'Soude artificielle, renseignements dd^{és} . . . Réponses négatives . . . 1810'; [7] 'La Chambre de commerce de Rouen . . . 1810'; [8] 'Les fabricants de soude de Rouen . . . 1810'; [9] Untitled dossier relating to the question of salt-tax exemption for chemical manufacturers, 1806–10; [10] 'Renseignements sur les quantités de sel employées à la fabrication de la soude en 1816, 1817, 1818'; [11] 'Pelletan . . . 1807'; [12] 'Pelletan . . . 1808'; [13] 'Pelletan . . . 1810'; [14] 'Pajot Descharmes . . . 1807'; [15] 'Montgolfier, Desormes & Clément . . . 1807'; [16] 'Descroizilles . . . 1807'; [17] 'Payen & Pluvinet . . . 1808'; [18] 'Bérard & Martin sollicitent . . . 1809'; [19] 'Bérard & Martin demandent . . . 1810'; [20] 'Bérard & Martin à Montpellier . . . 1809'; [21] 'Soude artificielle. Encouragements généraux . . . Questions proposées sur la suppression de ces encouragements . . . 1816, 1817 à 1821'.

$F^{12}2255$: *Manufacture of alum* (*1769–1820*)*, etc.*
Doss: [1] 'Clément-Desprez . . . 1783–1785'; [2] 'Bourboulon et Buffault . . . 1783'.

$F^{12}2299–2300$: *Bleaching, 1788–1838*
Doss: [1] 'Carondelet . . . An 10'; [2] 'Blanchiment à la vapeur. Divers départements' (an IX); [3] 'Blanchiment à la vapeur' (an IX-1809); [4] 'Blanchiment, perfectionnement du' (an II-1828); [5] Untitled dossier on Bonjour, 1788–90.

$F^{12}2415$: *Bleaching, finishing, dyeing, 1760–88*
Doss: [1] '10. Le Sr Bonjour' (1788); [2] '12. Le Sr Brisson . . . 1780'; [3] '14. Les Srs Levêque et Cie . . . 1788'.

$F^{12}2456$: *Salt and salt-works*
Doss: [1] 'Demandes d'exemption d'impôt sur le sel . . . 1808, 1809'.

$F^{12}2512$: *Customs duties, preparation of tariff of 1822*
Doss: [1] 'Produits chimiques'; [2] 'Pierres, terres et autres fossiles'.

$F^{12}6728$: '*Sociétés anonymes*'
Doss: [1] 'Compagnie des salines et produits chimiques du Plan d'Aren . . . 1819'.

$F^{12}7589$: *Petitions regarding commerce, 1814–15*
Doss: [1] 'Soudes et savons, 1814'.

$F^{12}7591$: *Tables of imports, an V* [*1796–7*]–*1810, 1815–21*

$F^{14}4234–4249$: *Mines and quarries, etc., 1740–1850* (16 cartons)
Doss: [1] 'Aisne'; [2] 'Aveyron'; [3] 'Oise'.

$F^{20}289^1$: *Salt-works, 1809–12*
Doss: [1] 'Bouches-du-Rhône'.

H 940 : Documents relating to 'cahier' of 1788
Doss: [1] 'Languedoc—cahier de 1788'.
H 1438 : Miscellanea on Picardy, 1709–90
26 AQ 1 : Manufacture des glaces et produits chimiques de Saint-Gobain, Chauny et Cirey (1695–1840)
Doss: [1] 'Renseignements divers'; [2] 'Renseignements divers, 3ᵉ botte'.

AS: ARCHIVES DE LA SEINE
D^3B^673 : *Register of partnerships and other commercial and industrial agreements, 18th century*
D^3B^679 : *Id.*
$D^{31}U^3$: *Partnerships and other legal agreements, 1800–71* (306 cartons)
$DM^{12}24$: *Industrial statistics, 1817–37*

BML: BIBLIOTHÈQUE MUNICIPALE DE LYON
[1] Fonds Coste 1281: MS by the astronomer J. J. de Lalande on history of Bourg and la Bresse, 1764–1806.

BMM: BIBLIOTHÈQUE MUNICIPALE DE MARSEILLE
Collection Volcy-Boze
[1] vol. xxii, fᵒ7: Arrêté of 24 Jan. 1804, on works of J. B. Michel; [2] vol. xxviii, fᵒ51: Chaptal *fils* to Segaud, director of Plan d'Aren works, 13 Jan. 1826.

BMR: BIBLIOTHÈQUE MUNICIPALE DE ROUEN
Collection Girardin
Letters: [1] no. 93: Clément-Desormes to Darcet, 24 June 1830; [2] no. 136: G. Holker to Girardin, 23 Nov. 1851; [3] no. 137: G. Holker to Girardin, 17 Feb. 1861.

BUM: BIBLIOTHÈQUE UNIVERSITAIRE DE MONTPELLIER
[1] Manuscript H561 bis, t. 3, 4ᵉ pièce: Concerning a complaint about fumes from Chaptal's factory, 1791.

CCM: ARCHIVES DE LA CHAMBRE DE COMMERCE DE MARSEILLE
[1] Doss. 'Soudes' (unclassified); [2] Register 'Mémoires nᵒ 2'; [3] Liasse H.135.

CNAMA: CONSERVATOIRE NATIONAL DES ARTS ET MÉTIERS, ARCHIVES
Ancien fonds X : 'Appareils et produits chimiques' (Due to be reclassified)
[1] Doss. no. 32, containing 'Mémoire sur la fabrication de l'acide vitriolique' by Scanegatty; [2] Doss. no. 41, 'Malherbe . . . an 3ᵉ'; [3] Doss. no. 45, 'Valentino . . . an 3ᵉ'; [4] Piece no. 49, note on making sulphuric acid without saltpetre (by Conté?).

CNAMB: CONSERVATOIRE NATIONAL DES ARTS ET MÉTIERS, BIBLIOTHÈQUE
[1] MS 8ᵒFa40: 'Conservatoire des arts et métiers. Chimie industrielle. Professeur M. Clément-Desormes. Journal des cours de 1825 à 1830', by J. M. Baudot, 3 vols.

INPI: ARCHIVES DE L'INSTITUT NATIONAL DE LA PROPRIÉTÉ INDUSTRIELLE
Doss: [1] 'Le Blanc, Nicolas'; [2] 'Carny'; [3] 'Mollerat. Soude'; [4] 'Clément. Acide sulfurique'; [5] 'Pelletan fils'; [6] 'Chamberlain'; [7] 'Gazzino et Cie. Décomposition du muriate de soude' (1813).

MC: MINUTIER CENTRAL
Étude VI
[1] 'Inventaire après décès' of J. B. L. L. Barréra, 29 Sept. 1820.
Étude X
[1] Mortgage loan raised by Chaptal *fils* on the former works at les Ternes and on his share in the la Folie works, 5 Nov. 1825.
Étude XV
[1] Revision of partnerships between Darcet's company and Jacquemart *frères*, 23 and 25 March 1811.
Étude XXIII
[1] Agreement between Leblanc and Dizé, 1 April 1790; [2] Agreement between Leblanc and Dizé, 15 Jan. 1791; [3] Partnership between Orleans, Leblanc, Dizé and Shée, 27 Jan. 1791; [4] Descriptions of processes of Leblanc and Dizé, originally deposited 27 March 1790, redeposited 4 March 1856 after examination by Dumas commission.
Étude XXVI
[1] Sale of a property at la Paille by Chaptal to Bérard and Martin, 19 fructidor an X (6 Sept. 1802); [2] Cession of la Paille works by Chaptal to Bérard, Martin & Cie, 19 fructidor an X (6 Sept. 1802); [3] Dissolution of partnership between Chaptal and P. Coustou, 22 Aug. 1808; [4] Partnership between Chaptal *fils* and Berthollet *fils*, 25 Aug. 1808; [5] Lease of les Ternes works by Chaptal to Chaptal *fils* and Berthollet *fils*, 26 Aug. 1808; [6] 'Procuration ... Chaptal *fils* ... à Mr Joseph Ricôme ... pour liquider la maison de commerce de M. Chaptal à Marseille', 22 May 1815; [7] Marriage of Chaptal *fils*, 28 March 1816; [8] 'Dépôt en suite de mariage du 28 mars 1816' by Chaptal *père* and *fils*, 13 April 1816 (appended to previous item); [9] 'Inventaire après décès' of Chaptal, 9 Aug. 1832.
Étude XLVIII
[1] Partnership between J. B. Peeters and L. Alban, 10 Aug. 1776.
Étude LV
[1] Partnership between A. J. Buffault, P. J. B. Buffault and L. Alban, 6 floréal an V (25 April 1797).
Étude CIII
[1] 'Inventaire après décès' of Louis Simon, 12 messidor an XI (1 July 1803); [2] 'Inventaire après décès' of Léonard Alban, 18 fructidor an XI (5 Sept. 1803).

PP: ARCHIVES DE LA PRÉFECTURE DE POLICE
[1] 'Registre des procès-verbaux des séances du Conseil de salubrité', 16 Nov. 1807–21 March 1823; [2] 'Rapports du Conseil de salubrité', Register 1 (1806–9); [3] Id., Register 2 (1810–11); [4] Id., Register 3 (1812–13); [5] Id., Register 4 (1814–15).

SG: ARCHIVES DE LA COMPAGNIE DE SAINT-GOBAIN
[1] Dossier C7^3: 'Charles-Fontaine, 1806, 07, 1816'.

II: PRINTED SOURCES

ACADÉMIE DES SCIENCES (1910–22). *Procès-verbaux des séances de l'Académie des sciences, tenues depuis la fondation de l'Institut jusqu'au mois d'août 1835* (10 vols.). Hendaye.

Académie des sciences, arts et belles-lettres de Dijon. Séance publique. Dijon (1816, etc.)

ADELSON, J. (1957). The early evolution of business organisation in France. *Business history review* **31**, 226–43.

ADMINISTRATION DES DOUANES (n.d.). *Tableau des marchandises étrangères importées en France, et des marchandises françaises exportées à l'étranger, pendant l'année 1818 (– 1820).* [Paris]. [BN: Lf¹⁵⁸68]

AIKIN, A. and AIKIN, C. R. (1807). *A dictionary of chemistry and mineralogy* (2 vols.). London.

ALBAN, L. and VALLET (1785). *Précis des expériences faites par MM. Alban et Vallet, et souscription proposée pour un cours de direction aérostatique* (29 pp.). Paris.

ALCOCK, T. (1827). *An essay on the use of chlorurets of oxide of sodium and of lime as powerful disinfecting agents.* London.

ALLEN, J. FENWICK (1907). *Some founders of the chemical industry* (2nd edn). London and Manchester.

Almanach du commerce de la ville de Paris, by DUVERNEUIL and J. DE LA TYNNA, Paris, an VI [1797–8]–1806; continued as *Almanach du commerce de Paris, des départemens de l'Empire français et des principales villes du monde,* by J. DE LA TYNNA (from 1819 by S. BOTTIN), 1807–38; continued as *Almanach-Bottin du commerce,* 1839–56; continued as *Annuaire et almanach du commerce . . . (Firmin Didot et Bottin réunis),* 1857–1903. [Complete set in BN; fragmentary collection in BM.]

Almanach historique et commercial de Marseille et du département des Bouches-du-Rhône, suivi du Guide marseillais, pour l'année de grâce 1817. Marseilles (1817). [BN: Lc³¹279]

ALYON (1805). Note sur la poudre de Tennante et Knox, chimistes écossais. *Ann. chim.* **53**, 341–5.

ANASTASI, A. (1884). *Nicolas Leblanc: sa vie, ses travaux, et l'histoire de la soude artificielle.* Paris.

ANCELON, E. A. (1879). Historique de l'exploitation du sel en Lorraine. *Mémoires de l'Académie de Metz,* année 59: 1877–8, 153–222.

ANDRÉ-FÉLIX, A. (1971). *Les Débuts de l'industrie chimique dans les Pays-Bas autrichiens.* Brussels.

Annales de chimie (96 vols.). Paris (1789–94; 1797–1815). Continued as *Annales de chimie et de physique.* Paris (1816, etc.)

Annales de l'industrie nationale et étrangère. Paris (1820, etc.)

Annales des arts et manufactures (62 vols.). Paris (1800–17).

Annales des mines. Paris (1816, etc.)

Annuaire statistique du département de la Seine-inférieure, pour l'année 1823 (2 vols.). Rouen (1823). [BM: 879.e.1]

Aperçu de l'extraction et du commerce des substances minérales en France avant la Révolution. *J. des mines,* No. 1, 55–92 (1794).

ARBELET, P. (1918). Prieur de la Côte-d'Or, ministre des munitions. *Revue bleue* **56,** 14–19, 76–80, 108–12. [Includes text of Prieur's MS 'Révélations sur le Comité de salut public'.]

Archives des découvertes et des inventions nouvelles. Paris (1809, etc.)

Archives parlementaires de 1787 à 1860. Recueil complet des débats législatifs et politiques des Chambres françaises. First series: 1787–99 (82 vols. [covering 1787–94]). Paris (1867–1913; 1961–(in progress)). Second series: 1800–60 (127 vols.). Paris (1862–1914).

ARMONVILLE, J. R. (1818). *Le Guide des artistes, ou Répertoire des arts et manufactures.* Paris. [A subject bibliography.]

ARTZ, F. B. (1966). *The development of technical education in France, 1500–1850.* Cambridge, Mass.

ARVERS (1817). Mémoire sur les manufactures d'acides. *Séance publique de la Société d'émulation de Rouen*, année 1817, 43–63.

AULARD, F. V. A. (ed.) (1889–1964). *Recueil des actes du Comité de salut public* (28 vols. + 3 index vols.). Paris.

Avis relatif à l'extraction de la soude du sulfate de soude. *J. des mines*, No. 7, 63–4 (1795).

BACQUIÉ, F. (1927). *Les Inspecteurs des manufactures sous l'ancien régime, 1669–1791*. Toulouse. (In series: *Mémoires et documents pour servir à l'histoire du commerce et de l'industrie en France* (ed. J. Hayem), vol. 11.)

BAKER, K. M. (1973). Politics and social science in eighteenth-century France: the 'Société de 1789'. In *French government and society 1500–1800: essays in honour of Alfred Cobban* (ed. J. F. Bosher), pp. 208–30. London.

BALARD, A. J. (1834). Recherches sur la nature des combinaisons décolorantes du chlore. *Ann. chim. phys.* **57**, 225–304.

BALLOT, C. (1915). Procès-verbaux du Bureau de consultation des arts et métiers. *Bulletin d'histoire économique de la Révolution*, année 1913, 15–160.

—— (1923). *L'Introduction du machinisme dans l'industrie française*. Paris.

BARDON, A. (1896). L'usine de vitriol de Saint-Julien de Valgagues. *Revue du Midi* **20**, 507–29.

BARGALLÓ, M. (1952). Ciento cincuenta aniversario de la 'Cristallotechnie' de Nicolas Leblanc. *Ciencia: revista hispano-americana de ciencias puras y aplicadas* **12**, 261–5.

BARKER, R. J. (1969). The Conseil général des manufactures under Napoleon (1810–1814). *French historical studies* **7**, 185–213.

BARKER, T. C., DICKINSON, R., and HARDIE, D. W. F. (1956). The origins of the synthetic alkali industry in Britain. *Economica* (N.S.) **23**, 158–71.

BARRAL, J. A. (1873). Éloge biographique de M. Anselme Payen. *Mémoires publiés par la Société centrale d'agriculture de France*, année 1873, 67–87.

BARRAL, P. (1964). *Les Périer dans l'Isère au XIX* ᵉ *siècle*. Paris.

'BASIL VALENTINE' (1678). *Basil Valentine his Triumphant chariot of antimony, with annotations of Theodore Kirkringius*. London.

BAUD, P. (1932a). *L'Industrie chimique en France: étude historique et géographique*. Paris.

—— (1932b). Les débuts de l'industrie chimique en France. *Annales de l'Université de Paris* **7**, 223–41.

—— (1933a). La manufacture de soude de Nicolas Leblanc. *Comptes rendus* **196**, 701–3.

—— (1933b). Les premières soudières françaises. *Comptes rendus* **196**, 1498–1500.

—— (1933c). John Holker et la fabrication de l'acide sulfurique en France, au XVIIIᵉ siècle. *Comptes rendus* **196**, 1797–1800.

—— (1934). Les origines de la grande industrie chimique en France. *Revue historique* **174**, 1–18.

BÉCHAMP, P. J. A. (1866). *Éloge historique de J. A. Chaptal*. Paris.

BECQUEY, F. L. (ed.) (n.d.). *Statistique des routes royales de France, 1824*. Paris.

BELLOT, DE LA RIVIÈRE, DESESSARTZ, and DE VALLUN [1774]. *Rapport fait à la Faculté de médecine . . . par MM. Bellot [etc.] . . . qu'elle avoit nommés . . . pour examiner le laboratoire du Sieur Charlard, & juger les inconvéniens qui peuvent résulter pour les maisons voisines, de la distillation d'eau-forte* (18 pp.). [Paris]. [AN, F¹²879, [1]]

BERGERON, L. (1975). *Banquiers, négociants et manufacturiers parisiens du Directoire à l'Empire* (2 vols.). Lille.

BERTHOLLET, C. L. (1785). Mémoire sur l'acide marin déphlogistiqué. *Obs. phys.* **26**, 321–5. [Read to Acad. Sci. 6 April 1785.]

—— (1788a). Mémoire sur l'acide marin déphlogistiqué. *Mém. Acad.*, année 1785,

276–95. [Text differs from previous item. Presented to Acad. Sci. 21 December 1785.]

—— (1788b). Observations sur quelques combinaisons de l'acide marin déphlogistiqué, ou de l'acide muriatique oxigéné. *Obs. phys.* **33**, 217–24.

—— (1789a). Suite des expériences sur l'acide sulfureux. *Ann. chim.* **2**, 54–72.

—— (1789b). Description du blanchiment des toiles & des fils par l'acide muriatique oxigéné, & de quelques autres propriétés de cette liqueur relatives aux arts. *Ann. chim.* **2**, 151–90.

—— (1790a). Additions à la description du blanchiment. *Ann. chim.* **6**, 204–9.

—— (1790b). Mémoire sur l'action que l'acide muriatique oxigéné exerce sur les parties colorantes. *Ann. chim.* **6**, 210–40.

—— (1791a). *Éléments de l'art de la teinture* (2 vols.). Paris.

—— (1791b). Réponse [to Haussmann]. *Ann. chim.* **11**, 250–60.

—— (an III [1795]). Description de l'art du blanchiment par l'acide muriatique oxigéné. *J. des arts et mans.* **1**, 192–255.

—— (1803). *Essai de statique chimique* (2 vols.). Paris.

—— and BERTHOLLET, A. B. (1804). *Éléments de l'art de la teinture, avec une description du blanchiment par l'acide muriatique oxigéné* (2nd edn; 2 vols.). Paris.

—— and DE DIETRICH (1789). Rapport fait à l'Académie royale des sciences le 10 juin 1789. *Ann. chim.* **3**, 46–54. [On Chaptal (1791a).]

BERTIN, DULAC, ROZIER, and PONSARD [1848]. *Mémoire des propriétaires et habitans du quartier Béchevelin, ville de la Guillotière, pour la suppression de la fabrique d'acide sulfurique de MM. Estienne et Jalabert* (11 pp.). La Guillotière. [AML, liasse 'I 1 Guillotière', [2].]

BEUGNOT (1835). Notice sur Alexandre de Fontenay. *Séance publique de la Société libre d'émulation de Rouen*, année 1834, 70–93.

Biographie universelle, ancienne et moderne (published by J. F. and L. G. MICHAUD; 85 vols. (56–85 forming *Supplément*)). Paris (1811–62).

Biographie universelle . . . Nouvelle édition (45 vols.). Paris [1854–65].

BIOLLAY, L. (1885). *Études économiques sur le XVIIIᵉ siècle: . . . L'Administration du commerce*. Paris.

—— (1886). *Études économiques sur le XVIIIᵉ siècle: Les Prix en 1790*. Paris.

BIREMBAUT, A. (1966). Quelques réflexions sur les problèmes posés par la conservation et la consultation des archives techniques françaises. *Archives internationales d'histoire des sciences* **19**, 21–102. [A valuable guide to sources.]

BLACHETTE, L. J. (1827). *Traité théorique et pratique du blanchiment des toiles de lin, de chanvre et de coton*. Paris. [SRL: CW 55*.]

BLAESSINGER, E. (1952). Dizé. In *Les Grandes figures du service de santé militaire. 3ᵉ série. Quelques grandes figures de la chirurgie, de la médecine et de la pharmacie militaires*, pp. 59–97. Paris.

BONDOIS, P. M. (1933). Le privilège exclusif au XVIIIᵉ siècle. *Revue d'histoire économique et sociale* **21**, 140–89.

BONJOUR, J. (1853). *Notice biographique sur Bonjour, J.-F., chimiste, . . . par son neveu*. Lons-le-Saunier.

BONNASSIEUX, P. and LELONG, E. (1900). *Conseil de commerce et Bureau du commerce, 1700–1791. Inventaire analytique des procès-verbaux*. Paris.

BOSC (1803). Observations sur le blanchiment à la vapeur. *Bull. Soc. enc.* **1**, 67–71.

BOSHER, J. F. (1970). *French finances, 1770–1795: from business to bureaucracy*. Cambridge.

BOSSI (1808). Sur le commerce et l'industrie du département de l'Ain. *Ann. des arts et mans.* **30**, 5–32.

BOTTÉE DE TOULMONT, J. J. A. and RIFFAULT DES HÊTRES, J. R. D. (1811). *Traité de l'art de fabriquer la poudre à canon* ... *Précédé d'un exposé historique sur l'établissement du service des poudres et salpêtres en France.* Paris.

BOTTIN, S. (1820). *Le Livre d'honneur de l'industrie française* ... *Ire partie.* Paris. [Catalogue of award winners at industrial exhibitions, 1798–1819.]

BOUCHARD, G. (1938). *Guyton-Morveau, chimiste et conventionnel (1737–1816).* Paris.

—— (1946). *Un Organisateur de la victoire: Prieur de la Côte-d'Or.* Paris.

BOUDET, F. H. (1852). Notice historique sur la découverte de la soude artificielle par Leblanc et Dizé. *J. de pharmacie* (3rd series) **22**, 99–115. (Reprinted in Pillas and Balland (1906), 89–119.)

BOUILLON-LAGRANGE, E. J. B. (1798). Notice des divers procédés connus pour l'extraction de la soude du sel commun, extraits du Rapport publié sur cet objet par le Comité de salut public. *J. de la Société des pharmaciens de Paris* **1**, 164–9.

—— (an VII [1798–9]). *Manuel d'un cours de chimie* (2 vols.). Paris.

—— (1812). *Manuel d'un cours de chimie* (5th edn; 3 vols.). Paris.

BOULOISEAU, M. (1968). *Le Comité de salut public (1793–1795)* (2nd edn). Paris.

BOUVET, M. (1928). L'accusation du tribunal révolutionnaire contre Schemel, apothicaire de Metz (1794). *J. de pharmacie d'Alsace-Lorraine* **60**, 315–17.

—— (1935). Les pharmaciens victimes de la Révolution: Nicolas Schemel. *Bulletin de la Société d'histoire de la pharmacie* **5**, 121–4.

—— (1953). L'évolution dans la fabrication des substances médicamenteuses: de l'artisanat à l'usine. *France-pharmacie* **6**, 131–41.

BOYSSE, E. (ed.) (1887). *L'Administration des menus. Journal de Papillon de la Ferté.* Paris.

BRAUDEL, F. and LABROUSSE, E. (eds.) (1970–7). *Histoire économique et sociale de la France* (3 vols. in 5). Paris.

BRAYER DE BEAUREGARD, J. B. L. (1824–5). *Statistique du département de l'Aisne* (2 vols.). Laon.

BRIAVOINNE, N. (1839). *De l'industrie en Belgique. Sa situation actuelle. Causes de décadence et de prospérité* (2 vols.). Brussels.

[BRISSON, A. F.] [1780]. *Instructions sur le blanchissage des toiles de chanvre et de lin.* [Lyons?] [Wrongly ascribed to Lavoisier in BN catalogue.]

Bulletin de la Société d'encouragement pour l'industrie nationale. Paris (1802, etc.)

Bulletin de pharmacie (6 vols.). Paris (1809–14). Continued as *Journal de pharmacie.* Paris (1815, etc.)

Bulletin des lois. Paris (1794, etc.)

Bulletin des sciences par la Société philomathique de Paris (3 vols.). Paris (1797–1805). Continued as *Nouveau bulletin des sciences* (3 vols.). Paris (1807–13). Continued as *Bulletin des sciences* (11 vols.). Paris (1814–24).

BURKE, J. G. (1966). *Origins of the science of crystals.* Berkeley.

CADET DE GASSICOURT, C. L. (1802). Note sur quelques nouveaux procédés anglois. *Bull. Soc. enc.* **1**, 37, 48.

—— (1803). *Dictionnaire de chimie* (4 vols.). Paris. [SRL]

—— (1804). Sur la fabrication de l'acide sulfurique. *Ann. des arts et mans.* **17**, 67–76.

—— (1805). Sur les moyens de suppléer à la construction de chambres de plomb dans la fabrication de l'acide sulfurique. *Ann. des arts et mans.* **21**, 80–3.

—— (1806). Rapport ... sur la manufacture de savons établie à Paris ... par M. J.-G. Decroos. *Bull. Soc. enc.* **5**, 137–40.

—— (1811). Sur l'emploi du résidu des soudes lessivées. *Bull. pharm.* **3**, 71–2.

CALVET, H. (1933). *L'Accaparement à Paris sous la Terreur: essai sur l'application de la loi du 26 juillet 1793.* Paris.

CAMPBELL, W. A. (1957). Peter Woulfe and his bottle. *Chemistry and industry,* 1957, 1182–3.

—— (1971). *The chemical industry.* London. (Longman's Industrial archaeology series.)

Carny *(Jean-Antoine-Allouard)* (16pp.). [Paris? (*c.* 1850?)]. [BN: Ln²⁷ 3564]

CARON, P. (ed.) (1925). *La Commission des subsistances de l'an II: procès-verbaux et actes.* Paris.

—— (1947). *Manuel pratique pour l'étude de la Révolution française* (new edn). Paris.

CARPENTER, K. E. (1972). European industrial exhibitions before 1851 and their publications. *Technology and culture* **13**, 465–86.

CARRIÈRE, C. [1973?]. *Négociants marseillais au XVIIIᵉ siècle: contribution à l'étude des économies maritimes* (2 vols.). [Marseilles].

CHABERT, A.(1945). *Essai sur les mouvements des prix et des revenus en France de 1798 à 1820.* [*Vol. 1:*] *Les prix.* Paris.

CHAGNON, A., COSTE, P., and LACOIN, M. (1945). *Les Débuts de la grande industrie chimique et la Société d'encouragement pour l'industrie nationale: Nicolas Leblanc, Chaptal* (38 pp.). Paris. [Containing: Chagnon, 'Historique du procédé Leblanc'; Coste, 'Le procédé Leblanc et l'industrie chimique au siècle dernier'; Lacoin, 'Chaptal, ministre de la production industrielle du Premier Consul'.]

CHAMBERLAIN, E. (1822). *Rapports sur les produits chimiques et autres, de M. Chamberlain* (4 pp.). Rouen. [BN]

CHANLAIRE, P. G. (1815). *Atlas national de la France en départemens, revu et augmenté en 1810* (2 vols.). Paris.

CHAPMAN, S. D. (1969). The Peels in the early English cotton industry. *Business history* **11**, 61–89.

CHAPTAL, J. A. C. (1781). Premier mémoire sur quelques établissemens utiles à la province de Languedoc. *Obs. phys.* **17**, 365–9.

—— (1782). Mémoire sur la décomposition du soufre par l'acide nitreux. *Obs. phys.* **21**, 148–53.

—— (1789*a*). Observations sur l'acide muriatique oxigéné. *Mém. Acad.,* année 1787, 611–16.

—— (1789*b*). Observations sur quelques phénomènes que nous présente la combustion du soufre. *Ann. chim.* **2**, 86–92.

—— (1790). *Élémens de chimie* (3 vols.). Montpellier.

—— (1791*a*). Observations sur la manière de former l'alun par la combinaison directe de ses principes constituans. *Mém. Acad.,* année 1788, 768–77.

—— (1791*b*). *Elements of chemistry* (trans. W. Nicholson; 3 vols.). London.

—— (1797). Analyse comparée des quatre principales sortes d'aluns connues dans le commerce; et observations sur leur nature et leur usage. *Ann. chim.* **22**, 280–96; additional note, **23**, 222–4.

—— (1798). Observations sur la nécessité et le moyen de cultiver la barille en France. *Ann. chim.* **26**, 178–87.

—— (1800*a*). Essai sur le perfectionnement des arts chimiques en France. *J. de physique* **50**, 217–33.

—— (1800*b*). Notice sur une nouvelle méthode de blanchir le coton. *J. de physique* **51**, 305–9.

—— (1801). Notice sur un nouveau moyen de blanchir le linge dans nos ménages. *Ann. chim.* **38**, 291–6.

—— (1807). *Chimie appliquée aux arts* (4 vols.). Paris.

—— (1819). *De l'industrie françoise* (2 vols.). Paris.

—— (1893). *Mes souvenirs sur Napoléon* (ed. A. Chaptal). Paris. [Besides the title work, this contains: 'La vie et l'oeuvre de Chaptal. Mémoires personnels rédigés par lui-même de 1756 à 1804, continués d'après ses notes, par son arrière-petit-fils jusqu'en 1832'.]

—— and VAUQUELIN, N. L. (1806). Rapport du mémoire sur l'alun, de MM. Desormes et Clément. *Ann. chim.* **57**, 327–33.

CHARPENTIER, P. (1890). *Le Papier.* Paris. (Forming part of E. Frémy, *Encyclopédie chimique,* vol. 10.)

CHASSAGNE, S. (1971). *La Manufacture de toiles imprimées de Tournemine-lès-Angers (1752–1820): étude d'une entreprise et d'une industrie au XVIIIᵉ siècle.* Paris.

CHEVALLIER, J. B. A. (1826). Note sur la préparation du chlorure de chaux destiné au blanchiment et à la désinfection. *J. de chimie médicale* **2**, 172–7.

—— (1829). *L'Art de préparer les chlorures de chaux, de potasse et de soude.* Paris.

CHOFFEL, J. (1960). *Saint-Gobain: du miroir à l'atome.* Paris.

CHRISTOPHE, R. (1971). L'analyse volumétrique de 1790 à 1860. Caractéristiques et importance industrielle. Évolution des instruments. *Revue d'histoire des sciences* **24**, 25–44.

CLACQUESIN, P. (1900). *Histoire de la communauté des distillateurs. Histoire des liqueurs.* Paris. [Bibliothèque Forney, Paris.]

CLAPHAM, R. C. (1868–71). An account of the commencement of the soda manufacture on the Tyne. *Transactions of the Newcastle Chemical Society* **1**, 29–45.

CLÉMENT, *see* DESORMES.

CLOUZOT, H. (1928). *Histoire de la manufacture de Jouy et de la toile imprimée en France* (2 vols.). Paris and Brussels.

—— and FOLLOT, C. (1935). *Histoire du papier peint en France.* Paris.

CLOW, A. and CLOW, N. L. (1952). *The chemical revolution: a contribution to social technology.* London.

COCHIN, A. (1866). *La Manufacture des glaces de Saint-Gobain de 1665 à 1865.* Paris.

COCHOIS, P. (1902). *Étude historique et critique de l'impôt sur le sel en France.* Paris.

COIGNET, J. (1894). *Notice historique sur l'industrie des produits chimiques à Lyon* (42 pp., numbered 93–134). Lyons.

COLE, A. H. and WATTS, G. B. (1952). *The handicrafts of France as recorded in the Descriptions des arts et métiers, 1761–1788.* Boston, Mass.

Collection officielle des ordonnances de police imprimée par ordre de M. le Préfet de police. Tome 1: 1800–1848. Paris (1880).

COLLINS, P. (1976). Johann Wolfgang Döbereiner and heterogenous catalysis. *Ambix* **23**, 96–115.

COMPAGNIE DE SAINT-GOBAIN (1965). *1665–1965.* Paris. (A tercentennial history.)

Comptes rendus hebdomadaires des séances de l'Académie des sciences. Paris (1835, etc.)

CONSEIL D'ÉTAT (1809). *Rapport et projet de décret relatif à une augmentation du prix des poudres et salpêtres.* (*M. le comte de Cessac, rapporteur. 2ᵉ rédaction*). Paris, 4 July. [BM: 5403.c.8(97)]

—— (1810). *Rapport et projet de décret relatif aux manufactures et établissemens qui répandent une odeur insalubre ou incommode.* (*M. le comte R. de Saint-Jean-d'Angely, rapporteur. 1ʳᵉ rédaction*) (36 pp.). Paris, 30 June. [BM: 5403.b.4(32)]

COQUEBERT, C. E. (1794). Histoire de la décomposition du sel marin, avec un extrait du rapport des citoyens Lelièvre, Pelletier, Darcet et Giroud, sur les moyens d'en extraire la soude avec avantage. *J. des mines,* No. 3, 29–90.

Correspondance des propriétaires et entrepreneurs de la manufacture de Monseigneur Comte d'Artois, pour les acides et sels minéraux, établie à Javel, près Paris, sur la fabrication en grand de l'alun-factice (62 pp.). [Paris?, 1780]. [BN: Vz.1733]

COSTAZ, C. A. (1802–3). Notice sur les brevets d'invention, et sur la législation qui y est relative. *Bull. Soc. enc.* **1**, 41–2.

—— (1818). *Essai sur l'administration de l'agriculture, du commerce, des manufactures et des subsistances, suivi de l'historique des moyens qui ont amené le grand essor pris par les arts depuis 1793 jusqu'en 1815.* Paris.

—— (1843). *Histoire de l'administration, en France, de l'agriculture, des arts utiles, du commerce, des manufactures, . . . terminé par l'exposé des moyens qui ont amené le grand essor pris par l'industrie française, depuis la Révolution . . . Troisième édition* (3 vols.). Paris.

COSTAZ, L. (1819). *Rapport du jury central sur les produits de l'industrie française.* Paris. [1819 exhibition.]

COURT, S. (1972). The *Annales de chimie,* 1789–1815. *Ambix* **19**, 113–28.

CREUZÉ DE LESSER, H. F. (1824). *Statistique du département de l'Hérault.* Montpellier.

CROSLAND, M. P. (1962). *Historical studies in the language of chemistry.* London.

—— (1967). *The Society of Arcueil: a view of French science at the time of Napoleon I.* London.

—— (1978). *Gay-Lussac: scientist and bourgeois.* Cambridge.

CROUZET, F. (1964). Wars, blockade, and economic change in Europe, 1792–1815. *J. economic history* **24**, 567–88.

—— (1966). Angleterre et France au XVIIIᵉ siècle. Essai d'analyse comparée de deux croissances économiques. *Annales: économies, sociétés, civilisations* **21**, 254–91. [English translation in: R. M. HARTWELL (ed.) (1967). *The causes of the industrial revolution in England.* London.]

CUVIER, G. (1837–8). *Histoire des progrès des sciences naturelles, depuis 1789 jusqu'à ce jour* (2 vols.). Brussels.

DALTON, J. (1810). *A new system of chemical philosophy. Part II.* Manchester.

DARCET, J., LELIÈVRE, C. H., PELLETIER, B. and GIROUD, A. [1794]. *Extrait du rapport présenté au Comité de salut public, par Darcet, Pelletier et Lelièvre; sur la fabrication de la soude* (12 pp.). [Paris]. [BM, BN. The BN has two edns (differing very slightly).]

—— [1794?] *District d'Angers . . . Instruction révolutionnaire des poudres et salpêtres de la République. Extrait du rapport . . . par Darcet, Pelletier et Lelièvre, sur la fabrication de la soude* (4 pp.). Angers. [BN. Same text as previous item.]

—— (an III [1794]). *Description de divers procédés pour extraire la soude du sel marin, faite en exécution d'un arrêté du Comité de salut public du 3 pluviôse, an 2* (80 pp.+11 pl.). Paris. [SRL: ZA 29* (copy now being transferred to BM); AN: F¹²2243; BN: 4°V.4210 (copy lacking title page and plates, and catalogued under the title of the report which forms the principal item: 'Rapport sur les divers moyens d'extraire avec avantage le sel de soude du sel marin').]

—— (1797*a*). Extrait d'un rapport sur les divers moyens d'extraire avec avantage la soude du sel marin. Par les citoyens Lelièvre, Pelletier, d'Arcet, et Alexandre Giroud. *Ann. chim.* **19**, 58–156.

—— ('1794' [1797*b*]). Rapport sur les divers moyens d'extraire avec avantage le sel de soude du sel marin. Par Lelièvre, Pelletier, d'Arcet et Giroud. Extrait par J. C. Delamétherie. *J. de physique* **2**, 118–34, 191–9, pl.

—— (1798). Rapport sur les divers moyens d'extraire avec avantage la soude du sel marin. In *Mémoires et observations de chimie de Bertrand Pelletier* (ed. C. Pelletier and J. Sédillot), vol. 2, pp. 144–234. Paris. [Reprints the report and its supplement from Darcet *et al.* (an III [1794]), but without the plates and possibly slightly edited.]

—— *See also*: Bouillon-Lagrange; Coquebert.

DARCET, J. P. J. (1817). Lettre . . . sur la substitution du nitrate de soude au nitrate de potasse, dans les arts. *Ann. chim. phys.* **6**, 206–7.

—— (1821). A MM. Gay-Lussac et Arago. *Ann. chim. phys.* **16**, 68–72. [On extraction of gelatine from bones.]

DARDEL, P. (1966). *Commerce, industrie et navigation à Rouen et au Havre au XVIIIᵉ siècle.* Rouen.

DAUMAS, M. (ed.) (1962–8). *Histoire générale des techniques* (3 vols.). Paris.

DAVY, H. (1812). *Elements of chemical philosophy.* London.

DAVY, R. (1955). *Contribution à l'étude des origines de la droguerie pharmaceutique et de l'industrie du sel ammoniac en France: l'apothicaire Antoine Baumé (1728–1804). Thèse . . . Faculté de pharmacie de Strasbourg.* Cahors. [Bibliothèque nationale et universitaire, Strasbourg]

DEBIDOUR, A. (ed.) (1910–17). *Recueil des actes du Directoire exécutif* (4 vols.). Paris.

DECROOS, G. (1821). *Traité sur les savons solides, ou manuel du savonnier et du parfumeur.* Paris. [CNAMB]

DE LA FOLLIE, L. G. (1774). Réflexions sur une nouvelle méthode pour extraire en grand l'acide du soufre par l'intermède du nitre, sans incommoder ses voisins. *Obs. phys.* **4**, 335–9.

—— (1777). Supplément d'expériences et observations concernant la fabrication de l'huile de vitriol. *Obs. phys.* **10**, 139–44.

—— (1819). Supplément aux expériences sur la fabrication de l'huile de vitriol. *Précis analytique des travaux de l'Académie . . . de Rouen, depuis sa fondation en 1744 jusqu'à l'époque de sa restauration, le 29 juin 1803,* **4**, 113–16.

DE LAGET, P. (1962). Septèmes. *Provincia. Revue . . . publiée par la Société de statistique, d'histoire et d'archéologie de Marseille et de Provence* **5**, 74–6.

DELAMÉTHERIE, J. C. (1789). Discours préliminaire. *Obs. phys.* **34**, 3–55.

—— (1809). Des manufactures de soude. *J. de physique* **69**, 421–5.

DELUMEAU, J. (1962). *L'Alun de Rome, XVe–XIXe siècle.* Paris.

DEMACHY, J. F. (1773). *L'Art du distillateur d'eaux-fortes.* [Paris]. (In series: *Descriptions des arts et métiers.)*

—— (1775). *L'Art du distillateur liquoriste.* (n.p.). (In series: *Descriptions des arts et métiers.)*

—— (1780). *Descriptions des arts et métiers . . . Nouvelle édition . . . Par J. E. Bertrand . . . Tome XII. Contenant L'Art du distillateur d'eaux-fortes, L'Art du distillateur liquoriste, & L'Art du vinaigrier; avec des notes & des additions par M. Struve.* Neuchâtel.

DE PEYRE, A. (1959). Lettres de Jean-Antoine Chaptal à son fils (1808–1816). *Revue du Gévaudan, des Causses et des Cévennes* (N.S.) No. 5, 68–73.

DEPITRE, E. (1914–19). Les prêts au commerce et aux manufactures, de 1740 à 1789. *Revue d'histoire économique et sociale* **7**, 196–217.

DERRIEN (1804). Observations sur la fabrication de l'acide sulfurique. *Ann. des arts et mans.* **18**, 280–5.

DESCHARMES, *see* PAJOT DESCHARMES

Description des machines et procédés spécifiés dans les brevets d'invention, de perfectionnement et d'importation dont la durée est expirée (93 vols. +2 index vols.; vol. 1 ed. C. P. Molard, vols. 2–22 ed. G. J. Christian). Paris (1811–74).

Description d'un nouvel appareil pour procurer à peu de frais et avec abondance aux blanchisseries le chlore qui leur est nécessaire. *Ann. ind.* **8**, 135–42. (1822).

Descriptions des arts et métiers, faites ou approuvées par messieurs de l'Académie royale des sciences (73 titles + supplements). Paris (1761–88).

—— *Nouvelle édition publiée avec des observations, & augmentée de tout ce qui a été écrit de mieux sur ces matières, en Allemagne, en Angleterre, en Suisse, en Italie. Par J. E. Bertrand* (19 vols.). Neuchâtel (1771–83).

DESCROIZILLES, F. A. H. (an III [1795]). Description et usages du berthollimètre. *J. des arts et mans.* **1**, 256–76.

—— (1806a). *Notices sur les alcalis du commerce.* Paris.

—— (1806b). Notices sur les alcalis du commerce. *Ann. chim.* **60**, 17–60. [A shortened version.]

—— (1809). Notices sur les alcalis du commerce. Seconde partie. On y a joint la description d'un appareil absorbant, utile aux nouvelles manufactures de soude. *Ann. chim.* **72**, 314–29.

—— (1818). *Notices sur l'alcali-mètre et autres tubes chimico-métriques . . . 2ᵉ édition, corrigée et augmentée.* Paris. [SRL]

DESMAREST, N. (1788). Papier (Art de fabriquer le). In *Encyclopédie méthodique. Arts et métiers mécaniques,* vol. 5, 463–595.

DESMIER DE SAINT-SIMON, E. J. A. (1843). *Description géologique du département de l'Aisne.* Paris.

DESORMES, C. B. and CLÉMENT, N. (1806). Théorie de la fabrication de l'acide sulfurique. *Ann. chim.* **59**, 329–39.

DICKINSON, H. W. (1939). The history of vitriol making in England. *Transactions of the Newcomen Society* **18**, 43–60. [On oil of vitriol.]

Dictionary of national biography (66 vols.). London (1885–1901).

Dictionary of scientific biography (ed. C. C. Gillispie; 14 vols.). New York (1970–6).

Dictionnaire chronologique et raisonné des découvertes, inventions, innovations, perfectionnemens, observations nouvelles et importations, en France, dans les sciences, la littérature, les arts, l'agriculture, le commerce et l'industrie, de 1789 à la fin de 1820 (17 vols.). Paris (1822–4).

Dictionnaire de biographie française. Paris (1933– (in progress)).

Dictionnaire technologique (22 vols. + 2 vols. plates). Paris (1822–35).

Dictionnaire universel de commerce, banque, manufactures, douanes, pêches, navigation marchande (2 vols.). Paris (1805).

DIEUDONNÉ, C. (1804). *Statistique du département du Nord* (3 vols.). Douai.

DIZÉ, M. J. J. (an X [1802]). *Précis historique sur la vie et les travaux de J. d'Arcet.* Paris.

—— (1803). Sur la décomposition des sulfures alcalins par les oxydes de plomb et de manganèse. *J. de chimie et de physique.* Reprinted in Pillas and Balland (1906), 212–19.

—— (1810). Mémoire historique de la décomposition du sel marin . . . suivi de quelques considérations sur l'importance de la fabrication de la soude artificielle en France. *J. de physique* **70**, 291–300.

DOIN, G. T. (1828). Dictionnaire des teintures. In *Encyclopédie méthodique. Manufactures, arts et métiers,* vol. 4.

DORVEAUX, P. (1929). L'invention de l'eau de Javel. *Bulletin de la Société d'histoire de la pharmacie* **17**, 286–7.

DOSSIE, R. (1758). *The elaboratory laid open.* London.

DUBLED, H. (1972). Un épisode de la lutte du Comtat-Venaissin contre l'administration française au XVIIIᵉ siècle: la question des poudres et salpêtres. *Rencontres,* No. 100, 3 pp. Carpentras.

DUBOIS, E. [1845?]. *Almanach statistique du canton de la Guillotière pour l'année 1845.* La Guillotière.

DUCHEMIN, P. P. (1890). *Petit-Quevilly.* Pont-Audemer.

—— (1893). *Sotteville-lès-Rouen et le faubourg Saint-Sever.* Rouen.

DUECKER, W. W. and WEST, J. R. (eds.) (1959). *The manufacture of sulfuric acid.* New York.

DUHAMEL DU MONCEAU, H. L. (1739). Sur la base du sel marin. *Mém. Acad.,* année 1736, 215–32.

DUJARDIN-SAILLY (1813). *Tarif chronologique des douanes de l'Empire françois* (8th edn). Paris. [Annual editions from 1806.]

DULIEU, L. (1950–2). Le chimiste Étienne Bérard, trésorier de l'École de pharmacie de Montpellier (1764–1839). *Revue d'histoire de la pharmacie* **10**, 40–4.

—— (1958). Le mouvement scientifique montpelliérain au XVIIIᵉ siècle. *Revue d'histoire des sciences* **11**, 227–49.

DUMAS, J. B. A. (1828–46). *Traité de chimie appliquée aux arts* (8 vols. + plates). Paris.

—— [1844?]. *Discours . . . prononcé aux funérailles de M. d'Arcet, le 5 août 1844.* Paris.

—— (1856). Rapport relatif à la découverte de la soude artificielle. *Comptes rendus* **42**, 553–78.

DUNOYER, C. (1842). Notice nécrologique sur Clément-Désormes. *Le Moniteur universel*, 14 Jan.

DUPIN, F. P. C. (1827). *Forces productives et commerciales de la France* (2 vols.). Paris.

DU PORTEAU (1790). Lettre . . . sur un procédé anglois pour faire l'acide vitriolique. *Obs. phys.* **37**, 227–8.

DUTIL, L. (1911). *L'État économique du Languedoc à la fin de l'ancien régime, 1750–1789*. Paris.

DUVAL, C. (1951). François Descroizilles, the inventor of volumetric analysis. *J. chemical education* **28**, 508–19.

DU VERDIER, P. (1972). Les procès-verbaux des séances de la Commission d'agriculture et des arts (an II–an III). *Bulletin d'histoire économique et sociale de la Révolution française*, 1970, 11–20.

EDELSTEIN, S. M. (1955). Two Scottish physicians and the bleaching industry: the contributions of Home and Black. *American dyestuff reporter* **44**, 681–4.

—— (1958). A Frenchman named O'Reilly: modern bleaching 150 years ago. *American dyestuff reporter* **47**, 253–7.

Encyclopédie, ou Dictionnaire raisonné des sciences, des arts et des métiers (ed. D. Diderot and J. le Rond D'Alembert; 17 vols. + 12 vols. plates). Paris (1751–77).

Encyclopédie méthodique. Arts et métiers mécaniques (8 vols. + 8 vols. plates). Paris (1782–91).

Encyclopédie méthodique. Chymie, pharmacie et métallurgie (6 vols. + 1 vol. plates). Paris (1786–1815).

Encyclopédie méthodique. Manufactures, arts et métiers (4 vols.). Paris (1785–1828).

Exhibition of the works of industry of all nations, 1851. Reports by the juries. London (1852).

Exposition publique des produits de l'industrie française. Catalogue des produits industriels qui ont été exposés au Champ-de-Mars pendant les trois derniers jours complémentaires de l'an VI . . . suivi du procès-verbal du jury (25 pp.). Paris (Vendémiaire an VII [1798]). [BM: F.548.(1)]

Exposition publique des produits de l'industrie française, an 10 [1802]. Procès-verbal des opérations du jury (72 pp.). Paris (Vendémiaire an XI [1802]). [BM: F.548.(3)]

Exposition de 1806. Rapport du jury sur les produits de l'industrie française, . . . précédé du procès-verbal des opérations du jury. Paris (1806).

FAGES, U. (1860). Industriels et inventeurs: Christophe Oberkampf. *Revue des deux mondes* **29**, 594–626.

FAUJAS DE SAINT-FOND, B. (1783–4). *Description des expériences aérostatiques de MM. Montgolfier, et de celles auxquelles cette découverte a donné lieu* (2 vols). Paris.

FAURE, H. (1889). *Histoire de la céruse, depuis l'antiquité jusqu'aux temps modernes, suivie d'un essai sur l'histoire du plomb*. Lille.

FAYET, J. (1960). *La Révolution française et la science, 1789–1795*. Paris.

FIGUIER, G. L. (1873–7). *Les Merveilles de l'industrie, ou Description des principales industries modernes* (4 vols.). Paris.

FODÉRÉ, F. E. (1813). *Traité de médecine légale et d'hygiène publique* (2nd edn; 6 vols.). Paris.

FOURCROY, A. F. (1800). *Système des connaissances chimiques* (10 vols.), Paris (1800) + Index, Paris (1802).

—— and LAVOISIER, A. L. (1957). Rapport sur le procédé de Bertholet, pour le blanchiement des toiles. Reprinted from *J. du Lycée des arts* (13 May 1793) in L. SCHELER, *Lavoisier et la Révolution française. I. Le Lycée des arts*, pp. 20–6. Paris.

—— and VAUQUELIN, N. L. (1797). Mémoire pour servir à l'histoire de l'acide sulfureux. *Ann. chim.* **24**, 229–309.

—— BERTHOLLET, C. L., and VAUQUELIN, N. L. (1808). Extrait d'un rapport fait à l'Institut . . . sur un mémoire de MM. Mollerat, concernant la carbonisation du bois en vaisseaux clos et l'emploi de différens produits qu'elle fournit. *Bull. Soc. enc.* 7, 175–80.

FOURCY, A. (1828). *Histoire de l'École polytechnique.* Paris.

FRÉMY, E. (1909). *Histoire de la Manufacture royale des glaces de France au XVIIe et au XVIIIe siècle.* Paris.

FRESNEL, A. (1866–70). *Oeuvres complètes* (ed. Henri de Senarmont, E. Verdet, and L. Fresnel; 3 vols.). Paris.

GARÇON, J. (1900). Les sources bibliographiques des sciences chimiques. *Congrès bibliographique international tenu à Paris du 13 au 16 avril 1898 sous les auspices de la Société bibliographique. Compte rendu des travaux,* vol. 2, pp. 161–85. Paris. [BM]

—— (1900–1). *Répertoire général ou dictionnaire méthodique de bibliographie des industries tinctoriales et des industries annexes, depuis les origines jusqu'à la fin de l'année 1896* (3 vols.). Paris. [An extensive subject bibliography, ranging over many branches of chemical industry.]

GAY-LUSSAC, J. L. (1807). Mémoire sur la décomposition des sulfates par la chaleur. *Mém. Soc. Arcueil* 1, 214–51.

GÉLÉBART, F. (1949). Le premier pharmacien-chef de la Pharmacie centrale de l'armée, Jérôme Dizé. *Revue d'histoire de la pharmacie* 13, 422–7.

GERBAUX, F. (1899). Les papeteries d'Essonnes, de Courtalin et du Marais de 1791 à 1794. *Le Bibliographe moderne* 3, 206–15.

—— (1903). La papeterie de Buges en 1794. *Le Bibliographe moderne* 7, 25–83.

—— and SCHMIDT, C. (eds.) (1906–37). *Procès-verbaux des Comités d'agriculture et de commerce de la Constituante, de la Législative et de la Convention* (4 vols. + Index). Paris.

GERBER, C. (1925–7). Le fabricant de produits chimiques Chaptal et la question du salpêtre. *Bulletin de la Société d'histoire de la pharmacie* 4, 185–94.

GILBERT, L. F. (1952). W. H. Wollaston MSS. at Cambridge. *Notes and records of the Royal Society* 9, 311–32.

GILLE, B. (1961). *Le Conseil général des manufactures. Inventaire analytique des procès-verbaux, 1810–1829.* Paris.

—— (1963). *Documents sur l'état de l'industrie et du commerce de Paris et du département de la Seine, 1778–1810, publiés avec une étude sur les essais d'industrialisation de Paris sous la Révolution et l'Empire.* Paris.

—— (1964). *Les Sources statistiques de l'histoire de France: des enquêtes du XVIIe siècle à 1870.* Paris and Geneva.

GILLISPIE, C. C. (1957*a*). The discovery of the Leblanc process. *Isis* 48, 152–70.

—— (1957*b*). The natural history of industry. *Isis* 48, 398–407.

—— (1965). Science and technology. In *The new Cambridge modern history. Volume IX. War and peace in an age of upheaval, 1793–1830* (ed. C. W. Crawley), pp. 118–45. Cambridge.

GIRARDIN, J. P. L. (1846). *Leçons de chimie élémentaire appliquées aux arts industriels* (3rd edn; 2 vols.). Paris.

GITTINS, L. (1966). The manufacture of alkali in Britain, 1779–1789. *Annals of science* 22, 175–90.

GODECHOT, J. (1968). *Les Institutions de la France sous la Révolution et l'Empire* (2nd edn). Paris.

GUERLAC, H. (1959). Some French antecedents of the chemical revolution. *Chymia* 5, 73–112.

—— (1961). *Lavoisier: the crucial year.* Ithaca, New York.

GUIFFREY, J. J. (1876). Les comités des assemblées révolutionnaires, 1789–1795. Le Comité de l'agriculture et du commerce. *Revue historique* 1, 438–83.

GUTTMANN, O. (1901). The early manufacture of sulphuric and nitric acid. *J. Society of chemical industry* **20**, 5–8.

GUYTON DE MORVEAU, L. B. (1801). *Traité des moyens de désinfecter l'air, de prévenir la contagion, et d'en arrêter les progrès*. Paris.

—— and CHAPTAL, J. A. C. (1805). Rapport demandé à la Classe des sciences physiques et mathématiques de l'Institut, sur la question de savoir si les manufactures qui exhalent une odeur désagréable peuvent être nuisibles à la santé. *Ann. chim.* **54**, 86–103.

H. P. [Henri Pillore] [1810]. *Quelques mots sur les dangers des fabriques nouvelles d'acides sulphuriques et de soudes factices, lorsqu'elles ne sont point à une distance sagement calculée des villes* (16 pp.). Rouen. [ADSM, 5 MP 1252, [1]]

HABER, L. F. (1969). *The chemical industry during the nineteenth century. A study of the economic aspect of applied chemistry in Europe and North America* (2nd impression). Oxford.

—— (1971). *The chemical industry 1900–1930*. Oxford.

HAHN, R. (1971). *The anatomy of a scientific institution: the Paris Academy of Sciences, 1666–1803*. Berkeley.

HALL, A. R. (1974). What did the Industrial Revolution in Britain owe to science? In *Historical perspectives: studies in English thought and society in honour of J. H. Plumb* (ed. N. McKendrick), pp. 129–51. London.

HALL, M. B. (1973). La croissance de l'industrie chimique en Grande-Bretagne au XIXe siècle. *Revue d'histoire des sciences* **26**, 49–68.

HALLADE, J. (1973). *Histoire de la soudière de Chauny. (Cie de Saint-Gobain, Péchiney-Saint-Gobain, Rhône-Progil). 1822–1972* (118 pp.). [Bichancourt, Chauny: the author].

HARDIE, D. W. F. (1950). *A history of the chemical industry in Widnes*. [Widnes?]

—— (1952). The Macintoshes and the origins of the chemical industry. *Chemistry and industry*, 1952, 606–13.

—— and PRATT, J. D. (1966). *A history of the modern British chemical industry*. Oxford.

HASSENFRATZ, J. H. (1792). Explications de quelques phénomènes qui paroissent contrarier les loix des affinités chimiques. *Ann. chim.* **13**, 3–38.

HAUSSMANN, J. M. (1791). Lettre . . . à C. L. Berthollet. *Ann. chim.* **11**, 237–50.

HAYES, R. (1949). *Biographical dictionary of Irishmen in France*. Dublin.

HELLANCOURT (1790). Lettre . . . à M. Lavoisier, sur le blanchissage des toiles en Beauvoisis, en Flandre & en Basse-Picardie. *Ann. chim.* **7**, 263–77.

HENDERSON, W. O. (1972). *Britain and industrial Europe, 1750–1870: studies in British influence on the industrial revolution in Western Europe* (3rd edn). Leicester.

HENKEL, J. F. (1757). *Pyritologia: or, a History of the pyrites*. London.

HENNEZEL D'ORMOIS, J. M. F. DE (1933). *Gentilshommes verriers de la Haute-Picardie. Charles-Fontaine*. Nogent-le-Rotrou.

HERBIN DE HALLE, P. E. (1803–4). *Statistique générale et particulière de la France* (7 vols. + *Atlas*). Paris.

HÉRICART DE THURY, L. E. F. (1819). *Rapport du jury d'admission des produits de l'industrie du département de la Seine à l'exposition du Louvre, comprenant une notice statistique sur ces produits*. Paris.

HERPIN (1839). Rapport . . . sur divers procédés de blanchissage du linge. *Bull. Soc. enc.* **38**, 38–54, 74–86.

HIGGINS, S. H. (1924). *A history of bleaching*. London.

Histoire de l'Académie royale des sciences, avec les Mémoires de mathématique et de physique, années 1699–1790 (93 vols.). Paris (1702–97).

HOLMES, F. L. (1974). *Claude Bernard and animal chemistry*. Cambridge, Mass.

HOMASSEL (an VII [1798–9]). *Cours théorique et pratique sur l'art de la teinture*. Paris.

—— (1807). *Cours théorique et pratique sur l'art de la teinture* (2nd edn; revised by E. J. B. Bouillon-Lagrange). Paris. [SRL]

HOME, F. (1756). *Experiments on bleaching.* Edinburgh.

—— (1762). *Essai sur le blanchiment des toiles.* Paris. [A translation of the previous work.]

HUET DE COETLIZAN, J. B. C. R. (an XII [1803–4]). *Recherches économiques et statistiques sur le département de la Loire-inférieure. Annuaire de l'an XI.* Nantes.

HYSLOP, B. F. (1965). *L'Apanage de Philippe Égalité, duc d'Orléans (1785–1791).* Paris.

IMPERIAL CHEMICAL INDUSTRIES LTD. (1955). *Sulphuric acid: manufacture and uses.* London.

Index biographique des membres et correspondants de l'Académie des sciences du 22 décembre 1666 au 15 novembre 1954. Paris (1954).

INSTITUT DE FRANCE (1810). *Rapports et discussions de toutes les classes de l'Institut de France, sur les ouvrages admis au concours pour les prix décennaux.* Paris. [BM: 733.g.16.(39–41)]

JANOT, J. M. (1952). *Les Moulins à papier de la région vosgienne* (2 vols.). Nancy.

JAUBERT, P. (1773). *Dictionnaire raisonné universel des arts et métiers; contenant l'histoire, la description, la police des fabriques et manufactures de France et des pays étrangers. Nouvelle édition, corrigée et . . . augmentée* (5 vols.). Paris.

JOURDAN, A. J. L., DECRUSY, and ISAMBERT, F. A. ([1822]–33). *Recueil général des anciennes lois françaises, depuis l'an 420 jusqu'à la révolution de 1789* (29 vols.). Paris.

Journal de commerce. Brussels (1759–62).

Journal de l'Empire. Paris (an XIII [1805]–1814).

Journal de la Société des pharmaciens de Paris (3 vols. in 2). Paris (ans V–VIII [1797–9]).

Journal de Normandie. Rouen (1785–91). Continued as *Journal de Rouen.* [ADSM]

Journal de physique, de chimie et d'histoire naturelle (53 vols.). Paris (1794–1823). (A continuation of *Observations sur la physique.*)

Journal de Rouen. Rouen (1791, etc.). [ADSM]

Journal des arts et manufactures. Publié sous la direction de la Commission exécutive d'agriculture et des arts (3 vols.). Paris (ans III–V [1795–7]). [BN]

Journal des mines. Paris (an III [1794]–1815). Continued as *Annales des mines.*

Journal du commerce. Paris (an III [1794–5]–1811; 1814, etc.). [BN]

JULIA (1804). *Mémoire sur la culture de la soude dans la ci-devant province du Languedoc. Ann. chim.* **49**, 267–85.

JULIA DE FONTENELLE, J. S. E. (1834). *Manuel complet du blanchiment et du blanchissage* (2 vols.). Paris.

JULLIANY, J. (1842–3). *Essai sur le commerce de Marseille* (2nd edn; 3 vols.). Marseilles and Paris.

KERSAINT, G. (1958). Sur la fabrique de produits chimiques établie par Fourcroy et Vauquelin 23, rue du Colombier, à Paris. *Comptes rendus* **247**, 461–4. [The works made fine chemicals. Kersaint has written on it further in: *Revue d'histoire de la pharmacie* **47** (1959), 25–30; *Revue générale des sciences pures et appliquées* **67** (1960), 93–102; *Revue des docteurs en pharmacie de France*, année xlvi, (N.S.) No. 43 (1960), 13–32.]

—— (1961). Sur l'usine de Chaptal aux Ternes. *Comptes rendus* **252**, 1407–9.

—— (1966). *Antoine François de Fourcroy (1755–1809), sa vie et son oeuvre.* Paris. (*Mémoires du Muséum national d'histoire naturelle*, N.S., ser. D, vol. 2.)

KIRK, R. E. and OTHMER, D. F. (eds.) (1963–72). *Encyclopedia of chemical technology* (2nd edn; 23 vols. + Index). New York.

KIRKBY, W. (1902). *The evolution of artificial mineral waters.* Manchester.

KUHLMANN (1827). Notice sur la fabrication de l'acide sulfurique. *Recueil des travaux de la Société des sciences, de l'agriculture et des arts de Lille*, année 1826–7, 120–6.

Kuhlmann, Établissements (1926). *Cent ans d'industrie chimique: les Établissements Kuhlmann, 1825–1925*. Paris.

Kuscinski, A. (1916–19). *Dictionnaire des Conventionnels*. Paris. (Reprinted Breuil-en-Vexin, 1973.)

Kuznetsov, V. I. (1966). The development of basic ideas in the field of catalysis. *Chymia* 11, 179–204.

L. [Le Bertre] [1810 or 1811]. *Quelques lignes de réponse, aux quelques mots de Mr H. P., sur les dangers des fabriques d'acide sulfurique et de soude* (16 pp.). Rouen. [ADSM, 5 MP 1252, [1]]

Labarraque, A. G. (1822). Extrait d'un mémoire de M. Labarraque sur l'art du boyaudier. *Bull. Soc. enc.* 21, 370–4.

—— (1826). Note sur la préparation des chlorures désinfectans. *J. de chimie médicale* 2, 165–72.

La Berge, A. F. (1975). The Paris Health Council, 1802–1848. *Bulletin of the history of medicine* 49, 339–52.

Labouchère, A. (1866). *Oberkampf (1738–1815)*. Paris.

Laboulaye, C. P. Lefebvre (1853–61). *Encyclopédie technologique. Dictionnaire des arts et manufactures* (2nd edn; 3 vols.). Paris.

—— (1867–70). *Dictionnaire des arts et manufactures* (3rd edn; 2 vols. + Complément). Paris.

Lacroix, A. (1863). *Historique de la papeterie d'Angoulême*. Paris.

Laferrère, M. (1952). Les industries chimiques de la région lyonnaise. *Revue de géographie de Lyon* 27, 219–56.

—— (1960). *Lyon, ville industrielle: essai d'une géographie urbaine des techniques et des entreprises*. Paris.

—— (1972). Le rôle de la chimie dans l'industrialisation à Lyon au XIXe siècle. In *L'Industrialisation en Europe au XIXe siècle: cartographie et typologie. Colloques internationaux du Centre National de la Recherche Scientifique. Sciences Humaines. Lyon, 7–10 octobre 1970* (ed. P. Léon, F. Crouzet, and R. Gascon), pp. 393–9. Paris.

Lalande, J. J. Le François de [1761]. *Art de faire le papier*. Paris. (In series: *Descriptions des arts et métiers*.)

—— (1803). *Bibliographie astronomique*. Paris.

Lambeau, L. (1912). *Histoire des communes annexées à Paris en 1859. Vaugirard*. Paris.

—— (1914). *Histoire des communes annexées à Paris en 1859. Grenelle*. Paris.

Landes, D. S. (1965). Technological change and development in Western Europe, 1750–1914. In *The Cambridge economic history of Europe* (ed. H. J. Habakkuk and M. Postan), vol. 6, pp. 274–601, 943–1007. Cambridge.

—— (1969). *The unbound Prometheus. Technological change and industrial development in Western Europe from 1750 to the present*. Cambridge.

Laurens (1808). Observations sur l'emploi des soudes dans les fabriques à savon de Marseille. *Ann. chim* 67, 97–106.

—— (1812). Recherches sur les savons du commerce, suivies de quelques observations relatives aux moyens de détruire les sulfures contenus dans les lessives des soudes artificielles. *Mém. Acad. Marseille* 9, année 1811, seconde partie, 1–33.

La Vitriolerie. *Rive gauche*, No. 42, 18–20. Lyons (1972).

[Lavoisier, A. L.] (1779). *L'Art de fabriquer le salin et la potasse, publié par ordre du roi, par les régisseurs-généraux des poudres et salpêtres*. Paris.

—— (1786). Observations sur la fabrique d'acides minéraux établie à Montpellier par M. Chaptal. In *Oeuvres*, vol. 6, 69–73.

—— (1789). *Traité élémentaire de chimie* (2 vols.). Paris.

—— (1862–93). *Oeuvres* (ed. J. B. Dumas and E. Grimaux; 6 vols.). Paris.

—— and Berthollet, C. L. (1789). Rapport d'un mémoire de M. Chaptal, sur quelques propriétés de l'acide muriatique oxigéné. *Ann. chim.* 1, 69–72.

LEBLANC, N. (1800). Mémoire[s] et rapports concernant la fabrication du sel ammoniac et de la soude. *J. de physique* **50**, 462–71. [Two memoirs by Leblanc, and reports on them by Fourcroy and Vauquelin, all presented to the *Lycée des arts*.]

—— (1802). *De la crystallotechnie, ou Essai sur les phénomènes de la cristallisation.* [Paris].

—— (1803). Observations sur la manière d'extraire la soude du sulfate de soude. *Bull. Soc. enc.* **1**, 75–6.

—— (1804). Observations sur la confection et l'usage de la soude. *Ann. chim.* **50**, 92–106.

LEBRETON, T. (1865). *Biographie rouennaise.* Rouen.

LECLAIR, E. (1901). *La Fabrication des acides forts à Lille avant 1790* (15 pp.). Poitiers. [Bibliothèque de la Faculté de médecine, Paris : 63383 (11), 169]

LEFROY (1810). Notice sur la décomposition du muriate de soude par les eaux provenant de la lixiviation des terres pyriteuses effleuries. *J. des mines* **27**, 231–6.

LELIÈVRE, *see* DARCET.

LELIVEC, H. (1806). Suite de la statistique des mines et usines du département du Mont-Blanc. *J. des mines* **20**, 407–502.

LEMAY, P. (1932). Berthollet et l'emploi du chlore pour le blanchiment des toiles. *Revue d'histoire de la pharmacie* **3**, 79–86.

—— (1933–4). La lessive de Berthollet et l'eau de Javel. *Revue d'histoire de la pharmacie* **4**, 99–103. [Also in *Le Courrier médical*, 1933, 343–4.]

—— (1949). Desormes et Clément découvrent et expliquent la catalyse. *Chymia* **2**, 45–9.

—— and OESPER, R. E. (1946). Claude Louis Berthollet (1748–1822). *J. chemical education* **23**, 158–65, 230–6.

LEMERY, N. (1697). *Cours de chymie* (9th edn). Paris.

LENOIR, H. (1922). *Historique et législation du salpêtre. Les pharmaciens et les ateliers révolutionnaires du salpêtre, 1793–1795. (Contribution à l'histoire de la pharmacie en Normandie).* Paris.

LE NORMAND, L. S. (1833–4). *Manuel du fabricant de papiers* (2 vols. + plates). Paris.

LÉRUE, J. A. DE (1875). *Notice sur Descroizilles (François-Antoine-Henri), chimiste, né à Dieppe, et sur les membres de sa famille* (76 pp.). Rouen.

LEUILLIOT, P. (1951–2). Notes sur les Haussmann et la manufacture du Logelbach (jusqu'en 1830). *Annuaire de la Société historique et littéraire de Colmar* **2**, 85–99.

—— (1959–60). *L'Alsace au début du XIXᵉ siècle. Essais d'histoire politique, économique et religieuse, 1815–1830* (3 vols.). Paris.

LEVASSEUR, E. (1900–1). *Histoire des classes ouvrières et de l'industrie en France avant 1789* (2nd edn; 2 vols.). Paris.

—— (1903). *Histoire des classes ouvrières et de l'industrie en France de 1789 à 1870* (2nd edn; 2 vols.). Paris.

LÉVY-LEBOYER, M. (1964). *Les Banques européennes et l'industrialisation internationale dans la première moitié du XIXᵉ siècle.* Paris.

LORGNA, A. M. (1786). Recherches sur l'origine du natrum ou alkali minéral natif. *Obs. phys.* **29**, 30–44, 161–71, 295–304, 373–86.

LOYSEL, P. (an VIII [1799–1800]). *Essai sur l'art de la verrerie.* Paris.

—— (1801). Mémoire sur le blanchiment de la pâte du papier. *Ann. chim.* **39**, 137–59.

LUCION, R. (1889). Einige Beiträge zur Geschichte des Ammoniaksoda-Processes. *Chemiker-Zeitung* **13**, 627. [On researches by the physicist Fresnel, 1811–12.]

LUNGE, G. (1879–80). *A theoretical and practical treatise on the manufacture of sulphuric acid and alkali, with the collateral branches* (3 vols.). London.

—— (1891–6). *A theoretical and practical treatise . . .* (2nd edn; 3 vols.). London.

MCCLOY, S. T. (1952). *French inventions of the eighteenth century.* [Lexington].

MCDONALD, D. (1960). *A history of platinum, from the earliest times to the eighteen-eighties.* London.

McKie, D. (1952). *Antoine Lavoisier: scientist, economist, social reformer.* London.

Macquer, P. J. (1766). *Dictionnaire de chymie* (2 vols.). Paris.

—— (1771). *A dictionary of chemistry* [trans. with notes by J. Keir] (2 vols.). London.

—— (1777). *A dictionary of chemistry* (2nd edn; 3 vols.). London.

—— (1778). *Dictionnaire de chymie* (2nd edn; 2 vols.). Paris.

Mactear, J. (1877). On the growth of the alkali and bleaching-powder manufacture of the Glasgow district. *Chemical news* **35**, 4–5, 14–17, 23–6, 35–6.

—— (1878). On the part played by carbon in reducing the sulphates of the alkalis. *J. Chemical Society* **33**, 475–87.

—— (1880–2). History of the technology of sulphuric acid. *Proceedings of the Philosophical Society of Glasgow* **13**, 409–27.

Madsen, E. R. (1958). *The development of titrimetric analysis till 1806.* Copenhagen.

Magnien (1786). *Recueil alphabétique des droits de traites uniformes, de ceux d'entrée et de sortie des cinq grosses fermes, de douane de Lyon et de Valence* (4 vols.). Avignon.

—— (1808). *Tarif des droits de douane et de navigation maritime de l'Empire français.* Paris. [Further edns 1811, 1815.]

—— and Deu de Perthes, L. J. (1809). *Dictionnaire des productions de la nature et de l'art, qui font l'objet du commerce de la France . . . et des droits auxquels elles sont imposées* (2 vols. in 3). Paris.

Maindron, E. (1881). *Les Fondations de prix à l'Académie des sciences. Les lauréats de l'Académie, 1714–1880.* Paris.

Malherbe, J. F. M., Gautherot, and Milet-Mureau (1795). Rapport sur la verrerie de Muntzdhal, ci-devant Saint-Louis. *J. des inventions et découvertes* [*J. du Lycée des arts*], No. 2, Vendémiaire an IV, 138–48.

Mannoury d'Ectot, Mis de [1856]. *Notice sur la vie et les travaux de Nicolas Le Blanc, inventeur des procédés de l'extraction de la soude du sel marin* (6 pp.). Paris.

Mannoury d'Ectot, M^{ise} de (1880). *Notice sur la vie et les découvertes de Nicolas Le Blanc, par sa petite-fille H. de Mannoury d'Ectot, née Nicolas Le Blanc* (12 pp., port.). Paris.

Marcus, A. (1887). *Les Verreries du Comté de Bitche. Essai historique.* Nancy.

Marion, M. (1923). *Dictionnaire des institutions françaises aux XVIIe et XVIIIe siècles.* Paris.

Markovitch, T. J. (1965). L'industrie française de 1789 à 1964. Sources et méthodes. *Cahiers de l'I.S.E.A.* [Institut de Science Économique Appliquée, Paris], Cahier AF4, No. 163, Juillet. (Series 'Histoire quantitative de l'économie française'.)

—— (1966a). L'industrie française de 1789 à 1964. Analyse des faits. *Cahiers . . .*, AF5, No. 173, Mai. [Statistics of production, etc.]

—— (1966b). L'industrie française de 1789 à 1964. Analyse des faits (suite). *Cahiers . . .*, AF6, No. 174, Juin.

—— (1966c). L'industrie française de 1789 à 1964. Conclusions générales. *Cahiers . . .*, AF7, No. 179, Novembre.

Marshall, A. E. (1927). The use of lead in sulphuric acid manufacture. *Transactions of the American Institute of Chemical Engineers* **20**, 149–59.

Masson, P. (1919). Le commerce de Marseille de 1789 à 1814. In *Marseille depuis 1789: études historiques*, vol. 1. Paris.

—— (ed.) (1926). *Les Bouches-du-Rhône. Encyclopédie départementale*, vol. 8, *L'Industrie.* Marseilles.

Masuyer (1824). [Letter on use of hypochlorites as disinfectants, and on extraction of gelatine from bones]. *J. de la Société des sciences, agriculture et arts du département du Bas-Rhin* **1**, 153–63.

Matagrin, A. (1925). *L'Industrie des produits chimiques et ses travailleurs.* Paris.

Mathias, P. (1972). Who unbound Prometheus? Science and technical change, 1600–

338 BIBLIOGRAPHY

1800. In *Science, technology and economic growth in the eighteenth century* (ed. A. E. Musson), pp. 69–96. London.

MATHIEZ, A. (1917). La mobilisation des savants en l'an II. *La Revue de Paris*, année 24, t.6, 542–65.

MATTHEWS, G. T. (1958). *The royal general farms in eighteenth-century France*. New York.

MATTHEWS, M. H. (1976). The development of the synthetic alkali industry in Great Britain by 1823. *Annals of science* **33**, 371–82.

MELLOR, J. W. (1922–37). *A comprehensive treatise on inorganic and theoretical chemistry* (16 vols.). London.

Mémoire pour les sieurs Mallez frères, fabricans de produits chimiques, à Septèmes; contre les opposans à l'autorisation qu'ils ont demandée d'établir, dans le vallon de Friguières, une fabrique de sulfate et de carbonate de soude, sous l'offre d'en condenser les vapeurs (59 pp.). Marseilles (1818). [BMM: Fonds de Provence, 3.762]

Mémoire des fabricants de soude de Marseille, en réfutation d'une pétition de quelques habitants du département de l'Aude, présenté à la Chambre des Pairs et à celle des Députés (33 pp.). Marseilles [1819]. [BN: Vp.1954. See further: AN, $F^{12}2243$, [7].]

Mémoires de l'Académie royale des sciences, see Histoire.

Mémoires de l'Institut national des sciences et des arts. Sciences mathématiques et physiques (14 vols.). Paris (an VI [1798]–1818).

Mémoires de physique et de chimie de la Société d'Arcueil (3 vols.). Paris (1807–17).

Mémoires des sociétés savantes et littéraires de la République française (2 vols.). Paris (1801–2).

Mémoires publiés par l'Académie de Marseille (12 vols.). Marseilles (an XI [1803]–1814).

MÉRIMÉE, J. F. L. (1817). Rapport . . . sur un grand vase de platine, présenté à la Société par MM. Cuoq et Couturier. *Bull. Soc. enc.* **16**, 33–6.

MICHELET, H. (1965). *L'Inventeur Isaac de Rivaz (1752–1828). Ses recherches techniques et ses tentatives industrielles*. Martigny.

MITCHELL, B. R. and DEANE, P. (1962). *Abstract of British historical statistics*. Cambridge.

MOILLIET, J. L. (1966). Keir's caustic soda process: an attempted reconstruction. *Chemistry and industry*, 1966, 405–8.

Moniteur universel (Le). Paris (1811, etc.). Previously *Gazette nationale, ou le Moniteur universel*. Paris (1789–1810).

MONTEIL, J. (1974). Un procès de pollution industrielle à Montpellier en 1791. *Histoire des sciences médicales* **8**, 825–7. Colombes. [On Chaptal's works.]

MULTHAUF, R. P. (1965). Sal ammoniac: a case history in industrialization. *Technology and culture* **6**, 569–86.

—— (1966). Heavy chemicals. In *The origins of chemistry*, Ch. 16. London.

—— (1971). The French crash program for saltpeter production, 1776–94. *Technology and culture* **12**, 163–81.

MUSPRATT, J. S. [1860]. *Chemistry, theoretical, practical, and analytical, as applied and relating to the arts and manufactures* (2 vols.). Glasgow.

MUSSON, A. E. (ed.) (1972). *Science, technology and economic growth in the eighteenth century*. London.

—— (1975). Continental influences on the Industrial Revolution in Great Britain. In *Great Britain and her world, 1750–1914: essays in honour of W. O. Henderson* (ed. B. M. Ratcliffe), pp. 71–85. Manchester.

—— and ROBINSON, E. (1969). *Science and technology in the Industrial Revolution*. Manchester.

NICOLAS (1797). Extrait d'un mémoire sur les salines nationales des départemens de la Meurthe, du Jura, du Doubs et du Mont-Blanc. *Ann. chim.* **20**, 78–188.

Note sur les blanchisseries berthollliennes. *Bull. Soc. enc.* **2**, 122–4. (1803).

Notice sur les manufactures et ateliers qui répandent une odeur insalubre ou incommode. *Bull. Soc. enc.* **13**, 66–71. (1814).

Notices sur les objets envoyés à l'exposition des produits de l'industrie française; rédigées et imprimées par ordre de S.E.M. de Champagny, Ministre de l'intérieur. An 1806 (349 pp.). Paris (1806). [BN: Vz.2233. Catalogued under the name of C. A. Costaz, who as head of the Ministry's *Bureau des arts et manufactures* wrote the introduction and was presumably the editor.]

Nouvelle biographie générale (ed. J. C. F. Hoefer; 46 vols.). Paris (1852–66).

Observations sur la physique, sur l'histoire naturelle et sur les arts (43 vols.). Paris (1773–93). Continued (and earlier popularly known) as the *Journal de physique*.

OESPER, R. E. (1942 and 1943). Nicolas Leblanc (1742–1806). *J. chemical education* **19** (1942), 567–72; **20** (1943), 11–20.

[O'NEILL, C.] (1876 and 1877). Materials for a history of textile colouring, no. 4 [–5]. *The textile colourist* **2** (1876), 295–303; **3** (1877), 3–14, 63–75. [On Home's *Experiments on bleaching.*]

O'REILLY, R. (1801). *Essai sur le blanchiment, avec la description de la nouvelle méthode de blanchir par la vapeur d'après le procédé du citoyen Chaptal.* Paris. [There is an abridged English translation in: *Philosophical magazine* **10** (1801), 97–111, 247–64, 299–317.]

[——] (1801). Manière d'extraire la soude du sel marin. *Ann. des arts et mans.* **5**, 131–9.

[——] (1802). Notice sur le blanchiment à la vapeur. *Ann. des arts et mans.* **10**, 189–92.

[——] (1807). Sur les papiers faits avec diverses substances végétales, et sur le blanchiment des pâtes de papier. *Ann. des arts et mans.* **25**, 173–82.

OUDIETTE, C. (1812). *Dictionnaire topographique des environs de Paris, dans lequel on trouve une nouvelle description des bourgs, villes, villages, renfermés dans cet espace; leur population, productions, industrie, commerce . . . .* Paris.

PADLEY, R. (1951–2). The beginnings of the British alkali industry. *University of Birmingham historical journal* **3**, 64–78.

PAILLET, J. [1820s?]. *Notice sur M. [J. P. J.] d'Arcet* (6 pp.). Paris.

PAJOT DESCHARMES, C. (1798). *L'Art du blanchiment des toiles, fils et cotons de tout genre.* Paris.

—— (1799). *The art of bleaching piece-goods, cottons, and threads, of every description* (trans. W. Nicholson). London.

—— ('1814' [1815]). Instruction sur l'emploi des soudes factices indigènes, en remplacement des soudes végétales étrangères, à l'usage des verriers travaillant en teinte blanche. *Bull. Soc. enc.* **13**, 253–78; also in *J. de phys.* **80** (1815), 317–44.

—— (1824). Note sur la première application faite en France du chlore (acide muriatique oxigéné) pour décolorer les toiles peintes et les toiles ordinaires teintes avant ou après le tissage. *Bull. Soc. enc.* **23**, 23–4.

—— (1826). Mémoire sur les moyens de remédier aux inconvéniens occasionnés par les vapeurs ou gaz délétères qui s'élèvent des fabriques de soude artificielle. *Ann. ind.* **21**, 262–97.

PALMER, R. R. (1970). *Twelve who ruled: the year of terror in the French Revolution* (2nd edn). Princeton.

PANCIER, F. (1937–8). Une famille picarde: les de Ribaucourt. *Bull. Société des antiquaires de Picardie* **37**, 37–68.

PARISEL, L. U. (1836). Revue des établissements industriels de Lyon. II. Rive gauche du Rhône. *Revue du lyonnais* **4**, 369–83.

PARKER, H. T. (1965a). French administrators and French scientists during the Old Regime and the early years of the Revolution. In *Ideas in history* (ed. R. Herr and H. T. Parker), pp. 85–109. Durham, N.C.

—— (1965*b*). Two administrative bureaus under the Directory and Napoleon. *French historical studies* **4**, 150–69.

PARKES, S. (1815). *Chemical essays, principally relating to the arts and manufactures of the British Dominions* (5 vols.). London.

PARMENTIER, A. A. (1807). Observations sur la lettre de M. Masuyer, médecin, concernant les fumigations du gaz acide muriatique oxigéné. *Ann. chim.* **64**, 268–72.

PARNELL, E. A. (1844). *Applied chemistry; in manufactures, arts, and domestic economy* (2 vols.). London.

PARTINGTON, J. R. (1961–70). *A history of chemistry* (4 vols.). London.

PAYAN, R. (1934). *L'Évolution d'un monopole: l'industrie des poudres avant la loi du 13 fructidor an V. Thèse pour le doctorat.* Paris.

PAYEN, A. (1825). Instruction sur les procédés à suivre dans l'emploi du chlorure de chaux dans le blanchiment des substances végétales. *Ann. ind.* **19**, 78–85.

—— (1826). Du chlorure de soude médicinal. *J. de chimie médicale* **2**, 513–20.

—— (1829–32). *Rapport du jury départemental de la Seine sur les produits de l'industrie admis au concours de l'exposition publique de 1827* (2 vols.). Paris.

—— (1855). *Précis de chimie industrielle* (3rd edn; 2 vols. (text + plates)). Paris.

—— (1866). Les industries chimiques au dix-neuvième siècle. II. La soude artificielle. *Revue des deux mondes* **62**, 958–83.

—— and CARTIER (1832). *Die Fabrikation der Schwefelsäure nach den neuesten französischen und englischen Methoden und Verbesserungen* (43 pp., pl.). Quedlinburg.

—— and CHEVALLIER, J. B. A. (1824). Essais sur les moyens de reconnaître la valeur réelle des soufres destinés à la fabrication de l'acide sulfurique. *J. de pharmacie* **10**, 500–2.

PAYEN, J. (1969). *Capital et machine à vapeur au XVIIIᵉ siècle: les frères Périer et l'introduction en France de la machine à vapeur de Watt.* Paris.

PÉCLET, J. C. E. (1823). *Cours de chimie.* Marseilles.

PÉLIGOT, E. M. (1888). *Notice biographique sur Jean Robert Bréant* (16 pp.). Bernay.

PELLETAN, P. (1810). Essai sur les moyens de retenir l'acide muriatique qui se dégage pendant la décomposition en grand du sel marin par l'acide sulfurique; et description d'un appareil propre à cet usage. *Ann. chim.* **75**, 176–93, pl.

[——] [1810]. *Réclamation sur l'article du rapport du jury [des prix décennaux], concernant la fabrication de la soude* (8 pp.). [Paris, *c.* Aug.–Oct. 1810]. [I am grateful to Dr. R. P. Multhauf for drawing my attention to this piece, of which there is a copy in the collections of the American Philosophical Society.]

—— (1822–4). *Dictionnaire de chimie générale et médicale* (2 vols.). Paris.

PELLETIER, *see* DARCET.

PENOT, A. (1831–2). *Statistique générale du département du Haut-Rhin* (2 vols. in 1). Mulhouse.

PERRAUD DE THOURY, E. (1853). *Notice biographique sur M. d'Arcet (Jean-Pierre-Joseph)* (15 pp.). Paris.

PERRET (1867). Note sur la substitution des pyrites au soufre dans la fabrication de l'acide sulfurique. *Ann. chim. phys.* (4th series) **11**, 479–83.

Pétition à Monsieur le comte de Villeneuve, préfet du département des Bouches-du-Rhône, pour le Sieur Aaron-Jacques-Charles-Pierre Bernard [et al.] . . . Contre: le Sieur Vidal fils, banquier et fabricant de soudes factices (41 pp.). Marseilles [1816]. [AMM, 23 F 1]

PEUCHET, J. and CHANLAIRE, P. G. (1810). *Description topographique et statistique de la France . . . avec la carte de chaque département* (3 vols.). Paris.

PICARD, C. (1865–7). *Saint-Quentin, de son commerce et de ses industries* (2 vols.). Saint-Quentin.

PIGEIRE, J. (1931). *La Vie et l'oeuvre de Chaptal (1756–1832). Thèse pour le doctorat.* Paris.

—— (1932). *La Vie et l'oeuvre de Chaptal (1756–1832)*. Paris. [Omits some of the footnote material of the 1931 edition.]

PIGEONNEAU, H. and FOVILLE, A. DE (1882). *L'Administration de l'agriculture au Contrôle général des finances (1785–1787)*. *Procès-verbaux et rapports*. Paris.

PILLAS, A. and BALLAND, J. A. F. (1906). *Le Chimiste Dizé, sa vie, ses travaux, 1764–1852*. Paris. [Pillas was Dizé's grandson.]

—— —— (1915). Le chimiste Dizé. *Bulletin des sciences pharmacologiques* **21**, 111–19, 214–22. [Presents material supplementary to their book.]

PINGRET, E. H. T. and BRAYER DE BEAUREGARD, J. B. L. (1821). *Monumens, établissemens et sites les plus remarquables du département de l'Aisne*. Paris. [Lithographs by Pingret with notes by Brayer.]

POUSSIER, A. (1924). *De l'origine du blanchiment du coton à Rouen* (10 pp.). Rouen.

POUTET, J. J. E. (1828). Traité des savons. In *Encyclopédie méthodique. Manufactures*, vol. 4. Paris.

Précis analytique des travaux de l'Académie des sciences, belles-lettres et arts de Rouen, depuis sa fondation en 1744, jusqu'à l'époque de sa restauration le 29 juin 1803 (5 vols.). Rouen (1814–21).

Précis analytique des travaux de l'Académie . . . de Rouen, pendant l'année 1804, etc. Rouen (1807, etc.)

PRIS, C. (1974). La Manufacture des glaces de Saint-Gobain avant la révolution industrielle. *Revue d'histoire économique et sociale* **52**, 161–72.

—— (1975). *La Manufacture royale des glaces de Saint-Gobain, 1665–1830* (3 vols.). Lille: Service de reproduction des thèses.

Procès-verbal de l'Assemblée de Nosseigneurs des États de la province de Languedoc. Montpellier (annual volumes up to 1789). [AN, H*748]

PULSIFER, W. H. (1888). *Notes for a history of lead, and an enquiry into the development of the manufacture of white lead and lead oxides*. New York.

QUEVAUVILLER, A. (1955). L'hygiène et la sécurité du voisinage industriel: les établissements classés. *Chimie et industrie* **73**, 1002–10.

RANC, A. (1923). La question des nitrates sous la Restauration. *Revue scientifique* **61**, 615–18.

Rapport des deux sociétés de médecine de Marseille et conclusions motivées pour les propriétaires opposans à l'établissement d'une fabrique de sulfate et de carbonate de soude au Vallon-de-Friguières (c^{ne} de Septèmes), contre les Srs Mallez frères, demandeurs en autorisation (33 pp.). Marseilles (1819). [BMM: Fonds de Provence, O,3622]

Rapport sur les effets de l'appareil condensateur, que le Sr B. Rougier a établi à sa fabrique de soude à Septèmes, adressé à M. le Comte de Villeneuve, préfet du département des Bouches-du-Rhône (48 pp.). Marseilles (1826). [BMM: Fonds de Provence, 3.761]

Recherches statistiques sur la ville de Paris et le département de la Seine; recueil de tableaux dressés et réunis d'après les ordres de Monsieur le Comte de Chabrol . . . préfet (6 vols.). Paris (1821–60). [BM. Catalogued under 'Seine'.]

RÉMOND, A. (1946). *John Holker, manufacturier et grand fonctionnaire en France au XVIIIᵉ siècle, 1719–1786*. Paris.

RIBERO, L. (n.d.). Le salicor, en 1810; histoire d'un échec. *Fédération historique du Languedoc méditerranéen et du Roussillon. 33ᵉ, 34ᵉ, et 36ᵉ Congrès (Bagnols, Narbonne, Lodève—1959, 1960, 1963)*, pp. 251–6. Montpellier. [BN. On the production of natural soda at Narbonne.]

RICHARD, C. (1922). *Le Comité de salut public et les fabrications de guerre sous la terreur*. Paris.

ROLANTS, E. (1922). Un cours de chimie à Lille avant la Révolution. *L'Écho médical du Nord*, 1922, 304–8.

ROUGET DE LISLE, A. A. (1851 and 1852). Notice historique, théorique et pratique sur le blanchissage du linge de toile, de la flanelle de santé et des divers vêtements. *Bull. Soc. enc.* **50** (1851), 22–41, 405–26, 591–600; **51** (1852), 227–44.

ROUGIER, B. (1812). Mémoire sur la fabrication de la soude artificielle. *Mém. Acad. Marseille* **10**, 57–117, 2 pl.

ROUQUETTE (1964–5). Jérôme Dizé, pharmacien en chef du premier Magasin général des pharmacies. *Revue d'histoire de la pharmacie* **12**, 411–18.

RUPP, T. L. (1798). On the process of bleaching with the oxygenated muriatic acid. *Memoirs of the Literary and Philosophical Society of Manchester* **5**, 298–313.

SADOUN-GOUPIL, M. (1974). Science pure et science appliquée dans l'oeuvre de Claude-Louis Berthollet. *Revue d'histoire des sciences* **27**, 127–45.

—— (1977). *Le Chimiste Claude-Louis Berthollet (1748–1822). Sa vie—son oeuvre.* Paris.

SAINT-LÉGER (1918–19). Les mémoires statistiques des départements pendant le Directoire, le Consulat et l'Empire. *Le Bibliographe moderne* **19**, 5–43.

SAUVAGEAU, C. (1920). *Utilisation des algues marines.* Paris.

SCAGLIOLA, R. [1943]. *Les Apothicaires de Paris et les distillateurs.* Clermont-Ferrand. [Bibliothèque de la Sorbonne: H.F.u.f.343(333)8°.]

SCHEELE, C. W. (1785). *Mémoires de chymie . . . tirés des Mémoires de l'Académie royale des sciences de Stockholm* (trans. Mme Picardet, Guyton de Morveau, *et al.*; 2 pts.). Dijon.

—— (1931). *The collected papers of Carl Wilhelm Scheele* (trans. L. Dobbin). London.

SCHELER, L. A. F. (1957). *Lavoisier et la Révolution française. I. Le Lycée des arts* (rev. edn). Paris.

—— (1961). Antoine-Laurent Lavoisier et Michel Adanson, rédacteurs de programmes des prix à l'Académie des sciences. *Revue d'histoire des sciences* **14**, 257–84.

SCHEURER-KESTNER, A. (1885). Nicolas Leblanc et la soude artificielle. *Revue scientifique* (3rd series) **9**, 385–96.

SCONCE, J. S. (ed.) (1962). *Chlorine: its manufacture, properties and uses.* New York.

SCOVILLE, W. C. (1950). *Capitalism and French glassmaking, 1640–1789.* Berkeley. (Reprinted New York, 1968.)

Séance publique de la Société libre d'émulation de Rouen (27 vols.). Rouen (1810–36).

Seconde exposition des produits de l'industrie française. Procès-verbal des opérations du jury (39 pp.). Paris (Vendémiaire an X [1801]). [BM: F.521.(11)]

SHÉE, H. (1797). Notes sur l'entreprise de manufactures de soude en France, par la décomposition du sel marin. *J. des arts et mans.* **3**, 284–99. [Notes addressed to the Darcet commission in 1794.]

SIMON, L. C. V. (1921). *Le Chimiste Descroizilles (François-Antoine-Henri), 1751–1825. Sa vie—son oeuvre. (Faculté de médecine et de pharmacie de Lille. Thèse pour le doctorat)* (87 pp.). Rouen. [Bibliothèque universitaire de Lille]

SINGER, C. J. (1948). *The earliest chemical industry: an essay in the historical relations of economics and technology illustrated from the alum trade.* London.

—— HOLMYARD, E. J., HALL, A. R., and WILLIAMS, T. I. (eds.) (1954–8). *A history of technology* (5 vols.). Oxford.

SMEATON, W. A. (1957). L.B. Guyton de Morveau (1737–1816): a bibliographical study. *Ambix* **6**, 18–34.

—— (1959). F.-J. Bonjour and his translation of Bergman's 'Disquisitio de attractionibus electivis'. *Ambix* **7**, 47–50.

—— (1962). *Fourcroy, chemist and revolutionary, 1755–1809.* Cambridge.

SOCIÉTÉ INDUSTRIELLE DE MULHOUSE (1902). *Histoire documentaire de l'industrie de Mulhouse et de ses environs au XIXe siècle* (2 vols.). Mulhouse.

SONOLET, J. and POULET, J. (1972). La dynastie médicale des Pelletan. *Semaine des hôpitaux de Paris* **48**, 3513–20.

SOREL, E. A. L. (1902–4). *La Grande industrie chimique minérale* (2 vols. (not numbered)). Paris.

Specification of the patent granted to Mr. Charles Tennant . . . for his method of using calcareous earths . . . instead of alkaline substances, for neutralising the muriatic acid gas used in bleaching . . . Dated January 23, 1798. *Repertory of arts and manufactures* **9**, 303–9. (1798).

Specification of the patent granted to Mr. Charles Tennant . . . for his invention for preparing the oxygenated muriates of calcareous earths . . . in a dry, undissolved, or powdery form . . . Dated April 13, 1799. *Repertory of arts and manufactures* **13**, 1–5. (1800).

STAHL, G. E. (1766). *Traité du soufre*. Paris.

SULIAC, J. (1950). La découverte de la soude artificielle. Dizé (1764–1852). *France-pharmacie* **3**, 211–12.

SZABADVARY, F. (1966). *History of analytical chemistry*. Oxford.

SZRAMKIEWICZ, R. (1974). *Les Régents et censeurs de la Banque de France nommés sous le Consulat et l'Empire*. Geneva.

Tableau des négociants, manufacturiers, fabricants et marchands de la ville de Rouen. 1808. [Rouen (1808)]. [BMR: Norm.118[1]]

Tableau des négociants, manufacturiers, fabricants et marchands de la ville de Rouen et des autres communes de l'arrondissement. Rouen (1817). [BMR: Norm.118[1]]

Tableaux 1° des substances minérales qui ont été importées de l'étranger ou exportées de France en 1816 et 1817; 2° des produits bruts des mines, minières, tourbières, sources salées et marais salans du royaume, en 1817. *Ann. des mines* **3**, 571–94. (1818).

Tarif chronologique (4 parts). [Paris] (1853–[4?]). Published by the Ministère de l'agriculture, du commerce et des travaux publics, in its series: *Annales du commerce extérieur. France. Législation commerciale*, No. 65, Jan. 1853 (1re partie. Matières animales); No. 102, July 1853 (2e partie. Matières végétales); No. 121, Feb. 1854 (3e partie. Matières minérales); and a fourth part, without number or date (4e partie. Fabrications). The copy consulted is bound in 2 vols. in the Bibliothèque de la Chambre de commerce de Paris (press-mark 200.271). The work lists, for each commodity, the import and export duties successively imposed, from 1791 to the time of publication.

TATON, R. (ed.) (1964). *Enseignement et diffusion des sciences en France au XVIIIe siècle*. Paris.

TAYLOR, F. S. (1957). *A history of industrial chemistry*. London.

TENNANT, E. W. D. (1947). The early history of the St. Rollox chemical works. *Chemistry and industry*, 1947, 667–73.

THENARD, L. J. (1813–16). *Traité de chimie élémentaire théorique et pratique* (4 vols.). Paris.

—— and ROARD, J. L. (1806). *Mémoire sur l'emploi comparé des aluns dans les arts* (22 pp.). Paris. (Issued with *Bull. Soc. enc.* **5** (1806–7).)

THÉPOT, A. (1966). Le système continental et les débuts de l'industrie chimique en France. *Revue de l'Institut Napoléon* **99**, 79–84.

THIÉRY, L. V. (1787). *Guide des amateurs et des étrangers voyageurs à Paris* (2 vols.). Paris.

THILLAYE, L. J. S. (1829). *Manuel du fabricant de produits chimiques* (2 vols.). Paris.

THOMAS, L. J. [1936]. *Montpellier, ville marchande. Histoire économique et sociale de Montpellier des origines à 1870*. Montpellier.

THOMSON, A. G. (1974). *The paper industry in Scotland, 1590–1861*. Edinburgh.

TINTHOIN, R. (1956). *Bicentenaire de la naissance de Chaptal (1756–1956)* (48 pp.). Mende. [An exhibition catalogue published by the Archives départementales de la Lozère.] [BN: 16°V. Pièce.943]

—— (n.d.) Chaptal créateur de l'industrie chimique française. In *Fédération historique du Languedoc méditerranéen et du Roussillon, XXXe et XXXIe Congrès, Sète-Beaucaire (1956–1957)*, pp. 195–206. Montpellier. [BN]

TISSEYRE, J. (1969). Les mines d'alun de Fontaynes. *Revue de Rouergue* **23**, 367–75.

TISSIER (1822). Notice sur les arts chimiques cultivés à Lyon. *Compte rendu des travaux de la Société d'agriculture, histoire naturelle et arts utiles de Lyon*, année 1821–2, 241–8.

TOOKE, T. and NEWMARCH, W. (1838–57). *A history of prices and of the state of the circulation, from 1792 to 1856* (6 vols.). London. (Reprinted London, 1928.)

TORAUDE, L. G. (1921). *Bernard Courtois (1777–1838) et la découverte de l'iode (1811).* Dijon.

TRESSE, R. (1952). Le Conservatoire des arts et métiers et la Société d'encouragement pour l'industrie nationale au début du XIXe siècle. *Revue d'histoire des sciences* **5**, 246–64.

Troisième exposition publique des produits de l'industrie française. Jours complémentaires an 10. Catalogue des productions industrielles qui seront exposées (48 pp.). Paris (Fructidor an X [1802]). [BM: F.548.(2)]

TUETEY, A. (ed.) (1917). *Correspondance du ministre de l'intérieur relative au commerce, aux subsistances, et à l'administration générale (16 avril–14 octobre 1792).* Paris.

URE, A. (1822). On the composition and manufacture of chloride or oxymuriate of lime (commonly called bleaching powder), and on the atomic weight of manganese. *Quarterly journal of science, literature and the arts* **13**, 1–23. [Translated in part in: *Bull. Soc. enc.* **21** (1822), 192–5; *Ann. ind.* **11** (1823), 88–94.]

—— (1839). *A dictionary of arts, manufactures, and mines.* London.

VAUQUELIN, N. L. (1797). Mémoire sur la nature de l'alun du commerce. *Ann. chim.* **22**, 258–79; additional note (1798), **25**, 107–8.

—— (1799). Notice sur un sel provenant de la manufacture du cit. Payen. *Ann. chim.* **32**, 296–306.

—— (1802). Note sur l'hydrosulfure de soude. *Ann. chim.* **41**, 190–3.

—— (1804a). Analyse comparée de différentes sortes d'aluns. *Ann. chim.* **50**, 154–72.

—— (1804b). Réflexions sur la décomposition du muriate de soude par l'oxide de plomb. *Mém. Inst.* **5**, 171–6.

—— (1814). Note sur une couleur bleue artificielle, analogue à l'outremer. *Ann. chim.* **89**, 88–91.

VIENNET, O. (1947). *Napoléon et l'industrie française. La crise de 1810–1811.* Paris.

VILLENEUVE-BARGEMENT, C. DE (1821–34). *Statistique du département des Bouches-du-Rhône* (4 vols. + *Atlas*). Marseilles.

VITALIS, J. B. (1823). *Cours élémentaire de teinture . . . et sur l'art d'imprimer les toiles.* Paris. [SRL]

WALLON, H. (1897). *Une page d'histoire locale: la bourse découverte et les quais de Rouen.* Rouen.

WEBB, K. R. (1956). J. A. C. Chaptal, Comte de Chanteloup (1756–1832). A pioneer of chemical industry. *Chemistry and industry*, 1956, 1443–5.

WEINER, D. B. (1974). Public health under Napoleon: the *Conseil de salubrité de Paris*, 1802–1815. *Clio medica* **9**, 271–84.

WELTER, F. (1910). Jean-Joseph Welter, professeur et chimiste-mécanicien. *Mémoires de l'Académie de Metz* (3rd series), année 37: 1907–8, 319–54.

WELTER, J. J. (1817). Sur la combinaison du chlore avec la chaux. *Ann. chim. phys.* **7**, 383–7.

—— and GAY-LUSSAC, J. L. (1820). Observations sur l'essai des soudes et des sels de soude du commerce. *Ann. chim. phys.* **13**, 212–21.

WOLFF, K. H. (1974). Textile bleaching and the birth of the chemical industry. *Business history review* **48**, 143–63.

YOUNG, A. (1931). *Voyages en France en 1787, 1788 et 1789* (trans. H. Sée; 3 vols.). Paris.

III: UNPUBLISHED DISSERTATIONS

BARRIELLE, G. L'implantation des industries chimiques dans la 'région des étangs' et 'l'affaire du Pourra', 1806–1848. Faculté des lettres et sciences humaines d'Aix-en-Provence, Diplôme d'études supérieures d'Histoire contemporaine, présenté sous la direction de M. le Professeur Guiral. Année 1966–1967. [I am grateful to Professor Guiral for his kindness in making a copy of this dissertation available to me. It deals with public health problems indirectly caused by the Plan d'Aren soda works after the Empire.]

SMITH, J. G. Studies of certain chemical industries in Revolutionary and Napoleonic France. Ph.D. thesis in History of Science, University of Leeds, 1970.

INDEX

More important references are distinguished by page numbers in bold type; numbers in italic indicate pages with illustrations. Abbreviations: mfg, manufacturing; mfr, manufacturer; mfre, manufacture; n, note; nr, near; ptnr, partner; Rev., Revolutionary; wks, works.

Plate 1. John Holker. (Bibliothèque municipale de Rouen)

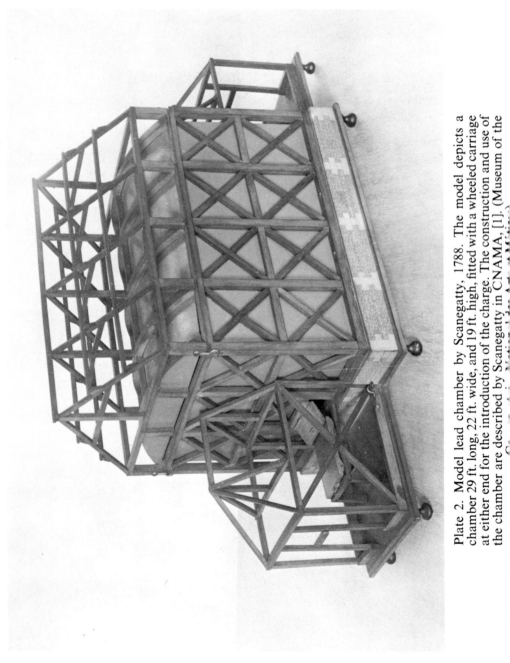

Plate 2. Model lead chamber by Scanegatty, 1788. The model depicts a chamber 29 ft. long, 22 ft. wide, and 19 ft. high, fitted with a wheeled carriage at either end for the introduction of the charge. The construction and use of the chamber are described by Scanegatty in CNAMA, [1]. (Museum of the

Plate 3. J. A. C. Chaptal as a young man. Portrait of unknown date by L. A. G. Bouchet (probably *c*. 1790 to judge from Chaptal's age). Chaptal is holding a book on dyeing and in the background is a dyeworks. (Musée Carnavalet. Photo: Charmet)

Plate 4. The la Paille chemical works in 1824. Sketch by Amelin.
(Bibliothèque municipale de Montpellier)

Plate 5. Chaptal as Minister of the Interior. Portrait by A. C. G. Lemonnier, 1808. (Faculté de Médecine, Paris. Photo: Lauros-Giraudon)

Plate 6. Traditional bleaching at Jouy, 1783. Detail from the famous cotton print *Les travaux de la manufacture*, designed for Oberkampf's works at Jouy by J. B. Huet. (Musée de l'Impression sur Étoffes, Mulhouse)

Plate 7. C. L. Berthollet. Lithograph after a portrait of the Revolutionary period. From F. S. Delpech, *Iconographie des contemporains* (2 vols., Paris, 1832), vol. 1. (British Library)

Plate 8. F. A. H. Descroizilles. Miniature by Judlin. (Musée des Beaux-Arts, Rouen. Photo: BMR)

Plate 9. Chlorine bleachery and sulphuric acid works of Dupuis at Saint-Quentin, 1821. From Pingret and Brayer (1821), Pl. 34. (British Library)

Plate 10. P. L. Athénas. (Bibliothèque Nationale, Cabinet des Estampes)

Plate 11. Nicolas Leblanc in middle life. Portrait of unknown date attributed to Gros (?). (Musée Carnavalet [photograph only—location of original unknown]. Photo: Charmet)

Plate 12. M. J. J. Dizé. Wax medallion by Pinson, 1793. (Pharmacie centrale des armées)

Plate 13. Louis Philippe Joseph, Duke of Orleans. From F. S. Delpech, *Iconographie des contemporains* (2 vols., Paris, 1832), vol. 2. (British Library)

Plate 14. Leblanc in later life. Statue erected in 1886 in the courtyard of the Conservatoire des Arts et Métiers. (Conservatoire National des Arts et Métiers)

Plate 15. Pierre Pelletan. (Musée d'Histoire de la Médecine, Paris. Photo: Charmet)

Plate 16. J. P. J. Darcet. (Wellcome Institute for the History of Medicine. By
courtesy of the Wellcome Trustees.)

Plate 17. An early soda works near Marseilles. Lithograph by Deroy
depicting a factory at les Goudes. In the background can be seen the fort of
Mont Rose. The date of the print is not known but we would guess it to have
been produced towards the mid-nineteenth century. The factory at les
Goudes seems to have been established between 1828 and 1837 (being listed
by Julliany (1842–3), vol. 3, 310, but not by Villeneuve-Bargement (1821–34),
vol. 4, 795). At the end of the 1830s it was a works of moderate size,
producing 2·4 million kg of soda a year. (Musée du Vieux-Marseille, Ancien
fond)